Data-Intensive Science

Chapman & Hall/CRC
Computational Science Series

SERIES EDITOR

Horst Simon
Deputy Director
Lawrence Berkeley National Laboratory
Berkeley, California, U.S.A.

AIMS AND SCOPE

This series aims to capture new developments and applications in the field of computational science through the publication of a broad range of textbooks, reference works, and handbooks. Books in this series will provide introductory as well as advanced material on mathematical, statistical, and computational methods and techniques, and will present researchers with the latest theories and experimentation. The scope of the series includes, but is not limited to, titles in the areas of scientific computing, parallel and distributed computing, high performance computing, grid computing, cluster computing, heterogeneous computing, quantum computing, and their applications in scientific disciplines such as astrophysics, aeronautics, biology, chemistry, climate modeling, combustion, cosmology, earthquake prediction, imaging, materials, neuroscience, oil exploration, and weather forecasting.

PUBLISHED TITLES

PETASCALE COMPUTING: ALGORITHMS AND
APPLICATIONS
Edited by David A. Bader

PROCESS ALGEBRA FOR PARALLEL AND DISTRIBUTED
PROCESSING
Edited by Michael Alexander and William Gardner

GRID COMPUTING: TECHNIQUES AND APPLICATIONS
Barry Wilkinson

INTRODUCTION TO CONCURRENCY IN
PROGRAMMING LANGUAGES
**Matthew J. Sottile, Timothy G. Mattson, and
Craig E Rasmussen**

INTRODUCTION TO SCHEDULING
Yves Robert and Frédéric Vivien

SCIENTIFIC DATA MANAGEMENT: CHALLENGES,
TECHNOLOGY, AND DEPLOYMENT
Edited by Arie Shoshani and Doron Rotem

INTRODUCTION TO THE SIMULATION OF DYNAMICS
USING SIMULINK®
Michael A. Gray

INTRODUCTION TO HIGH PERFORMANCE
COMPUTING FOR SCIENTISTS AND ENGINEERS
Georg Hager and Gerhard Wellein

PERFORMANCE TUNING OF SCIENTIFIC APPLICATIONS
**Edited by David Bailey, Robert Lucas, and
Samuel Williams**

HIGH PERFORMANCE COMPUTING: PROGRAMMING
AND APPLICATIONS
John Levesque with Gene Wagenbreth

PEER-TO-PEER COMPUTING: APPLICATIONS,
ARCHITECTURE, PROTOCOLS, AND CHALLENGES
Yu-Kwong Ricky Kwok

FUNDAMENTALS OF MULTICORE SOFTWARE
DEVELOPMENT
**Edited by Victor Pankratius, Ali-Reza Adl-Tabatabai,
and Walter Tichy**

INTRODUCTION TO ELEMENTARY COMPUTATIONAL
MODELING: ESSENTIAL CONCEPTS, PRINCIPLES, AND
PROBLEM SOLVING
José M. Garrido

COMBINATORIAL SCIENTIFIC COMPUTING
Edited by Uwe Naumann and Olaf Schenk

HIGH PERFORMANCE VISUALIZATION:
ENABLING EXTREME-SCALE SCIENTIFIC INSIGHT
**Edited by E. Wes Bethel, Hank Childs,
and Charles Hansen**

CONTEMPORARY HIGH PERFORMANCE COMPUTING:
FROM PETASCALE TOWARD EXASCALE
Edited by Jeffrey S. Vetter

DATA-INTENSIVE SCIENCE
Edited by Terence Critchlow and Kerstin Kleese van Dam

Data-Intensive Science

Edited by
Terence Critchlow
Kerstin Kleese van Dam

CRC Press
Taylor & Francis Group
Boca Raton London New York

CRC Press is an imprint of the
Taylor & Francis Group, an **informa** business

A CHAPMAN & HALL BOOK

CRC Press
Taylor & Francis Group
6000 Broken Sound Parkway NW, Suite 300
Boca Raton, FL 33487-2742

First issued in paperback 2016

© 2013 by Taylor & Francis Group, LLC
CRC Press is an imprint of Taylor & Francis Group, an Informa business

No claim to original U.S. Government works

Version Date: 20130405

ISBN 13: 978-1-138-19968-2 (pbk)
ISBN 13: 978-1-4398-8139-2 (hbk)

Library of Congress Cataloging-in-Publication Data

Data-intensive science / editors, Terence Critchlow and Kerstin Kleese van Dam.
    pages cm -- (Chapman & Hall/CRC computational science series)
    Includes bibliographical references and index.
    ISBN 978-1-4398-8139-2 (alk. paper)
    1. Science--Data processing. I. Critchlow, Terence. II. Van Dam, Kerstin Kleese.

Q183.9.D35 2013
502.85--dc23   2012048693

Visit the Taylor & Francis Web site at
http://www.taylorandfrancis.com

and the CRC Press Web site at
http://www.crcpress.com

For my beloved Jodi,
You are the kindest, gentlest person I know. Your constant support has made many things possible, including this book. You are my rock, my inspiration, my partner. I am lucky—and thankful—to have you with me on this great journey. With whom else could I watch polar bears fall asleep? Or spend hours quietly waiting for a leopard to stretch? Or share tea and scones? Or simply marvel at the beauty all around? You have shown me over and over again that anything can happen when we are together.
Thank you.
Always,
Terence

Contents

Editors

Terence Critchlow is currently the chief scientist for the Scientific Data Management Group in the Computational Sciences and Mathematics Division of the Pacific Northwest National Laboratory (PNNL). He earned his BSc from the University of Alberta in 1990, and his MS and PhD in computer science from the University of Utah in 1992 and 1997, respectively. He worked at Lawrence Livermore National Laboratory (LLNL) from 1997 to 2007, spending time as a postdoctoral, individual contributor, and finally a principal investigator. Dr. Critchlow led several projects while at LLNL, including data management efforts supporting the Advanced Simulation and Computing program and several programs for the Department of Homeland Security, including the Biodefense Knowledge Center. He joined the PNNL in April 2007 as a chief scientist within the Computational Sciences and Mathematics Division. In this role, he has continued to lead projects on data analysis, data dissemination, and workflow systems. Dr. Critchlow is currently a senior member of both the Institute of Electrical and Electronics Engineers (IEEE) and the Association for Computing Machinery (ACM). His research interests are focused in the areas of large-scale data management, metadata, data analysis, online analytical processing, data integration, data dissemination, and scientific workflows.

Kerstin Kleese van Dam (recipient of the British Female Innovators and Inventors Silver Award) is currently an associate division director and lead of the Scientific Data Management Group at the Pacific Northwest National Laboratory in the United States. Her prior positions include director of computing at the Biomedical Sciences Faculty at the University College London, United Kingdom; IT program manager and lead of the Scientific Data Management Group at the Science and Technology Facilities Council in the United Kingdom; high-performance computing specialist at the

German Climate Computing Center (DKRZ); and software developer at INPRO, a research institute of the German Automotive Industry. Kerstin has led collaborative data management efforts in scientific disciplines such as molecular science (e-Minerals), materials (e-Materials, MaterialsGrid), climate (PRIMA, NERC Data Services, DOE BER Climate Science for a Sustainable Energy Future), biology (DOE Bio Knowledgebase Prototype Project, Integrative Biology), and Experimental Facilities (ICAT). Her research is focused on data management and analysis in extreme scale environments.

Contributors

Isha Arkatkar
North Carolina State University
Raleigh, North Carolina

Alkyoni Baglatzi
School of Rural and Surveying
    Engineering
National Technical University of
    Athens
Athens, Greece

and

Institute for Geoinformatics
University of Münster
Münster, Germany

Ian Bird
European Organization for
    Nuclear Research (CERN)
Geneva, Switzerland

Daniel Brat
Department of Biomedical
    Informatics and Department
    of Pathology & Laboratory
    Medicine
Center for Comprehensive
    Informatics
Emory University
Atlanta, Georgia

Peer-Timo Bremer
University of Utah
Salt Lake City, Utah

and

Lawrence Livermore National
    Laboratory
Livermore, California

Sharath Cholleti
Department of Biomedical
    Informatics and Department
    of Pathology & Laboratory
    Medicine
Center for Comprehensive
    Informatics
Emory University
Atlanta, Georgia

Lee Cooper
Department of Biomedical
    Informatics and Department
    of Pathology & Laboratory
    Medicine
Center for Comprehensive
    Informatics
Emory University
Atlanta, Georgia

Terence Critchlow
Computational Sciences and
    Mathematics Division
Pacific Northwest National
    Laboratory
Richland, Washington

Eli Dart
Energy Sciences Network (ESnet)
Lawrence Berkeley National
    Laboratory
Berkeley, California

Ian T. Foster
Argonne National Laboratory
Chicago, Illinois

Geoffrey Fox
Indiana University
Bloomington, Indiana

Damian D. G. Gessler
University of Arizona
Tucson, Arizona

Zhenhuan Gong
North Carolina State University
Raleigh, North Carolina

David Gutman
Department of Biomedical
    Informatics and Department
    of Pathology & Laboratory
    Medicine
Center for Comprehensive
    Informatics
Emory University
Atlanta, Georgia

Attilay Gyulassy
University of Utah
Salt Lake City, Utah

Tony Hey
Microsoft Research
Redmond, Washington

Mark F. Horstemeyer
Department of Mechanical
    Engineering
Mississippi State University
Starkville, Mississippi

Bill Howe
Department of Computer Science
    and Engineering
eScience Institute
University of Washington
Seattle, Washington

John Jenkins
North Carolina State University
Raleigh, North Carolina

William Johnston
Energy Sciences Network (ESnet)
Lawrence Berkeley National
    Laboratory
Berkeley, California

Bob Jones
European Organization for
    Nuclear Research (CERN)
Geneva, Switzerland

Cliff Joslyn
Pacific Northwest National
    Laboratory
Richland, Washington

Tomi Kauppinen
Department of Media
    Technology
School of Science
Aalto University
Aalto, Finland

and

Institute for Geoinformatics
University of Münster
Münster, Germany

Carsten Keßler
Institute for Geoinformatics
University of Münster
Münster, Germany

Scott Klasky
Oak Ridge National
    Laboratory
Oak Ridge, Tennessee

Kerstin Kleese van Dam
Pacific Northwest National
    Laboratory
Richland, Washington

Jun Kong
Department of Biomedical
    Informatics and Department
    of Pathology & Laboratory
    Medicine
Center for Comprehensive
    Informatics
Emory University
Atlanta, Georgia

Tahsin Kurc
Department of Biomedical
    Informatics and Department
    of Pathology & Laboratory
    Medicine
Center for Comprehensive
    Informatics
Emory University
Atlanta, Georgia

Sriram Lakshminarasimhan
North Carolina State University
Raleigh, North Carolina

Michael Lautenschlager
German Climate Computing
    Centre
Hamburg, Germany

Bryan Lawrence
British Atmospheric Data Center
Harwell Oxford, United
    Kingdom

D. Lecarpentier
CSC—IT Center for Science
Espoo, Finland

Eric S. Myra
Department of Atmospheric,
    Oceanic and Space
    Sciences
University of Michigan
Ann Arbor, Michigan

Kanchana Padmanabhan
North Carolina State University
Raleigh, North Carolina

Tony Pan
Department of Biomedical
Informatics and Department
of Pathology & Laboratory
Medicine
Center for Comprehensive
Informatics
Emory University
Atlanta, Georgia

Valerio Pascucci
Pacific Northwest National
Laboratory
Richland, Washington

Paulo Pinheiro
Pacific Northwest National
Laboratory
Richland, Washington

J. Reetz
RZG—Garching Computing
Centre of the Max Planck Society
Garching, Germany

Rob Ross
Argonne National Laboratory
Lemont, Illinois

Joel Saltz
Department of Biomedical
Informatics and Department
of Pathology & Laboratory
Medicine
Center for Comprehensive
Informatics
Emory University
Atlanta, Georgia

Nagiza F. Samatova
North Carolina State University
Raleigh, North Carolina

and

Oak Ridge National
Laboratory
Oak Ridge, Tennessee

Eric Schendel
North Carolina State University
Raleigh, North Carolina

Neil Shah
North Carolina State University
Raleigh, North Carolina

Ashish Sharma
Department of Biomedical
Informatics and Department
of Pathology & Laboratory
Medicine
Center for Comprehensive
Informatics
Emory University
Atlanta, Georgia

Eric G. Stephan
Pacific Northwest National
Laboratory
Richland, Washington

Brian Summa
University of Utah
Salt Lake City, Utah

F. Douglas Swesty
Department of Physics and
  Astronomy
State University of New York at
  Stony Brook
Stony Brook, New York

George Teodoro
Department of Biomedical
  Informatics and Department
  of Pathology & Laboratory
  Medicine
Center for Comprehensive
  Informatics
Emory University
Atlanta, Georgia

Anne Trefethen
Oxford University
Oxford, United Kingdom

Karin Verspoor
University of Colorado
Boulder, Colorado

Fusheng Wang
Department of Biomedical
  Informatics and Department
  of Pathology & Laboratory
  Medicine
Center for Comprehensive
  Informatics
Emory University
Atlanta, Georgia

Patrick Widener
Department of Biomedical
  Informatics and Department
  of Pathology & Laboratory
  Medicine
Center for Comprehensive
  Informatics
Emory University
Atlanta, Georgia

Dean N. Williams
Lawrence Livermore National
  Laboratory
Livermore, California

P. Wittenburg
Max Planck Institute for
  Psycholinguistics (MPI-PL)
Nijmegen, the Netherlands

CHAPTER 1

What Is Data-Intensive Science?

Terence Critchlow and Kerstin Kleese van Dam

CONTENTS

Today, we are living in a digital world where scientists often no longer interact directly with the physical object of their research but do so via digitally captured, reduced, calibrated, analyzed, synthesized and, at times, visualized data. Advances in experimental and computational technologies have led to an exponential growth in the volume, variety, and complexity of data and although the deluge is not happening everywhere in an absolute sense, it is in a relative one. Science today is data intensive. Data-intensive science has the potential to transform not only how we do science but also how quickly we can translate scientific progress into complete solutions, policies, decisions, and ultimately economic success.

Critically, data-intensive science touches some of the most important challenges we are facing today. Consider a few of the grand challenges outlined by the U.S. National Academy of Engineering: make solar energy economical, provide energy from fusion, develop carbon sequestration methods, advance health informatics, engineer better medicines, secure cyberspace, and engineer the tools of scientific discovery. Arguably, meeting any of these challenges requires the collaborative effort of transdisciplinary teams and also significant contributions from enabling data-intensive technologies. Indeed, for many of them advances in data-intensive research will be the single most important factor in developing successful and timely solutions. Simple extrapolations of how we currently interact with and utilize data and knowledge are not sufficient to meet this need. Given the importance of these challenges, a new, bold vision for the role of data in science, and indeed how research will be conducted in a data-intensive environment, is evolving.

1.1 A VISION FOR THE FUTURE OF SCIENCE

Mainstream data-intensive science has so far focused on addressing the basic needs of the community. The 2008 update to the road map of the European Strategy Forum on Research Infrastructures (ESFRI) identifies, for the first time, not only the need for leading-edge experimental and computational facilities to drive future scientific progress but also the importance of an underpinning e-infrastructure consisting of integrated communication networks, distributed grids, high-performance computing (HPC), and digital repositories components. ESFRI further states that data in their various forms (from raw data to scientific publications) will need to be stored, maintained, and made available and openly accessible to all scientific communities. They are placing a new emphasis on digital repositories as places to capture and curate scientific data for the good of both science and the economy. Similarly, Gordon Bell, in *The Fourth Paradigm* (Hey et al. 2009, p. xiii), describes three pillars of data-intensive science as "capture, curation, and analysis." Intellectual and technological progress in these areas has particularly been driven by centers of excellence, large-scale long-term infrastructure projects, and organizations with visionary leadership and an in-depth understanding of data-intensive sciences. Some key examples for international centers and projects are UK Data Curation Centre (http://www.dcc.ac.uk/), the SciDAC Scientific Data Management Center (*Scientific Data Management*, Shoshani, 2009), the Earth Systems Grid and its international partners (The Earth System Grid: Supporting the Next Generation of Climate Modeling Research, Bernholdt et al. 2005), e-Infrastructure for Large Scale Experimental Facilities (ICAT: Integrating Infrastructure for Facilities Based Science, Flannery et al. 2009), and the Biomedical Informatics Research Network (http://www.birncommunity.org/). These projects have clearly demonstrated the potential of data-intensive science technologies; however, the report *Data-Intensive Research Theme* (Atkinson et al. 2010, p. 1) notes, "Current strategies for supporting it demonstrate the power and potential of the new methods. However, they are not a sustainable strategy as they demand far too much expertise and help in addressing each new data-intensive task." This and other recent publications (Strategy for a European Data Infrastructure, Koski et al. 2009; Data Intensive Science in the Department of Energy, Ahrens et al. 2010) clearly show the community consensus that more generalized, easy-to-use solutions need to be developed to make possible a more widespread use of these basic data-intensive technologies. Thought leaders are also pointing out that although current developments of infrastructure surrounding the management of data continue to be important, it is time to go beyond these basic approaches and focus on the data itself—developing the means to transform data into an infrastructure in its own right.

Increasingly, societal challenges such as securing a sustainable energy future, improving health, or mitigating and adapting to climate change require the timely and effective integration of research results from and the development of new insights that are shared between transdisciplinary research teams. New insights need to be effectively communicated not only between scientists but also to both industry innovation drivers and policy and decision makers to drive progress in addressing these challenges. Research infrastructures of the future will therefore need to enable the effective access and exchange of knowledge in transdisciplinary teams across different levels of theory, different research methodologies, different disciplines, and different levels of experience. Understanding the drivers and hurdles of collaboration and knowledge exchange in innovative research communities and addressing these effectively within an IT infrastructure will be critical for the success of any such infrastructure.

For example, policy and decision making, in response to climate change, will require both economic and environmental trade-offs. Decisions about allocating scarce water across competing municipal, agricultural, and ecosystem demands is just one of the challenges in this scenario, along with decisions regarding competing land use priorities such as biofuels, food, and species habitats. To be able to address this challenge, policy makers will look toward inter- and multidisciplinary science teams to provide the required underpinning information. These teams might need to couple regional climate models with a regional model of human decision making that considers energy and land use from a socioeconomic perspective. They might also want to integrate this system with detailed models of agriculture and energy infrastructures so that they can see in detail how the future landscape might change. In this way, researchers provide decision and policy makers with evaluations of different options. For example, a plan to boost renewable energy in a region could evaluate the likely locations for growing biofuel crops or the potential for wind resources in the future, taking into account both the future climate (will there be enough water, enough wind?) and the socioeconomic context (How many people will there be? What will the regional demand for energy be? What other resources are available for energy?). One can assume that particular effort will be extolled on evaluating the uncertainty associated with such modeling as well as the underpinning data, ensuring that the quality and uncertainty of the results is appropriate for the level of decision making required.

As the aforementioned examples demonstrate, the ability to transfer, utilize, and synthesize knowledge across scales, techniques, and disciplines will crucially influence the speed at which necessary complex insights can be acquired and translated into tangible policies and solutions. Equally important will be furthering collaboration and vigorous scientific engagement and exchange between scientists and other stakeholders, helping to break down traditional boundaries and gathering people around a common cause. Data as an infrastructure and a tool for scientific discovery can play a pivotal role. It can help to effectively validate or reject theories and scenarios within and across scientific domains, integrate and transfer knowledge in data-rich environments with diverse stakeholders, and facilitate collaboration around common interests, allowing for experimenting in data.

Enabling experimenting in data would allow scientists to develop complex in situ scenarios in which they can collaborate as teams, ad hoc or asynchronous (reusing previous results), to test their hypotheses by using predominantly digital sources. Such scenarios might include exchanging, combining, and manipulating data from multiple, very heterogeneous sources in real time and analyzing these with a combination of integrated methods from different domains; the results might be utilized to evaluate a hypothesis directly or be integrated to drive other experimental or computational research scenario evaluations. Data as an infrastructure would enable scientists to contribute their own expertise relevant to the context of an experiment, synthesize their findings with contributions from others, and subsequently enrich their own research through the new impulses gained by working with scientists from other domains. Therefore, experimenting in data would permit multidisciplinary teams to gain faster and more in-depth insights into the chosen problem space by pooling their combined knowledge to provide more complete information to all, thus increasing the speed and quality of each of their individual and combined research.

For such a new approach to be useful to researchers, it is important to take into account that scientific research and collaboration is very much context and trust driven and that personal contact and in-depth knowledge of technologies and people are critical factors in the uptake of new research methodologies. When the utilization of a new technique or discipline is considered, the existing mental model of the current research domain is extended as knowledge grows, linking and placing new experiences in context to existing knowledge and understanding. The new generation of multidisciplinary data needs to emulate and support this process as effectively as possible by providing information and environments tailored to individual researchers and their experience and expertise. Data as an infrastructure would enable experimenting in data across scales, techniques, and disciplines not only by providing the raw information required but also by engendering trust through providing the means to vet the contribution of others and control the utilization of one's own research by others; furthermore, it would encourage the integration of different insights by adopting principles from cognitive, behavioral, motivational, and educational sciences in its implementation and delivery.

Providing scientists with a rich tapestry of information regarding current research processes would enable queries to be formulated with ontology support weighted by personal experience (research subject) to restrict answers to relevant results. Answers could be drawn from a multitude of sources: from raw and derived data; publications enhanced by auxiliary information from blogs, wikis, web pages, Twitter, RSS feeds; and other social networking sources categorized in terms of evaluated links. Where possible, data sets would not only reference the authors but also provide access to their publication lists (those linked to the data set highlighted), curricula vitae and, similar to LinkedIn or Xing, information on whether one knows them, is linked to them via other collaborators, or has used their work before and had rated it as good material or not. If there are further questions, one would be able to contact them via various instant media if the person has allowed this.

Scientists need to be able to easily choose and vet available types of data (produced by suitable techniques, from reputable sources, i.e., repositories or latest research results) weighted by appropriateness for this type of experiment (preferred type), and its quality and appropriate level of uncertainty. If an identified data source is too big or too complex for immediate appraisal, directed data mining tools should be offered, as well as tools for the extraction of required subsets or features for further analysis. Enabling the combination of multisource data will require scalable techniques to identify and specify which features are contributed from a particular data source to the overall experiment, for example, chemical composition, structure, particular reaction determination, environmental interaction, property refinement, or a higher level abstraction of particular features to be fed to next higher process level. Researchers will need access to adaptable, scalable analysis methods that can deal not only with the complexity and volume of data in data-intensive research environments but also deliver the results in real or near real time where the results are required to influence time-critical decisions or fast interchange of information, that is, between simulation codes or people.

Data-intensive technologies have become an essential foundation in many different domains for sustainable progress in research and innovation. To address society's most pressing challenges, the community will need to move beyond one-off solutions and make the usage of data-intensive technologies a mainstream scientific research methodology on par with observational, experimental, and computational techniques. Data as an infrastructure needs to become ubiquitous and transcend levels of theory, research techniques, and disciplinary boundaries, thus acting as both an integrator and a facilitator of collaborative working. Realizing this vision will depend on technological advances and even more so on our ability to take the research community along on the journey, helping them to welcome and actively participate in the change and realize the opportunities it will offer.

These are daunting goals that have been espoused for over a decade; however, the advancement and convergence of several domains indicate that the community has reached such a maturity level of thought and technology that this goal may become achievable over the coming decade.

1.2 ORGANIZATION OF THIS BOOK

Over the past five years, the amount of information, in particular digital information, has increased exponentially. Chapter 2, "Where Does All the Data Come From?" frames the opportunities that are presented by data-intensive science and discusses why science is becoming data intensive at this point in history. It also describes some of the key sources of this information—from images and videos to sensitive scientific instruments and to blogs and social networks—and how all of this information can be used to make significant advances in science, if only we can figure out how to use it effectively.

The remainder of this book is organized into three sections, roughly corresponding to a high-level data-intensive-science workflow: (1) understanding the science problems, (2) understanding current capabilities, and (3) moving from current challenges to solutions. Although we believe this is the most intuitive presentation, given the breadth of material being presented it is unlikely that reading the book cover to cover, in order, would be an approach well suited to most reader's motives. Thus, we encourage the reader to skip those chapters that fall outside their current motivation and focus on those that are most relevant to them, returning to the other chapters as interest dictates. To aid in this process, the rest of this chapter provides a high-level introduction to each of the remaining chapters.

1.2.1 Section 1: Data-Intensive Grand Challenge Science Problems

Over the past several years, a number of groups, including the National Academy of Engineering, have identified grand challenge problems facing scientists and engineers around the world. While addressing these problems will have a global impact, solutions are years away at best and the next set of challenges are likely to be even harder to solve. Because of the complexity of questions being asked, meeting these challenges requires large, multidisciplinary teams to work closely together for extended periods of time. Enabling this new type of science, involving distributed teams that need to collaborate despite vastly different backgrounds and interests, is the cornerstone of data-intensive science. This section presents four disciplines, each facing their own grand challenge problems, selected to highlight the breadth of the problems being faced. Each chapter shows a vision for the future of the respective discipline and describes how technology is expected to enable it.

- Chapter 3, "Large-Scale Microscopy Imaging Analytics for In Silico Biomedicine": over the past few years, advances in medical treatments have been developing hand in hand with IT. Pharmacies regularly use software solutions to minimize the potential for drug interactions, x-rays are digital (not film), and doctors use robotics to perform surgeries that would be impossible for an unaided human to perform. From genetically customized medications to secure but accessible health records and from computer-aided surgery to computer-aided monitoring of patients, advances in medicine and health care will be closely tied to technology in the foreseeable future. This chapter looks at how data-intensive science will continue the health-care revolution.

- Chapter 4, "Answering Fundamental Questions about the Universe": understanding the universe, how it started, our place in it, and whether or not we are alone has been a quest for humankind for thousands of years. Beyond philosophical questions, however, astrophysics provides keen insights into practical problems such as nuclear and fusion energy. Currently, there is significant disconnect between the observational and computational branches of the science, without significant collaboration or data sharing occurring between different science teams. As this chapter outlines, in the future these teams will be active collaborators, with observation feeding computational experiments that in turn identify new events to look for.

- Chapter 5, "Materials of the Future: From Business Suits to Space Suits": to create novel materials that have super properties, experimentally validated and numerically verified multiscale theoretical/computational models need to be created as trial-and-error methods will not produce the next genre of materials. To provide a fundamental understanding for directing and controlling matter starting at the quantum, atomic, and molecular levels to achieve novel physical and thermo-chemo-mechanical properties, particularly under extreme environments, one must also be driven by the requirements at the largest length scales in terms of the boundary value problem requirements. The idea is to develop experimentally validated computation enabling technologies to allow the design and synthesis of new, revolutionary forms of structures with tailored properties particularly when characterizing and quantifying the structure–property relations at different length scales is usually far away from equilibrium. As described in this chapter, a cyberinfrastructure that hosts the repositories of codes, models, data, and tutorials is required to facilitate these advancements.

1.2.2 Section 2: Case Studies

Although the focus of this book is on the future of data-intensive science, it is important to realize that significant work is already underway attempting to realize this future. This section briefly highlights three projects that have made significant strides in advancing the state of the art in data-intensive science. This is not intended to be a survey of current projects or of the field at large but rather to provide some insight into current best practices.

- Chapter 6, "Earth System Grid Federation: Infrastructure to Support Climate Science Analysis as an International Collaboration": for decades, the climate research community has sought a predictive understanding of the Earth's climate system (i.e., atmosphere, oceans, land, and sea ice) and has applied this understanding to assessments of anthropogenic impacts on climate. To support such investigations, a global, community-based data infrastructure capable of managing access to and analysis of numerical model outputs, observations, and climate data reanalyses is needed. To this end, the Earth System Grid (ESG) project, funded by the U.S. Department of Energy's Office of Science, has created a federated, distributed production infrastructure for the interchange of data and information. The international organization Global Organization for Earth System Science Portals has adopted and expanded these concepts to implement a global system known as the ESG Federation, which is described in Chapter 6.

- Chapter 7, "Data-Intensive Production Grids": the high-energy physics community regularly pushes the boundaries of both physics and computer science in its experiments. The EGEE the LHC international computing infrastructure is no exception, with the goal of the facility to identify new particles and increase our understanding of matter and energy. Chapter 7 will explore the use of grid infrastructure to support these demanding environments, paying particular attention to authentication issues, communication networks, and the rapid analysis of extremely large sensor data.

- Chapter 8, "EUDAT: Toward a Pan-European Collaborative Data Infrastructure": science and research communities from a wide range of scientific fields are faced with increasingly large amounts of relevant data that stem from new sources such as powerful new sensors and scientific instruments used in analyses, experiments, and observations as well as growing volumes of data from simulations and the digitization of library resources. This chapter describes EUDAT, a new European initiative that will deliver a collaborative data infrastructure with the capacity and capability for meeting future researchers' needs in a sustainable way. Its design will reflect a comprehensive picture of the data service requirements of the research communities in Europe and beyond. This will become increasingly important over the next decade as we face the challenges of massive expansion in the volume of data being generated and preserved and the complexity of that data and the systems required to provide access to it.

1.2.3 Section 3: From Challenges to Solutions

This section is the heart of the book and describes how we expect to move into a research environment where data-intensive science is standard practice. Each of the eight chapters in this section focuses on one or more components in the data-intensive science environment and how they need to work together to provide the infrastructure required to enable community-scale scientific collaborations.

- Chapter 9, "Infrastructure for Data-Intensive Science: A Bottom-Up Approach": the distributed nature of data-intensive science requires effective and efficient data transfer capabilities in order to support international collaborations. This chapter presents a bottom-up view of the infrastructure needed to efficiently move large data sets based on the authors' years of operational experience as infrastructure providers engaged in the design, deployment, and operation of networks tailored for science applications. Networks built for science differ in several ways from those built for other applications. However, just as networks are critical for the successful operation of a modern business enterprise, so too are they critical for data-intensive science. This chapter describes the key aspects of network infrastructure that are critical for successful data-intensive science.

- Chapter 10, "A Posteriori Ontology Engineering for Data Driven Science": science has a rich tradition in categorical knowledge management. This continues today in the generation and use of ontologies. Unfortunately, the link between hard data and ontological content is predominately qualitative, not quantitative. The usual approach is to construct ontologies of qualitative concepts and then annotate the data to the ontologies. This process has seen great value, yet it is laborious and predominantly unquantified as to how well the ontologies are managing and organizing the full information content of the data. An alternative approach is the converse: use the data itself to quantitatively drive ontology creation. Under this model, one generates ontologies at the time they are needed, allowing them to change as more data influences both their topology and concept space. This chapter outlines two approaches to achieve this, the first using the mathematical approach of formal concept analysis and the second using the semantic web approach of the web ontology language.

- Chapter 11, "Transforming Data into the Appropriate Context": once relevant data is identified, it must be transformed into the right context for it to be usable. These transformations crucially rely on both the structure and the semantics of the data. Top-down solutions to this data integration problem—global ontologies, schemas, standards—are incomplete. There will always be data that is not "born into compliance" that needs to be integrated post hoc. Consider search engines: Google and Bing do not refuse to index sources that are not compliant with HTML 5.0 and neither should science refuse to consider, say, Earth science data that is not compliant with Open Geospatial Consortium (OGC) standards. This chapter explores these issues, emphasizing bottom-up, "pay-as-you-go" approaches that can provide minimal services (i.e., cataloging) over all data and more sophisticated services (i.e., structured query) over well-structured, metadata-equipped data.

- Chapter 12, "Bridging the Gap between Scientific Data Producers and Consumers: A Provenance Approach": data-intensive science often requires computational processing to occur on grids, across distributed services, and in parallel on HPC machines. This creates a challenge when recording the derivation of digital objects such as data sets produced by distributed scientific processes. However, maintaining this information is critical for ensuring the reproducibility of experiments and providing sufficient context to make the experimental results useful to others. This chapter explores the current state of the art in maintaining data provenance.

- Chapter 13, "In Situ Exploratory Data Analysis for Scientific Discovery": multidisciplinary teams must interactively engage in analysis, exploration, and experimentation against data. Exploratory analysis against heterogeneous; distributed; and, in some cases, large, complex data will require new parallel/distributed analytical methods, algorithms, and architectures. In many cases, applications can be easily parallelized by constructing indexes across massive numbers of distributed commodity processors and using informatics techniques such as MapReduce to support data-intensive computation. However, many scientific applications cannot be formulated in this way. Recent collaborative work by the scientific computing and social media communities in extending traditional mathematical and statistical tools, that is, algebraic operations, spectral methods, and numerical techniques, will be essential. As outlined in this chapter, enabling scientists to perform in situ experimentation, in which a hypothesis can be effectively tested, will require advances in data analysis to fully utilize the vast amount of multimodal data available.

- Chapter 14, "Interactive Data Exploration": sometimes, scientists are not exactly sure about what they are looking for. Instead of trying to prove or disprove a hypothesis, they are simply trying to understand the available information. In these cases, a guided interactive exploration of the data may be more useful than a predefined analysis. The sheer amount of information available requires advanced tools to identify relevant information while filtering out the rest. Although methods have been found to search certain types of information sources, for example, Google for the World Wide Web, we do not have the means to search everything at once and do so in a meaningful and timely way. This chapter explores approaches for effective interaction with these diverse data sets.

- Chapter 15, "Linked Science: Interconnecting Scientific Assets": once new insight is obtained, it needs to be effectively disseminated to the broad research community so that others may build on the results. The increasing use of digital documents means that not only can the final paper be disseminated but also the data and analysis workflows can be effectively distributed. This will greatly enhance the value of the results since the process will have much greater transparency. Since data-intensive science is inherently an in silico analysis, this dissemination must ensure scientific reproducibility, a key tenant of science. Furthermore, as observational data ceases to have a physical representation, it is critical that this information (e.g., recordings of supernovae) is available to future generations. Ensuring that this is the case is a significant challenge given the pace of technological evolution (who can still read Microsoft Word 1.0 documents?). This chapter explores approaches for sharing and preserving digital scientific records.

The short, final chapter in this book (Chapter 16) summarizes the common themes presented in the previous sections, reiterating the key messages and placing the recurrent themes in context.

REFERENCES

Ahrens, J., Hendrickson, B., Long, G., Miller, S., Ross, R., and Williams, D. 2010. Data Intensive Science in the Department of Energy: Case Studies and Future Challenges. info.ornl.gov/sites/publications/Files/Pub31432.pdf (accessed February 7, 2012).

Atkinson, M., Kersten, M., Szalay, A., and van Hemert, J. 2010. *Data-Intensive Research Theme*. http://www.esi.ac.uk/files/esi/Theme15-proposal.pdf (accessed February 7, 2013).

Bernholdt, D., Bharathi, S., Brown, D., Chanchio, K., Chen, M., Chervenak, A., Cinquini, L., et al. 2005. The earth system grid: Supporting the next generation of climate modeling research. *Proceedings of the IEEE*, 93(3): 485–495.

Flannery, D., Matthews, B., Griffin, T., Bicarregui, J., Gleaves, M., Lerusse, L., Downing, R., et al. 2009. ICAT: Integrating data infrastructure for facilities based science. e-Science'09. In *Proceedings of Fifth IEEE International Conference on e-Science*. December 9–11, 2009, Oxford, U.K: IEEE, pp. 201–207.

Hey, T., Tansley, S., and Tolle, K. (eds.). 2009. The *Fourth Paradigm: Data-Intensive Scientific Discovery*. Redmond, WA: Microsoft Research.

Koski, K., Gheller, C., Heinzel, S., Kennedy, A., Streit, A., and Wittenburg, P. 2009. Strategy for a European data infrastructure. White paper. PARADE—Partnership for Accessing Data in Europe. http://www.csc.fi/english/pages/parade (accessed February 7, 2013).

Shoshani, A. and Rotem, D. (eds.). 2009. *Scientific Data Management: Challenges, Existing Technology, and Computational Science* Series. Boca Raton, FL: Chapman & Hall/CRC Press.

CHAPTER 2

Where Does All the Data Come From?

Geoffrey Fox, Tony Hey, and Anne Trefethen

CONTENTS

2.1 INTRODUCTION

The data deluge [1] is all around us, and this book describes the impact that this will have on science. Data are enabling new discoveries using a new—the fourth [2]—paradigm of scientific investigation. This chapter provides an overview of the diverse nature of the data driving the fourth paradigm with the data's richness of size, variety of characteristics, and need for computational processing. New fields are being born from, for example, the study of tweets and the acceleration pace of changes in previously quiescent ice sheets. Other fields such as earthquake prediction cry out for new observations to improve forecasting. Areas such as genomics and the search for fundamental particles at the Large Hadron Collider (LHC) drive the main stream of the field with many petabytes of data derived from advanced instruments. The deluge and its impact are pervasive.

Data born digital comes in many sizes and shapes and from a vast range of sources—even in our daily lives, we create a massive amount of digital information about ourselves and our lives. We do this by shopping with credit cards, using online systems, social networking, sharing videos, capturing traffic flow, measuring pollution, using security cameras, and so on. In addition, in health care, we now routinely create digital medical images as part of an "electronic health record." In this chapter, after looking briefly at the explosion of digital data and devices that are impacting our daily lives, we shall focus on some examples of the scientific data deluge. We anticipate that some of the tools, techniques, and visualizations that are now being used by scientists to explore and manage their challenging data sets will find their way into business and everyday life—just as was done with creation of the web by Tim Berners-Lee and the particle physics community. In our focus on scientific data, it is important not to ignore the digital aspects of the social sciences and the humanities, which, although physical artifacts still play an important role in research, are becoming increasingly digitally driven.

Scientific research creates its data from observations, experimentation, and simulation, and, of course, the outputs of such research include files and databases as well as publications in digital form. These textual publications increasingly need to be linked to the underlying primary data and this links the discussion of everyday data including web pages in Section 2.2 with the scientific data in Sections 2.3 through 2.5. Sections 2.6 and 2.7 note that all these data are useless unless we can label, sustain, and process it. Data are naturally refined at each step of the analysis, and rather than distinguishing different levels such as observation (raw data), data, information, knowledge, and wisdom, we use the generic term "data" in this chapter.

2.2 AN EXPLOSION OF DATA IN OUR LIVES

A recent report by McKinsey Global Institute [3] indicates that there are 30 billion pieces of content shared on Facebook every month, that the U.S. Library of Congress had collected 235 TB of data by April 2011, and that 15 out of 17 sectors in the United States have more data stored per company than stored by the U.S. Library of Congress. The future only sees this content increasing, with a projected growth of 40% in global data generated by commerce and individuals per year. The McKinsey report estimates that globally, enterprises stored more than 7 exabytes (EB) of new data on disc drives in 2010, comprising data generated through interactions with a customer base and data supporting the provision of services through the Internet. Individuals stored a similarly impressive amount—more than 6 EB of new data on home and hand-held devices.

The report also shows that there are over 30 million networked sensors deployed in the transportation, industrial, retail, and utilities sectors and that this number is increasing by more than 30% per year. We will return to the issue of sensor data later as we look in more detail at the Smart Grid and oceanographic examples of data collection.

It is a truism to say that the Internet has changed everything—today our lives are often as much digital as physical. Our collaborations and friendships are as likely to be virtual as being based on real, physical, face-to-face encounters and the management of our lives—be it banking, house utilities, health, or car insurance—is increasingly dominated by networked systems and online commerce.

The August 2011 Verisign report [4] indicates that by the end of second quarter 2011, there will be over 215 million domain name registrations—an increase of 8.6% over the previous year. It notes that the largest top-level domains in terms of base size were, in order, .com, .de (Germany), .net, .uk (United Kingdom), .org, .info, .nl (Netherlands), .cn (China), .eu (European Union), and .ru (Russian Federation). Figure 2.1 shows the breakdown.

Using this information, The Next Web [5] estimates the number of pages on the web to be between 42 and 121 billion, a 21% increase from 2008.

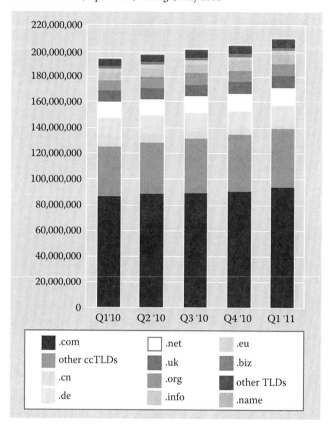

Total domain name registrations
Source: Zooknic, April 2011; Verisign, May 2011

FIGURE 2.1 **(See color insert.)** Breakdown of top-level domains over five quarters. (Verisign, The Domain Name Industry Brief, *The Verisign Domain Report*, August, 2011, 8, 3.)

There is an estimate of 45–50 billion for the indexed World Wide Web [6]. Verisign also reports that in the last decade, the number of Internet users has increased by 500%. It is notable, however, that this growth is not homogenous globally. Certain international regions are exploding in their use of the Internet. For example, a decade ago, Africa had less than 5 million Internet users; it now has more than 100 million. The report also estimates that in 2010, less than 40% of Internet users were English speaking.

Social use of the Internet is generating content constantly and in ever-increasing amounts. Search Engine Watch [7] includes some interesting statistics on YouTube, the video sharing community site. In 2010, more than 13 million hours of video were uploaded to the site; in 2 months (60 days), more videos were uploaded than had been created in six decades by the three major networks (ABC, CBS, and NBC). By May 2011, more than 48 hours of video were being uploaded per minute, and YouTube had surpassed three billion views per day.

The GreenPeace report, "How clean is your data?" [8] demonstrates that 1.2 zettabytes (ZB) of digital information has been generated by tweets, Facebook—where over 30 billion pieces of content are shared each month, e-mails, YouTube, and other social data transfers. The use of these social network and related tools is beautifully illustrated by JESS3 [9] in Figure 2.2.

Flickr, the photo-sharing site, now hosts over four billion images. These images are generally family photographs, holiday snaps, and the like. However, the increasing demand for digital image storage is a growing

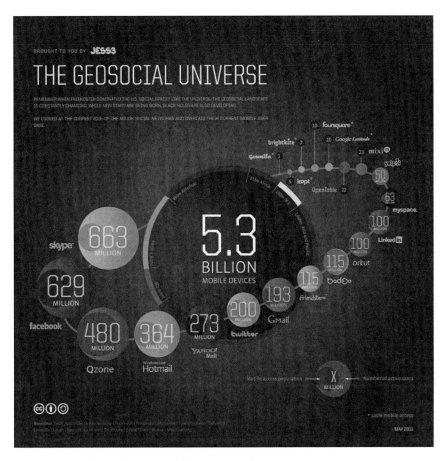

FIGURE 2.2 **(See color insert.)** The geosocial universe.

concern as we turn to health care and the medical images that are now a standard part of our health care systems. Medical images are created in many forms and for a broad range of diagnostic reasons. They include magnetic resonance imaging, digital mammography, positron emission tomography (PET), and x-ray computed tomography (CT). Together, these images amount to over 20,000 TB of data; a single CT study consists of sixty-four $512 \times 512 \times 16$ bit images and can correspond to as much as tens or even hundreds of megabytes. In the case of mammography, the size of digital data collected is approximately 200 MB per examination.

Ninety-six percent of radiology practices in the United States are film-less, and Table 2.1 illustrates the annual volume of data across the types of diagnostic imaging [10]; this does not include cardiology, which would take the total to over 10^9 GB (an exabyte).

The image repositories required to store medical images are more complex than those required to store photos on Flickr—the data are more complex, there are often diverse user access requirements, and a need to search and analyze the data in collections according to particulars of an individual or to a specific disease type. The National Cancer Institute maintains a survey of biomedical imaging archives [10], but the terabytes listed there are only the tip of the iceberg.

There are also other considerations addressing the management of medical images. In the United States, providers must meet Health Insurance Portability and Accountability Act regulations that require a data backup plan and disaster recovery plans. BridgeHead Software, that

TABLE 2.1   Types and Features of Radiology Data

Modality	Part B non-HMO (million)	All Medicare (million)	All Population (million)	Per 1000 Persons	Ave Study Size (GB)	Total Annual Data Generated in GB
CT	22	29	87	287	0.25	21,750,000
MR	7	9	26	86	0.2	5,200,000
Ultrasound	40	53	159	522	0.1	15,900,000
Interventional	10	13	40	131	0.2	8,000,000
Nuclear medicine	10	14	41	135	0.1	4,100,000
PET	1	1	2	8	0.1	200,000
X-ray, total incl. mammography	84	111	332	1,091	0.04	13,280,000
All diagnostic radiology	174	229	687	2,259	0.1	68,700,000

describes itself as a Healthcare Storage Virtualization company, found the top IT spending priorities of hospitals for 2010 were disaster recovery (44%), picture archiving and communication systems (38%), and digitizing paper records (35%) [11]. In addition, each state in the United States has its own medical record retention rules, typically for a minimum of 7 years.

Our actions in everyday life are captured by sensors of many kinds—pollution sensors in cities, cctv throughout many public places, and increasingly, by new technologies such as Smart Grids for electricity [12]. Smart Grids are a relatively new approach to energy management. A Smart Grid is generally an energy network that incorporates information technology to allow real-time management of the energy generation and distribution using two-way communication between generators and end-users. In the United Kingdom, the Department of Energy and Climate Change have set about rolling out smart meters across the United Kingdom [13] and it is anticipated that by 2019, 50 million smart meters will have been installed. The likely data generated and collected from these meters includes 12 readings per hour at 50 bytes per reading, which, together with the system monitoring tools, will create gigabytes of data each. The challenges for the Smart Grid are not only technical—how to store and analyze such data—but also raise issues of security and policy. How will the data be kept private? How long will it be kept for? How much data should be kept locally versus centrally? And of course the telecommunications networks required to implement such capabilities are not generally readily available. However, the move to Smart Grids is global and as microgeneration of electricity increases through local renewable energy schemes, the requirements for Smart Grid technologies will only increase [14].

Finally, an IDC report [15] titled "The 2011 Digital Universe Study: Extracting Value from Chaos" estimates that in 2011 the total volume of information created and replicated "will surpass 1.8 ZB (1.8 trillion gigabytes)—growing by a factor of 9 in just five years." The report estimates that 75% of this digital content will be created by individuals. The report also estimates that the number of files is growing even faster than the information itself and attributes this rapid growth to the increasing number of embedded systems now generating data.

2.3 RESEARCH DATA FROM OBSERVATIONS

2.3.1 Astronomy: The Square Kilometer Array

At the present time, hundreds of astronomers, computer scientists, and technology engineers across the globe are designing the next-generation radio telescope—the square kilometer array (SKA) [16]. It is anticipated that construction of the first phase of the telescope will begin in 2016 with the full telescope completed and in operation by 2022. The decision where to locate the SKA will be made in 2012, and it is likely be located in either Australia or South Africa, in a desert so as to have little or no interference, but will be a collaborative effort involving over 50 groups in 19 countries.

The present design [17–20] has a combination of aperture arrays in the core and up to 3000 phased array feeds on dishes giving a collecting area of approximately 1 km², with receptors extending out to a distance of 3000 km from the center of the telescope (Figure 2.3). The SKA will have a sensitivity of more than 50 times that of existing telescopes and 10,000 times the survey speed. It is intended to provide data to answer fundamental questions about gravitation and magnetism, galaxy formation, and even the question of life on other planets. The design of the SKA is developing through studies based on the science requirements and on a number of SKA "Pathfinder" projects that provide experience of design options and technology capability considerations.

The SKA provides an enormous information technology challenge with a typical data rate from each dish antenna on the order of 100 GB·s⁻¹ aggregating to over 100 TB·s⁻¹ [18] and with a need for exaflop-scale computation [21] for postprocessing. The IT infrastructure required to support the science at the SKA will range from real-time capability to transport and analyze the data at these high-data rates together with the capacity to store and "publish" the data for later analysis and interpretation by the global astrophysics community. The computational systems will likely

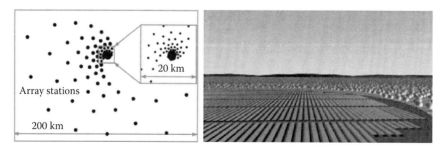

FIGURE 2.3 Possible configuration of SKA and artist's impression of the SKA core. (The Square Kilometre Array, *Precursors, Pathfinders and Design Studies* [accessed 2011 November 4], Available from: http://www.skatelescope.org/the-organisation/precursors-pathfinders-design-studies/.)

range from specifically designed field-programmable gate array–like units to exascale computing systems and Cloud data centers. The communications infrastructure will range from high-bandwidth intra-chip and inter-chip connections on a high-performance computing system to trans-oceanographic data transmission optical fibers supporting data rates of at least 100 Gb·s^{-1}. The SKA will succeed or not depending on both the physical implementation of the telescope design and the software infrastructure that will enable it. The software infrastructure required to realize this information technology challenge has been identified as >2000 person year task [22], but even this may not take full account of the complexity of the task.

2.3.2 Astronomy: The Large Synoptic Survey Telescope

The large synoptic survey telescope (LSST) [23] is the most ambitious survey currently planned in the optical part of the electromagnetic spectrum. The LSST is driven by four main science themes: probing dark energy and dark matter, taking an inventory of the solar system, exploring the transient optical sky, and mapping the Milky Way. It will be a large, wide-field ground-based telescope designed to obtain multiple images covering the sky that is visible from Cerro Pachón in Northern Chile. The current LSST design has an 8.4 m (6.7 m effective) primary mirror, a 9.6 deg^2 field of view, and a 3.2 gigapixel camera. This will allow about 10,000 deg^2 of sky to be covered using pairs of 15 second exposures twice per night, every three nights on average. The system is designed to yield high image quality as well as high astrometric and photometric accuracy. The total survey area will include 30,000 deg^2 and will be imaged multiple times in six bands covering the wavelength range 320–1050 nm. The project is scheduled to begin the regular survey operations before the end of this decade. About 90% of the observing time will be devoted to a deep-wide-fast survey mode that will uniformly observe an 18,000 deg^2 region about 1000 times (summed over all six bands) during the anticipated 10 years of operations. These data will result in databases including 10 billion galaxies and a similar number of stars and will serve the majority of the primary science programs [24,25].

In terms of numbers, LSST will handle 15 TB of raw scientific image data each night. The final image data archive is estimated to have around 200 PB of data. It is estimated that the project will require a sustained petaflop/s computing capability and, of course, significant local processing power in Chile and very high bandwidth connection to the U.S.

archive site. The project plans to use SciDB, a novel open source database system that is optimized for scientific data management of "big data" and for "big analytics" [26].

2.3.3 Earth Observation Data

NASA's Earth Observing System Data and Information System (EOSDIS) [27] manages and distributes data products through its Distributed Active Archive Centers (DAACs). Each DAAC processes, archives, documents, and distributes data from NASA's past and current research satellites and field programs. Each center serves one or more specific Earth Science subdiscipline and provides appropriate data products, data information, and services for its community. In Europe, the European Space Agency (ESA) [28] plays a similar role to EOSDIS and oversees an Earth observation program. For EOSDIS, the growth in the archive is around 1 PB/year. For ESA, including missions such as Envisat, Cryosat, and AATSR, the data volumes are comparable. Space missions from Japan and India contribute around 0.5 PB/year and aircraft missions, including LIDAR, probably account for another 0.5 PB/year. New missions, still in the planning stage, that are expected to launch in the next 5–10 years are likely to generate another 1 PB/year.

If we add up all these sources, we arrive at around 4 PB/year for Earth observation data, depending on the precise definitions (i.e., storage or distribution). For example, EOSDIS distributes 3.62 PB/year now—made up of 4200 data products, >400 million product distributions, and over one million users.

2.3.4 Oceanographic Data

Present ocean sensors tend to generate relatively low volumes of data and typically do so intermittently—during cruises or delivery to shore-side systems. All told, this probably only amounts to tens of gigabytes per day. However, there will be a transformational change with the deployment of real-time ocean observatories.

The Ocean Observatory Initiative (OOI) [29] is a long-term, National Science Foundation-funded program to provide 25–30 years of sustained ocean measurements to study climate variability, ocean circulation and ecosystem dynamics, air–sea exchange, seafloor processes, and plate–scale geodynamics [25]. The OOI will enable powerful new scientific approaches for exploring the complexities of Earth–ocean–atmosphere interactions and accelerate progress toward the goal of understanding, predicting, and managing our ocean environment. The Observatory is planned to be a networked infrastructure of science-driven sensor systems to measure the physical, chemical, geological, and biological variables in the ocean and seafloor. When complete, the OOI will be one fully integrated system collecting data on coastal, regional, and global scales. As a result, the data volumes are expected to increase dramatically and transform ocean science from being a data-impoverished branch of science to one having an abundance of data.

2.3.5 Earthquake Science

Fortunately, for society if not science, large earthquakes are infrequent and so the study of earthquakes is observational data-limited compared to other fields. Major quakes occur all over the world, and it is unrealistic to have substantial sensor networks deployed in most of these regions. Further, the quasiperiodicity of earthquakes implies that historical data is very important and we cannot increase that. Simulations can forecast damage and perhaps the aftershocks of an earthquake, but the most important capability—forecasting new quakes—is essentially entirely observational. Typically, one uses patterns (in time series) to forecast the future with simulations useful to check if a particular informatics approach is valid in an ensemble of simulated earthquakes. Important types of data include the following:

1. Catalogs of earthquakes with position and magnitude

2. Geometry of earthquake faults

3. Global positioning data (GPS) recording time-dependent positions

4. Synthetic aperture radar inferograms (InSAR) recording changes in regions over time

The first two types of data are gathered carefully with the recording of earthquakes and associated field analysis. This is small in size and only growing slowly but clearly of very high value.

For the GPS data, there are currently fewer than 10,000 GPS stations recording data at intervals varying between second and a day. Well-known GPS networks are the Southern California Integrated GPS Network, the Bay Area Regional Deformation Network in Northern California, and the Plate Boundary Observatory from University Navstar Consortium (UNAVCO) [30].

The InSAR data could become voluminous but currently totals some 350 images (each covering around 10,000 km^2) and is only 2 TB in size [31]. These data comes from uninhabited aerial vehicles (Uninhabited Aerial Vehicle Synthetic Aperture Radar [32] from Jet Propulsion Laboratory) or satellites (WInSAR from UNAVCO [30]). The situation could be revolutionized by the approval of the Deformation Ecosystem and Dynamics of Ice–Radar (DESDynI-R) Mission recommended in the Earth Science Decadal Survey [33]. DESDynI would produce around a terabyte of data per day, but the mission has not so far been approved and so is many years away from a possible launch. These InSAR data are analyzed (as by QuakeSim [34,35] for recent earthquakes) to find rates of changes, which are then used in simulations that can lead to better understanding of fault structures and their slip rates.

2.3.6 Polar Science

Another interesting case is polar science, which we illustrate with the work of the Center for Remote Sensing of Ice Sheets (CReSIS) [36] led by Kansas University that is pioneering new radar and unmanned autonomous vehicles (UAVs) to be used to study ice sheets [37].

For the project, multiple expeditions fly instruments that collect data including the following:

1. Ice thickness and internal layering from radar and seismics, and SAR images of ice-bed interface

2. Bed topography generated from ice thickness and surface elevation

3. Time series of change in surface elevation from airborne and satellite altimeters

4. Time series of surface velocity from repeat-pass satellite images, in situ GPS measurements, and aerial photos

5. Bed characteristics such as temperature, wetness, and sediment from seismics and radar

The spring 2011 CReSIS expedition took 80 TB of data in 2 months. After traditional processing with fast Fourier transforms, radar images are produced along multiple flight lines, as illustrated in Figure 2.4. Then, image processing is needed to identify the top and bottom of an ice sheet. Initially students performed this, but recently it has been automated with an image analysis tool developed at Indiana University [38]. The deployment of UAVs rather than current Orion and DC-8 conventional aircraft will increase data gathering capability by allowing continuous operation. There are also more complex data such as snow deposits showing the annual layers and revealing historical snow deposition.

The glacier-bed data illustrated in Figure 2.5 is fed into simulations that aim to understand the effect of climate change on glaciers. Note that

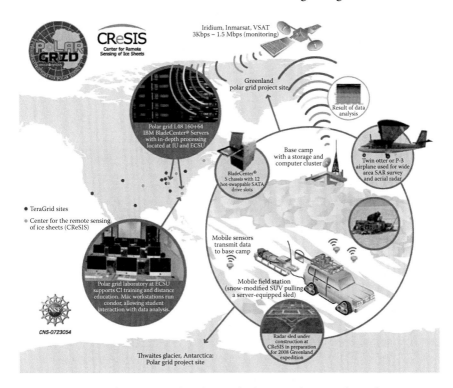

FIGURE 2.4 Architecture of PolarGrid data analysis Cyberinfrastructure. (PolarGrid NSF MRI funded partnership between Indiana University and Elizabeth City State University to acquire and deploy the computing infrastructure needed to investigate urgent problems in glacial melting. *Home Page* [accessed 2010 March 1], Available from: http://www.polargrid.org.)

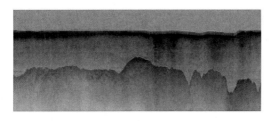

FIGURE 2.5 Radar imagery from CReSIS with top and bed of ice sheet determined.

gathering of data is complicated by the paucity of electrical power and poor Internet connectivity to the polar regions. The use of graphics processing unit is an interesting technology to potentially deliver lower power data processing. The data are gathered on removable discs mounted in a storage array connected to just one or two servers with rugged laptops as personal machines.

2.4 DATA FROM EXPERIMENTATION AND INSTRUMENTS

2.4.1 Particle Physics: The Large Hadron Collider

The LHC in Geneva is the highest energy particle accelerator ever constructed and operates in a 17 mi. tunnel around the CERN Laboratory near Geneva [39]. Two proton beams can be accelerated to energies of up to 7 TeV and collided to produce a spectacular spray of particles. A major goal is to find a key ingredient of the Standard Model of particle physics, the Higgs Boson. The LHC is currently operating at half of its design energy and plans to go to the full energy of 7 TeV per beam in 2014.

The LHC hosts four major experiments: Atlas, CMS, Alice, and LHCb. The first two experiments each record around 100 events per second with each event about 1.5 MB in size. These 100–450 events are selected in real time from the eventual 10^9 collisions (events) occurring every second at LHC. The experimental detectors contain 150 million sensors that record data 40 million times per second (each read out contains over 20 overlapping events). The reduction of a factor of 4×10^5 in data size is achieved with a multistage trigger [40,41]. Having an effective trigger is a major part of design and selection of an experiment. The trigger is based on detecting "unusual events" with signatures of high transverse momentum and interesting particles (leptons rather than baryons or mesons) being produced. The multistage trigger includes an initial hardware selection (giving a factor of about 400) followed by a software refinement executing on a dedicated cluster, which for CMS has 7000 cores. The software used in this final "higher level trigger" is a stripped down version of the basic analysis software and must reduce the Terabit/second input from the hardware trigger by about another factor of 1000. Alice is a heavy ion experiment to investigate collisions of lead nuclei in the LHC and has larger events and data rates. The LHCb experiment is lower in both respects than Atlas and CMS.

The LHC produces some 15 PB of data per year of all varieties with the exact value depending on duty factor of accelerator (which is reduced simply to cut electricity cost but also when there is malfunctioning of one or more of the many complex systems) and the experiments. The raw data produced by experiments are processed on the LHC Computing Grid [42], which has some 200,000 Cores arranged in a three-level structure. Tier-0 is CERN itself, Tier 1 are national facilities, and Tier 2 are regional systems. For example, one LHC experiment (CMS) has 7 Tier-1 and 50 Tier-2 facilities [43,44].

The initial data are analyzed in detail to find the parameters of the particles produced in the event and to disentangle the ~20 collisions in each event [45]. This analysis is often iterative as one improves the many calibration constants for the myriad of detector sensors. Detailed summaries of each event or reconstructed data are produced, which yields about half the size of the raw data, that is, ~0.75 Mb, with this process taking an average of around 15 minutes for each event. The experiments also create simple "analysis object data" (AOD) that provides a trade-off between event size and complexity of the available information to optimize flexibility and speed for analyses. An AOD (~0.1 MB) is 5% of size of the raw data but with enough information for a physics analysis including this event. The other 95% of raw data would be preserved elsewhere as it would be necessary if, for example, the physics quantities were to be recalculated with a reinterpretation or recalibration of the raw data. Finally, there are tags, about 2 kB per event that have enough information to select events for a physics analysis that could be performed with the AOD containing more details.

This analysis chain from raw data → reconstructed data → AOD and TAGS → Physics is performed on the multitier LHC Computing Grid. Note that every event can be analyzed independently so that many events can be processed in parallel apart from some concentration operations such as those to gather entries in a histogram. This implies that both Grid and Cloud solutions work with this type of data with currently Grids being the only implementation.

2.4.2 Photon Sources

The European Synchroton Radiation Facility in Grenoble uses x-ray radiation to study fields as diverse as protein crystallography, earth and materials science, as well as certain areas of physics and chemistry [46]. The beamlines are planned to be upgraded to increase the present rate of data production, currently around 1.5 TB per day, by two or three orders of magnitude in 10 years.

The European X-Ray Free Electron Laser Project (European XFEL) is an international project with 14 participating countries that is located near Hamburg in Germany [47]. Free electron lasers generate high-intensity electromagnetic radiation by accelerating electrons to relativistic speeds in a linear accelerator. By 2015, the European XFEL will produce high-intensity x-ray pulses at intensities much brighter than those produced by conventional light sources. The data rates and data volumes generated by advanced facilities such as the European XFEL and the LCLS at Stanford in the United States [41] will exceed those at conventional synchrotron light sources by at least an order of magnitude. Data rates are likely to be of the order of 7 TB per hour, depending on the experiment.

These futuristic projections need to be tempered with reality. At present, the UK Diamond Light Source is storing approximately 200 TB in 88 m files, increasing to a petabyte by 2014 [48]. For these facilities, it is often the number of files that is a challenge, rather than the total data volume.

2.4.3 Neutron Scattering and the Long Tail

ISIS is the UK's national neutron scattering facility, which draws several thousand visiting scientists a year from the United Kingdom and all over the globe to do experiments ranging from the safety of welds in aircraft engines to magnetic domains on hard discs, the provenance of ancient weapons, the structure and interactions of drugs, and the design of shampoos. There are over 25 different instruments at ISIS, which support different types of experiments generating up to a few terabytes of raw data in a day, with the facility running 100–200 days a year and producing in total up to a quarter of a petabyte of data in a year.

The large quantity of raw data does not necessarily need to be moved very far, however. Consider one instrument, the SANS2d, which is a small-angle scattering instrument. Neutrons are transported to the sample, where they are scattered at small angle and measured. One experimentalist [49] actually uses a white beam (neutrons from ~1.5 to 16 Å) and a time of flight mode with a pulsed source. Small-angle scattering gives information about distance correlations that are "large" on the atomic scale, that is, from about 8 to 400 Å. This class of experiments determines the structure of medium- to large-size biomolecules. This use of SANS2d can generate a gigabyte of raw data in a day, but these data largely record details that are not important for downstream analysis but are only important in the reduction of data down to the X–Y form in which it is normally used. Raw data files amounting to hundreds of megabytes are reduced to a single file of a few tens or maybe hundreds of kilobytes, and it is only that large because it is stored as a text file. The actual core data are probably less than a kilobyte. In the data information knowledge wisdom (DIKW) terminology, the information, I, is this final kilobyte.

ISIS has in-house and visiting users. An experiment, usually lasting a few days, will involve running hundreds of samples, which in turn generate hundreds of data files. These are transferred, usually via a pen drive, to a visiting scientist's laptop to be taken home for further analysis. The in-house experimentalist [49] has, therefore, accumulated perhaps 50 GB of files in tens of directories, each relating to a specific experiment or project. The analysis process usually involves taking each of the files from the experiment, fitting them to a model (or using the data to generate a model), often manually or in a semiautomated fashion. This probably doubles the quantity of files and volume on the local hard disc. External storage or additional compute capacity is rarely needed.

This analysis will probably get summarized in a table, which is then used to draw conclusions about how the experimental parameters affected the system being studied. The composition of a mixture, temperature, and pH are all common types of variables that are then graphed out and the changes compared to some physical theory. Alternately, a whole series of measurements might be used to generate one model, generally a structure, of the system under study. This often involves a lot of exploration of different methods for working with the data, building test models, and comparing the results from a range of different methods.

Ultimately, the results of the experiment are usually condensed into one figure, a graph or a model structure, that is presented in a published paper. The data are rarely if ever made available beyond a representation in some graphs, and the history—the provenance—of the analysis is almost never recorded. The replication of results would rely on obtaining the data from the authors and refitting the models suggested to it. In practice, this almost never happens, and critique of results is generally based on design of experiments rather than the analysis itself.

This example illustrates the "long tail science" [50,51]. This captures the reality that there are not, as in the LHC experiments, thousands of scientists collaborating on a single "big science" experiment but rather many thousands of individual scientists each doing their own experiment. Nevertheless, the total data involved for such "long tail science" approaches the petascale.

2.4.4 European Bioinformatics Institute

Life science is one of the best examples of the data deluge and we start off giving examples from the European Bioinformatics Institute (EBI), which, within Europe, is the primary host for bioinformatics data, and curates and shares data from throughout Europe and beyond [52]. EBI is an academic research institute located in Hinxton near Cambridge (United Kingdom) and is part of the European Molecular Biology Laboratory (EMBL) [53]. The EBI annual report [54] shows the status of the several databases within the EBI. For example, the European Nucleotide Archive had an accumulation rate of more than 500,000 bases per second. The data in Figure 2.6, taken from the annual report [54], show the growth across other databases hosted at the EBI and the story is much the same.

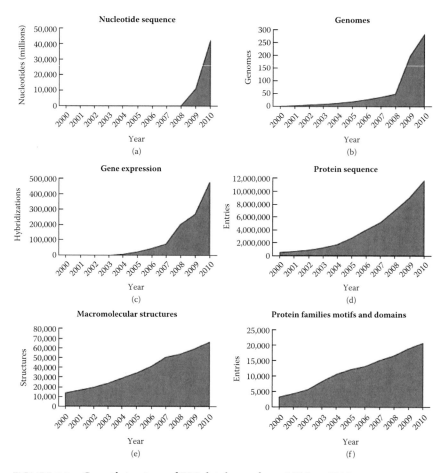

FIGURE 2.6   Growth in sizes of EBI databases from 2000 to 2010.

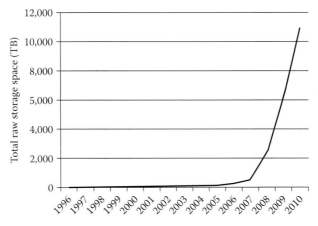

FIGURE 2.7   Storage requirements of the EBI from 1996 to 2010.

The increase in physical storage requirements of the Institute from 1996 to 2010 is shown in Figure 2.7. The EBI, as part of the Ensembl project, is now presenting data and services in the Cloud [55]—in this case, in the Amazon Cloud.

2.4.5 High-Throughput Gene Sequencing

The discussion of radiology in Section 2.2 shows that the life science data deluge impacts both research and our lives. Genomics is a field that directly spans this classification and here we look at the future growth of genomic data by studying the expectation that genome measurements for individuals will become an integral part of personal medicine. We will use one current sequencing instrument, the Illumina HiSEQ [56] machine, to quantify the analysis [57]. Each of these instruments generates 10^8 reads, each roughly 100 nucleotides long, each day. Each nucleotide is defined by 2 bits. It takes 100–10,000 cores to use the Blast algorithm to compare with data from a central database in one day. Each read is distilled from a coverage of 50–100 times as much data including duplicates.

If we take a unit as a human genome with 3×10^9 nucleotides or 6×10^9 bits, we can see more graphically the implications of this rate of data production. Each day one Illumina machine sequences 10^{10} nucleotides, equivalent to 3.3 human genome units per day. If we assume that today there are around 1000 Illumina machines deployed worldwide (500 in the United States), each capable of sequencing 3300 human genomes per day, then this amounts to 2×10^{13} bits per day corresponding to about 7 petabits of data per year (700 petabits per year including the extra factor of 100 from duplication of the sequence fragments that need to be included in the alignment process).

Measuring the genome of every new born would constitute ~11,000 human genomes per day for the United States and 200,000 human genomes per day for the world. Doing this on an ongoing basis—say 50 times in the lifetime of every human—yields 5×10^6 genomes measured per day for the world. This is 30 petabits per day or 10 exabits per year.

Thus, we see that it would require the capability equivalent to 1.5 million present-day Illuminas to measure the total human genomic data and 1.5×10^8 to 1.5×10^{10} continuously running present-day computational cores to perform a simple Blast analysis on the generated data [58–60]. Genomic data are notable for the intense computing effort associated with the data. This aspect is highlighted by the National Institutes of Health (NIH) observation that the cost of generating sequences has decreased over a factor of 100 more than the cost of computing over the last 3 years [61]. Note that NIH recently announced closure of a petabyte database [62] as they could not support it. Thus, building scalable computing and storage infrastructure for genomics is challenging.

2.5 DATA FROM SIMULATIONS

2.5.1 Data from Weather and Climate Simulations

At a September 2008 meeting involving 20 climate modeling groups from around the world, the World Climate Research Programme's [63] Working Group on Coupled Modelling [64] agreed to promote a new set of coordinated climate model experiments. These experiments comprise the fifth phase of the Coupled Model Intercomparison Project (CMIP5) [65]. CMIP5 will notably provide a multimodel context for the following:

1. Assessing the mechanisms responsible for model differences in poorly understood feedbacks associated with the carbon cycle and with clouds

2. Examining climate "predictability" and exploring the ability of models to predict climate on decadal time scales

3. Determining why similarly forced models produce a range of responses

The CMIP5 will involve the global production and analysis of several petabytes of data. The Program for Climate Model Diagnosis and Intercomparison [66], with responsibility for archival for CMIP5, has established the global "Earth System Grid Federation" (ESGF) [67] of data producers and data archives to support CMIP5. ESGF will provide a set of globally synchronized views of globally distributed data—including some large cache replicas, which will need to persist for (at least) several decades. ESGF will also stress the international networks, as well as the data archives themselves—but significantly less than would have been the case of a centralized archive. Developing and deploying the ESGF has exploited good will and best efforts, but future developments are likely to require more formalized architecture and management.

ESGF was born out of a number of initiatives to handle diverse, distributed data access for the climate community: In the United States, the Earth System Grid (ESG [68]); in the United Kingdom, the NERC DataGrid [69]; and in Germany, the Collaborative Climate Community Data Processing Grid (C3-Grid [70]). However, the dominant contribution has been that of the ESG. As a consequence, the ESGF architecture is currently a more mature version of the original ESG, extended and modified by both the code and experiences of the other partners.

There are five key information classes that underpin the ESGF [71–73]: the data themselves; the "data metadata," which exist within the data files; the "model and experiment metadata" created externally and ingested into the ESGF system; the "quality metadata" (which describes intrinsic checks on data fidelity rather than the extrinsic scientific quality); and "federation metadata" (to support user management and system deployment).

ESGF exploits this information using four major components: data nodes, gateways, federation metadata services (to support authentication and authorization), and data services to be deployed adjacent (or on) the data nodes.

2.5.2 Data from Petascale and Exascale Simulations

In studies [74–76] of the requirements of applications in exascale computing including high-energy physics, climate, nuclear physics, fusion, nuclear energy, basic energy sciences, biology, and national security, it was reported that exascale applications will generate terabytes of data per second making them one of the largest future sources of data. For exascale computing, this means a new design for memory, I/O, operating system, and software systems. It is likely that exascale simulations will behave as observational science does in terms of reducing the data created before seeking to store any results. This will require a new integrated pipeline from data creation to off-line storage.

These simulation data are produced for two important reasons. One is to provide checkpointing for restart and the second is for visualization and analysis of simulation results. As machines grow in performance and number of cores, the data produced by simulations naturally scale in size. However, the associated challenges grow even more since the mean time between failures (MTBF) of total system is reduced and the compute performance of high-end supercomputers will grow much faster than the disc I/O bandwidth. There have been several studies of these issues recently [77,78] as part of an examination of next-generation exascale systems. The latter will have up to a billion concurrent processes (perhaps arranged as a thousand threads on each of a million nodes) compared to largest simulations today on hundreds of thousands of cores. A new area of study is emerging of processing simulation data in parallel on the nodes of a supercomputer, and the Adaptable IO system is an interesting approach [79–81].

A study of fusion simulations [82] identified need to output 2 GB of data per simulated time step for each core in the parallel simulation. For "just" a million cores, this corresponds to 2 PB of data per time step requiring an aggregate I/O rate of 3.5 TB/second for a 10-minute time step with a simulation of one billion cells and one trillion particles. An exascale simulation might be 100 times this rate. These data rates are clearly much larger than those associated with observational data although checkpoint data, for example, do not need to be stored in perpetuity and can be overwritten, and perhaps visualization data will be analyzed in place (by parallel algorithms on same nodes as the simulation) and reduced in size before permanent storage.

2.6 DATA CONTEXT AND CURATION

Data without context are of little or no value. It matters where data have come from and how they has been processed up to that point. This is the "data provenance" that must be associated with the data for it to be usable by other researchers. Digital information is being generated in large quantities each day and, depending on the source, the data come with a variety of characteristics and issues—ranging from questions about the appropriateness of the metadata and semantics to describe it to the integrity and completeness of the data. Finally, all data come with a cost for keeping it and pragmatic choices must be made about what data to keep and for how long—since it is impractical to retain all the data we are now generating. In some fields, such as particle physics and astronomy, the raw data rates are now so large that only a selection of the data can be retained for future analysis.

The preparation of data for possible reuse and preservation is the process of "data curation." This includes such things as data-cleansing, to check the integrity of the data, and the adding of metadata—data about the data—to document what the data are, how the data have been collected and what format has been used, and so on, and perhaps adding higher-level annotations or semantic information by using ontologies or community-agreed vocabularies. All of these things are necessary to facilitate not only the "findability" of data by search engines but also to allow the possibility of reuse and the creation of meaningful "scientific mashups" of different data sets. Unfortunately, scientists are only human and the process of adding value to their data sets for others to use often receives scant reward. It is, therefore, not surprising that, all too often, the data curation and preservation procedures used will be imperfect or inadequate and that some important data will end up effectively being lost to future researchers. At least a part of our future occupation will likely be something like a "digital archaeologist"—trying to make sense of old data by piecing together the fragments of an inadequately documented historic past.

An interesting cautionary tale is that of the digital Domesday Book project in the United Kingdom [83,84]. After William the Conqueror won the Battle of Hastings and took control of what later became the kingdom of England, he decided to take a census of his new realm. The result was the Domesday Book completed in 1086, which can still be seen in the National Archives in London. In 1986, 900 years after the original census, the BBC produced a television program to celebrate this anniversary. One result of this project was an interactive video documentary implemented for the BBC microcomputer. At the launch of the UK Digital Preservation Coalition in London in 2002, broadcaster Lloyd Grossman called attention to the danger that the rapid evolution of computer media and recording formats would lead to the irretrievable loss of valuable historic records, and he gave the Domesday Book video-discs as an example. His remark sparked a heroic effort by many people to rescue the digital images, text, and video of the BBC Domesday interactive-video project. Eventually, after a lot of work by some exceptionally dedicated individuals, a version was produced in modern formats and that works on a modern PC.

2.7 DATA ARCHITECTURE FOR E-SCIENCE

The data deluge is changing the nature of computing in science and the architecture of systems designed to support it. Several important parameters differ between systems designed to support data analysis and those aimed at simulation. These include the following:

- The ratio of disc (I/O) bandwidth to instruction execution rate (Amdahl's I/O number [85,86]).

- The bandwidth and connection between the source of data and the computing system.

- The nature of data—size and dynamic structure. Is it an instrument or sensor generating a time series or a repository for simulation data?

A traditional computer system is often organized as shown in Figure 2.8.

Figure 2.8 shows a three-level data hierarchy with typically temporary data stored on cluster nodes, a shared set of files, and a backend archival storage. The shared files are shown in figure as either managed by computers in hosted storage or dedicated (SAN/NAS/etc.) storage. The shared file

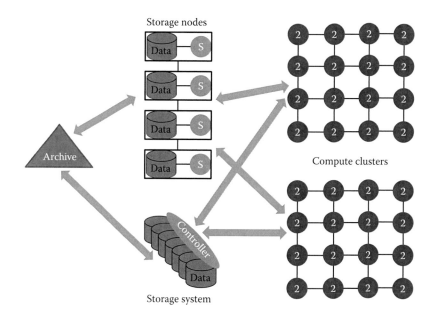

FIGURE 2.8 Computing system with a three-level storage hierarchy supporting multiple clusters where "the work" gets done. Each computer C has its own local disc not shown. S represents a storage noted.

system for scientific computing may support high-performance distributed file systems such as Lustre or GPFS. This architecture has been used for both data- and simulation-intensive work with good success. There are many attractive features of this architecture including the separation of concerns—storage and its backup are managed separately from the possibly large number of clusters supported, computers and storage can be separately upgraded, and a single storage system (and single copy of a data item) supports all computing venues. There is an obvious problem in data-intensive applications that the bandwidth between the computer and the data system components may be too small. Note that clusters typically have bisection bandwidths that are very large and scale up with system size. However, the link between the storage and the compute subsections is typically implemented using a static number of interconnects (perhaps some number of gigabit or 10 gigabit Ethernet connections). Even simulation systems face the same issues [77,78] at the largest scales when programs output data (for visualization) at volumes that overwhelm the connection to shared storage. Note that important technologies such as MPI-IO are built around this model.

An alternative architecture shown in Figure 2.9 addresses this issue by using data parallel file systems (DPFS) such as Google File System (MapReduce) [87], HDFS (Hadoop) [88], Cosmos (Dryad) [89–91], and Sector [92] with compute-data affinity optimized for data processing. This design was motivated by Internet applications but has seen little practical use outside that area.

Here, we have a simpler architecture with a uniform array of computer nodes with (large) local discs. User files are broken up into blocks, which are replicated several times and spread across different nodes and clusters. This architecture allows one to support "bringing the computing to the data" [2,93]. Archival storage is not necessary—all copies can be stored on spinning discs. The copies should be designed that some are near each other to support local computing, whereas at least one should be "far off" to provide a safe backup. Note that the discs and compute nodes within a cluster are linked to the scalable cluster interconnect and so good performance in fetching data from disc does not require the computing to be on the node where the relevant data are stored, but rather on a node with a high-performance (cluster) interconnect to data.

The architecture of Figure 2.9 supports several data management systems including both the important NOSQL developments [94,95] with constructs such as Bigtable [96], SimpleDB [97] in Amazon, and Azure

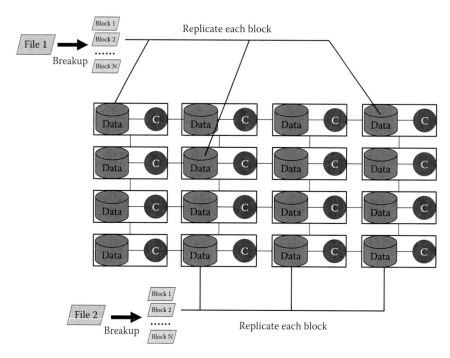

FIGURE 2.9 Data parallel file systems showing discs attached to compute nodes with files broken into blocks and stored across multiple computers with replications for fault tolerance.

Table [98] and the databases such as the SciDB and GrayWulf projects [26,86], which essentially collocate data and associated processing. NOSQL technologies emphasize distribution and scalability while their support of simple tables is interesting given that tables are clearly important in science as illustrated by the VOTable standard in astronomy [99] and the popularity of Excel [100]. However, there does not appear to be substantial experience in using tables outside clouds. It seems likely that tables will grow in importance for scientific computing, and academic systems could support this using two Apache projects: Hbase [101] for BigTable and CouchDB [102] for a document store. Another possibility is the open source SimpleDB implementation M/DB [103].

There are research issues both in data parallel file systems themselves and in their integration with programming models [104] and runtimes [105] such as MapReduce [106] and its iterative extensions [107–109]. A key change—illustrated by the Purlieus [110] project—is that the scheduling problem now is one of both data and computing rather than the usual approaches that just schedule computing tasks. Here, there is an important issue about locating user files. In the architecture of Figure 2.8, one only needs to place files in the shared file system to allow access for all clusters and applications. In Figure 2.9, one has to be more careful and place the data on or "near" (in terms of scalable connectivity) the compute systems that will be allocated to users of the files. There does not seem any experience in supporting the architecture of Figure 2.9 in a complicated heterogeneous multiuser environment. The problem is easier if one just has a few clusters as is, in fact, used in clouds for Internet search. Data parallel file systems on a grid of many (small) clusters seem difficult to use. More research is clearly needed here on multiuser environments in real data centers using data parallel file systems with multiple clusters.

Although DPFS originated in the cloud (Internet search) arena, commercial clouds tend to use a rather different object store model as seen in Amazon, Azure, and the open source OpenStack system. Here, one assigns a set of nodes to be storage servers as in the top of the middle layer of Figure 2.8, but rather than the full distributed file semantics of Lustre, one supports a simple object model. Objects have containers and metadata with operations such as get, put, update, delete, and copy objects. Again, there is little experience with this in scientific computing arena. The Simple Cloud APIs [111] for file storage, document storage services, and simple queues could help in providing a common environment between academic and commercial clouds. There is also some interesting work [112] involved relating these different file systems, so one can run applications—possibly with performance degradations—however, the data are stored. This is closely related to data movement as one needs to change between storage modes and possibly use a replica system as one does for data grids [113].

A traditional approach to scientific data establishes a repository that stores the data and metadata of a given experiment or set of experiments. This has and will play a critical role but is often inadequate for the common case of enormous amounts of data requiring enormous computing as emphasized in Section 2.5. We really need the data archives to be attached to an appropriate compute resource as it is often impractical for individual researchers to download their data to home compute resources.

2.8 CONCLUSIONS

Data are almost everywhere, large and growing. We have given examples of many sources of data: related to people—as in the web or medical data—scaling with the seven billion population of the planet; Big Science instruments such as the LHC with a few experiments or other, smaller scale instruments that support long tail science with a multitude of independent scientists; other examples such as genomics lie somewhere in between these two extremes. Whether it is the pleasing parallel Internet of individual web servers or the concentrated electronics of a giant telescope, the data deluge is only possible because of Moore's law—the electronics needed to gather and process data is continuing to get smaller and more powerful. Since this data must be transported to be useful, the ever increasing intra-planet communication bandwidth is an essential driver of the data deluge and the fourth paradigm of its analysis [2].

We have also emphasized that data (interpreted broadly as any component of the DIKW pipeline) come from many sources. Figure 2.10 shows that data are passed through a set of (filter) services as they go along the DIKW pipeline. As in Section 2.5, the source of data need not be an instrument or sensor; it can be a supercomputer or, in fact, any service, grid, or cloud. Section 2.6 emphasizes the importance of metadata and sustainability. In Section 2.7, we emphasized the need to reexamine the architectures used to support data-intensive science with distributed dynamic

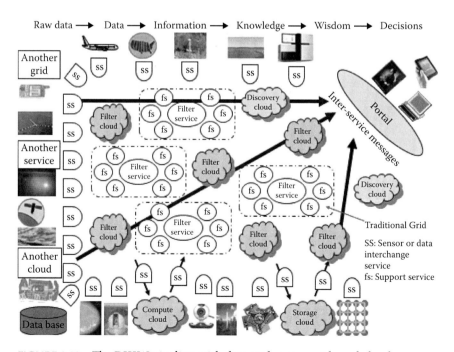

FIGURE 2.10 The DIKW pipeline with data and sensor grids and clouds.

data and major compute tasks associated with the data analytics. How do we bring the computing to the data?

Data have arrived; we need to learn how to use them for the benefit of the society.

ACKNOWLEDGMENTS

The authors thank Mark Abbott, Jeff Dozier, David Giaretta, Neil Geddes, Bryan Lawrence, Cameron Neylon, and Tony Tyson for their useful input and advice.

REFERENCES

1. T. Hey and A. Trefethen. The Data Deluge: An e-Science Perspective, Chapter 36. In *Grid Computing: Making the Global Infrastructure a Reality*, Fran Berman, Geoffrey Fox, and Tony Hey, John Wiley and Sons, Ltd. Chichester, U.K.

2. J. Gray, T. Hey, S. Tansley, and K. Tolle. *The Fourth Paradigm: Data-Intensive Scientific Discovery*. 2010 [accessed October 21, 2010]; Available from: http://research.microsoft.com/en-us/collaboration/fourthparadigm/.

3. McKinsey Global Institute. *Big Data: The Next Frontier for Innovation, Competition, and Productivity*. 2011 May [accessed November 3, 2011]; Available from: http://www.mckinsey.com/mgi/publications/big_data/index.asp.

4. Verisign. The Domain Name Industry Brief. *The Verisign Domain Report*, August, 2011. 8(3). [accessed January 19, 2013]; Available from: http://www.verisigninc.com/assets/domain-name-report-august2011.pdf.

5. TNW The Next Web. *24 Editors & Writers Across Europe, Australia, Middle East, Africa, Asia and North & South America: One of the top 10 Most Influential Blogs in the World*. [accessed November 3, 2011]; Available from: http://thenextweb.com/.

6. *Daily Estimated Size of the World Wide Web*. [accessed November 4, 2011]; Available from: http://www.worldwidewebsize.com/.

7. Search Engine Watch. *New YouTube Statistics: 48 Hours of Video Uploaded Per Minute; 3 Billion Views Per Day*. [accessed November 4, 2011]; Available from: http://searchenginewatch.com/article/2073962/New-YouTube-Statistics-48-Hours-of-Video-Uploaded-Per-Minute-3-Billion-Views-Per-Day.

8. Greenpeace International. *How Dirty Is Your Data? A Look at the Energy Choices That Power Cloud Computing*. April 2011.

9. JESS3. *A Creative Agency that Specializes in Data Visualization*. [accessed November 3, 2011]; Available from: http://jess3.com/.

10. NIH National Cancer Institute NCI. *CIP (Cancer Imaging Program) Survey of Biomedical Imaging Archives*. [accessed November 4, 2011]; Available from: https://wiki.nci.nih.gov/display/CIP/CIP+Survey+of+Biomedical+Imaging+Archives.

11. BridgeHead Software. *Data Management Healthcheck Survey Reveals Health IT's Top Spending Priorities.* June 14, 2010 [accessed November 4, 2011]; (BridgeHeadSoftwareistheHealthcareStorageVirtualization(HSV)company) Available from: http://www.ehealthserver.com/research-and-development /535-data-management-healthcheck-survey-reveals-health-its-top-spending -priorities.

12. AoL Energy. *Keeping the Lights on: Data Overload and Its Impact on Storage in the Smart Grid.* October 7, 2011 [accessed November 4, 2011]; Available from: http://energy.aol.com/2011/10/07/keeping-the-lights-on-data-overload-and -its-impact-on-storage-i/.

13. UK Department of Energy and Climate Change. *Smart Metering Implementation Programme.* March 2011 [accessed November 4, 2011]; Available from: http://www.decc.gov.uk/assets/decc/Consultations/smart -meter-imp-prospectus/1475-smart-metering-imp-response-overview.pdf.

14. UK Office of Gas and Electric Markets ofgem. *Smart Metering Implementation Programme: Data Privacy and Security.* July 27, 2010 [accessed November, 4 2011]; Available from: http://www.ofgem.gov.uk/e-serve /sm/Documentation/Documents1/Smart%20metering%20-%20Data%20 Privacy%20and%20Security.pdf.

15. IDC. *The 2011 Digital Universe Study: Extracting Value from Chaos.* June 2011 [accessed November 4, 2011]; Available from: http://www.emc .com/collateral/demos/microsites/emc-digital-universe-2011/index.htm.

16. The Square Kilometre Array. *Home Page.* [accessed November 4, 2011]; Available from: http://www.skatelescope.org/.

17. The Square Kilometre Array. *Precursors, Pathfinders and Design Studies.* [accessed November 4, 2011]; Available from: http://www.skatelescope.org /the-organisation/precursors-pathfinders-design-studies/.

18. P. J. Hall, R. T. Schilizzi, P. E. F. Dewdney, and T. J. W. Lazio. The Square Kilometre Array (SKA) Radio Telescope: Progress and Technical Directions. *The Radio Science Bulletin (URSI)*, September, 2008. (326). [accessed January 19, 2013]; Available from: http://www.skatelescope.org/uploaded/54709_ PHall_RSB_reprint.pdf.

19. R. T. Schilizzi, P. Alexander, J. M. Cordes, P. E. Dewdney, R. D. Ekers, A. J. Faulkner, B. M. Gaensler, P. J. Hall, J. L. Jonas, and K. I. Kellerman. *Preliminary Specifications for the Square Kilometre Array.* 2007 December. SKA Memo 100 [accessed January 19, 2013]; Available from: http://www .skatelescope.org/uploaded/5110_100_Memo_Schilizzi.pdf.

20. P. E. Dewdney, P. J. Hall, R. T. Schilizzi, and T. J. L. W. Lazio. The Square Kilometre Array. *Proceedings of the IEEE*, August, 2009. 97(8): 1482–1496. [accessed January 19, 2013]; Available from: http://www.skatelescope.org /uploaded/8388_Dewdney_IEEE.pdf.

21. T. J. Cornwell and Ger van Diepe. *Scaling Mount Exaflop: From the Pathfinders to the Square Kilometre Array.* 2008 [accessed November 4, 2011]; Available from: http://www.atnf.csiro.au/people/tim.cornwell/publications /MountExaflop.pdf.

22. A. J. Kemball and T. J. Cornwell. *A Simple Model of Software Costs for the Square Kilometre Array.* 2004 SKA Memorandum 51. [accessed January 19, 2013]; Available from: http://www.skatelescope.org/uploaded/8648_51_memo_Kemball.pdf.

23. Large Synoptic Survey Telescope. *Home Page.* 8.4m wide-field telescope facility. [accessed November 4, 2011]; Available from: http://www.lsst.org.

24. R. Kolb. *The Large Synoptic Survey Telescope (LSST).* 2005 A Whitepaper prepared for the Dark Energy Task Force Committee.

25. Large Synoptic Survey Telescope. *Community Science with LSST.* 2011 January 10–11. AAS Meeting in Seattle, WA, 9–13 January 2011 [accessed November 4, 2011]; Available from: http://www.lsst.org/lsst/news/aas_217.

26. *SciDB: Open Source Data Management and Analytics Software for Scientific Research.* [accessed May 25, 2011]; Available from: http://www.SciDB.org.

27. NASA. *The Earth Observing System Data and Information System (EOSDIS)* [accessed November 4, 2011]; Available from: http://earthdata.nasa.gov/.

28. European Space Agency (ESA). *Home Page.* [accessed November 4, 2011]; Europe's gateway to space Available from: http://www.esa.int.

29. NSF. *Ocean Observatories Initiative (OOI).* [accessed November 4, 2011]; Available from: http://www.oceanobservatories.org.

30. UNAVCO. *A Non-Profit Membership-Governed Consortium, Facilitating Geoscience Research and Education Using Geodesy.* [accessed November 4, 2011]; Available from: http://www.unavco.org/.

31. Alaska Satellite Facility. *Downlinks, Archives, and Distributes Satellite-Based Data to Scientific Users Around the World.* [accessed November 4, 2011]; Available from: http://www.asf.alaska.edu/.

32. NASA JPL UAVSAR. *Reconfigurable, Polarimetric L-Band Synthetic Aperture Radar SAR.* [accessed November 4, 2011]; Available from: http://uavsar.jpl.nasa.gov/.

33. DESDynl. *A dedicated U.S. InSAR and LIDAR Mission Optimized for Studying Hazards and Global Enviromental Change. Science Mission Providing Important Observations for Land Surface Change and Hazards (Surface Deformation), and Climatic Variables (Terrestrial Biomass and Ecosystem Structure and Ice Dynamics).* [accessed November 4, 2011]; Available from: http://desdyni.jpl.nasa.gov/.

34. QuakeSim Cyberinfrastructure supporting Earthquake Science. *Home Page.* [accessed March 21, 2010]; Available from: http://quakesim.jpl.nasa.gov/.

35. A. Donnellan, M. Pierce, D. McLeod, J. Parker, J. Rundle, L. Grant, R. Al-Ghanmi, G. Fox, and R. Granat. Integrating GPS, InSAR, and UAVSAR Date into the QuakeSim Computational Environment. In *IEEE Aerospace Conference.* March 5–12, 2011. Big Sky, Montana, USA. [accessed January 19, 2013]; Available from: http://grids.ucs.indiana.edu/ptliupages/publications/Integrating%20InSAR%20and%20UAVSAR%20date%20into%20the%20QuakeSim%20Computational%20Environment.docx.

36. Center for the Remote Sensing of Ice Sheets (CReSIS). *CReSIS Home Page.* [accessed December 31, 2009]; Available from: https://www.cresis.ku.edu/.

37. PolarGrid NSF MRI funded partnership between Indiana University and Elizabeth City State University to acquire and deploy the computing infrastructure needed to investigate urgent problems in glacial melting. *Home Page*. [accessed March 1, 2010]; Available from: http://www.polargrid.org.

38. D. Crandall, Indiana University. *Image Processing to Determine Ice Sheet Bed from Radar Data*, Personal Communication to, Geoffrey Fox. July, 2011.

39. J. Strickland. *How the Large Hadron Collider Works*. How Stuff Works article. [accessed November 4, 2011]; Available from: http://science.howstuffworks .com/science-vs-myth/everyday-myths/large-hadron-collider3.htm.

40. C. Foudas. *Overview of the LHC Triggers and Plans for LHC Start-up*. April 13, 2007 [accessed November 4, 2011]; Available from: http://www.docstoc .com/docs/42244651/Overview-of-the-LHC-Triggers-and-Plans-for-LHC.

41. M. Mozer. Triggers for New Physics at the LHC. *Journal of Physics: Conference Series*, 2009. 171(012101). DOI:10.1088/1742-6596/171/1/01210. http:// iopscience.iop.org/1742-6596/171/1/012101/pdf/1742 -6596_171_1_012101 .pdf.

42. Large Hadron Collidor Computing Grid. *Worldwide LHC Computing Grid Technical Site*. [accessed November 4, 2011]; Available from: http://lcg.web .cern.ch/LCG/.

43. Iberian Grid Collaboration. *The CMS Iberian Computing Sites Performance in the Advent of the LHC Era*. May 24–27, 2010 [accessed November 4, 2011]; Available from: http://www.ibergrid.eu/2010/pdfs/wednesday/3.% 20The%20 CMS%20Iberian%20Computing%20Sites%20performance%20in%20 the%20advent%20of%20the%20LHC%20era.pdf.

44. CMS LHC EXperiment. *CMS Computing Model* [accessed November 4, 2011]; Available from: https://twiki.cern.ch/twiki/bin/view/CMSPublic /WorkBookComputingModel.

45. I. Bird. Computing for the Large Hadron Collider. *Annual Review of Nuclear and Particle Science*, November, 2011. 61: 99–118. DOI:10.1146 /annurev-nucl-102010-130059

46. European Synchrotron Radiation Facility (ESRF). *Home Page for Joint Facility Supported and Shared by 19 European Countries*. [accessed November 4, 2011]; Available from: http://www.esrf.eu/.

47. The European X-Ray Free Electron Laser Project XFEL. *Home Page*. [accessed November 4, 2011]; Available from: http://www.xfel.eu/.

48. UK Diamond Light Source. *Home Page of UK's National Synchrotron Facility*. [accessed 2011 November 4]; Available from: http://www.diamond.ac.uk/.

49. Cameron Neylon, Senior Scientist, Science and Technology Facilities Council, Didcot, United Kingdom. *Neutron Scattering as Long Tail Science*, Personal Communication to, Tony Hey. October 30, 2011.

50. P. Bryan Heidorn. Shedding Light on the Dark Data in the Long Tail of Science. *Library Trends*, 2008. 57(2, Fall 2008): 280–299.

51. Peter Murray-Rust. *Big Science and Long-tail Science*. January 29, 2008 [accessed November 3, 2011]; Available from: http://blogs.ch.cam.ac.uk/pmr/2008/01/.

52. EMBL-EBI. *European Bioinformatics Institute*. [accessed November 3, 2011]; Available from: http://www.ebi.ac.uk/.

53. EMBL. *European Molecular Biology Laboratory*. [accessed November 3, 2011]; Available from: http://www.embl.org/.
54. EMBL-EBI. *EMBL-EBI's 2010 Annual Scientific Report*. [accessed November 3, 2011]; Available from: http://www.ebi.ac.uk/Information /Brochures/pdf/Annual_Report_2010_low_res.pdf.
55. EBI. *Using the Ensembl Data Hosted on Amazon AWS*. [accessed November 3, 2011]; Available from: http://www.ensembl.org/info/data/amazon_aws.html.
56. illumina. *HiSeq 2000*. [accessed November 4, 2011]; Available from: http:// www.illumina.com/systems/hiseq_2000.ilmn.
57. H. Tang, Indiana University. *Data Sizes for Genomics*, Personal Communication to Geoffrey Fox. May, 2011.
58. L. Stein. The Case for Cloud Computing in Genome Informatics. *Genome Biology*, 2010. 11(5): 207. [accessed January 19, 2013]; Available from: http:// genomebiology.com/2010/11/5/207.
59. E. Pennisi. Will Computers Crash Genomics? *Science*, February 11, 2011. 331(6018): 666–668. [accessed January 19, 2013]; Available from: http:// www.sciencemag.org/content/331/6018/666.short.
60. M. Schatz. *Cloud Computing and the DNA Data Race*. 2011 [accessed June 26, 2011]; Keynote Presentation at 3DAPAS/ECMLS workshops at HPDC June 8 2011 Available from: http://schatzlab.cshl.edu/presentations/2011-06-08 .HDPC.3DAPAS.pdf.
61. NIH. *DNA Sequencing Costs: Data from the NHGRI Large-Scale Genome Sequencing Program*. [accessed May 20, 2011]; Available from: http://www .genome.gov/sequencingcosts/.
62. U. G. Thomas. *NCBI to End Support for Sequence Read Archive as Federal Purse Strings Tighten*. February 18, 2011 [accessed August 27, 2011]; Available from: http://www.genomeweb.com/informatics/ncbi-end-support-sequence -read-archive-federal-purse-strings-tighten.
63. WCRP World Climate Research Programme. *Home Page of WCRP Facilitating Analysis and Prediction of Earth System Variability and Change for use in an Increasing Range of Practical Applications of Direct Relevance, Benefit and Value to Society*. [accessed November 4, 2011]; Available from: http://www.wcrp-climate.org/.
64. WCRP/CLIVAR Working Group on Coupled Modelling. *Home Page*. [accessed November 4, 2011]; Available from: http://www.clivar.org/.
65. WCRP World Climate Research Programme. *CMIP5 Coupled Model Intercomparison Project* [accessed November 4, 2011]; Available from: http:// cmip-pcmdi.llnl.gov/cmip5/.
66. PCMDI Program for Climate Model Diagnosis and Intercomparison. *Home Page for PCMDI with Mission to Develop Improved Methods and Tools for the Diagnosis and Intercomparison of General Circulation Models (GCMs) that Simulate the Global Climate*. [accessed November 4, 2011]; Available from: http://www-pcmdi.llnl.gov/.
67. ESGF Earth System Grid Federation. *Home page for ESGF, which is a Non-Profit Organization Formed by Participates in the GO-ESSP Collaboration to Bring their Knowledge and Experience to Bear on Critical Earth System*

Federations in the Dissemination of Climate Data and Related Products. [accessed November 4, 2011]; Available from: http://esg-pcmdi.llnl.gov/esgf.

68. ESG Earth Systems Grid. *Home Page for ESG with a Gateway and Two Major Projects ESGF (Earth System Grid Federation) and ESG-CET (U.S. Earth System Grid Center for Enabling Technologies).* [accessed November 4, 2011]; Available from: http://www.earthsystemgrid.org.

69. NERC Natural Environment Research Council. *NERC Data Grid Home Page.* [accessed November 4, 2011]; Available from: http://ndg.nerc.ac.uk/.

70. Collaborative Climate Community Data and Processing Grid (C3Grid). *Home Page.* [accessed November 4, 2011]; Available from: https://verc.enes .org/c3web.

71. METAFOR. *Home Page for Common Information Model (CIM) for climate data and the models that produce it.* [accessed November 4, 2011]; Available from: http://en.wikipedia.org/wiki/METAFOR.

72. B. Lawrence. *Quality Control, Documentation, and Long Term Support (of the CMIP5 Archive) (& a dose of the IPCC-DDC).* February 2011. "Managing Data to Support the Assessment Process" session at GEOSS Support for IPCC assessment, Geneva. [accessed November 5, 2011]; Available from: http://home.badc.rl.ac.uk/lawrence/static/2011/02/01/qc4ipcc.pdf.

73. K. Taylor. CMIP5 Handling of Model Output. 2011 October 6. In *4th Annual Meeting of the Integrated Assessment Modeling Consortium (IAMC), Austria Trend Hotel Savoyen Rennweg 16, A-1030 Vienna, Austria.* [accessed November 4, 2011]; Available from:http://www.iiasa.ac.at/Research/ENE /IAMC/mtg11/agenda.html Available from: http://www.iiasa.ac.at/Research /ENE/IAMC/mtg11/Taylor_CMIP5_data_archive.pdf.

74. M. Snir, W. Gropp, and P. Kogge. *Exascale Research: Preparing for the Post-Moore Era.* June 19, 2011. [accessed January 19, 2013]; Available from: http://www.ideals.illinois.edu./bitstream/handle/2142/25468 /Exascale%20Research.pdf.

75. U.S. Department of ENERGY. *ASCR News & Resources: ASCR Program Documents.* 2011 [accessed November 3, 2011]; Available from: http:// science.energy.gov/ascr/news-and-resources/program-documents/.

76. DOE OASCR *Workshop on Exascale Data Management, Analysis, and Visualization.* 2011 [accessed November 3, 2011]; Available from: http:// www.olcf.ornl.gov/event/exascale-2011/.

77. C. S. Chang. Needs for Extreme Scale DM, Analysis and Visualization in Fusion Particle Code XGC. *Center for Plasma Edge Simulation CPES at Exascale Data Management, Analysis, and Visualization,* Feb. 22–23, 2011, Houston, TX. [accessed July 23, 2011]; Available from: http://www.olcf.ornl .gov/wp-content/uploads/2011/01/CSChang_Fusion_final.pdf.

78. J. Chen and J. Bell. *Combustion Exascale Co-Design Center.* 6th International Exascale Software Project Workshop, San Francisco, CA, April 6–7, 2011. [accessed July 15, 2011]; Available from: http://www.exascale.org/mediawiki /images/3/38/Talk29-Chen.pdf.

79. *ADIOS Adaptable IO System* [accessed July 16, 2011]; Available from: http://adiosapi.org/index.php5?title = Main_Page.

80. J. Lofstead, F. Zheng, S. Klasky, and K. Schwan. Adaptable, Metadata rich IO Methods for Portable High Performance IO. In *Proceedings of the 2009 IEEE International Symposium on Parallel and Distributed Processing*. 2009, IEEE Computer Society. Rome, Italy. pp. 1–10.

81. Y. Tian, S. Klasky, H. Abbasi, J. Lofstead, R. Grout, N. Podhorszki, Q. Liu, Y. Wang, and W. Yu. EDO: Improving Read Performance for Scientific Applications Through Elastic Data Organization. In *IEEE Cluster*. September 26–30, 2011. Austin, TX.

82. *Scientific Grand Challenges: Fusion Energy Sciences and the Role of Computing at the Extreme Scale*. U. S. Department of Energy, March 18–20, 2009. Washington DC. [accessed July 15, 2011]; Available from: http://science.energy.gov/~/media/ascr/pdf/program-documents/docs/Fusion_report.pdf.

83. The Domesday Book Online. *Site to Enable Visitors to Discover the History of the Domesday Book, to Give an Insight into Life at the Time of its Compilation, and Provide Information and Links on Related Topics*. [accessed November 4, 2011]; Available from: http://www.domesdaybook.co.uk/.

84. J. Darlington, A. Finney, and A. Pearce. Domesday Redux: The Rescue of the BBC Domesday Project Videodiscs. *Ariadne Web magazine*, July, 2003(36). [accessed January 19, 2013]; Available from: http://www.ariadne.ac.uk/issue36/tna/.

85. G. Bell, J. Gray, and A. Szalay. Petascale Computational Systems: Balanced CyberInfrastructure in a Data-Centric World (Letter to NSF Cyberinfrastructure Directorate). *IEEE Computer*, January, 2006. 39(1): 110–112. Available from: http://research.microsoft.com/en-us/um/people/gray/papers/Petascale%20computational%20systems.doc.

86. Y. Simmhan, R. Barga, C. van Ingen, M. Nieto-Santisteban, L. Dobos, N. Li, M. Shipway, A. S. Szalay, S. Werner, and J. Heasley. GrayWulf: Scalable Software Architecture for Data Intensive Computing. In *42nd Hawaii International Conference on System Sciences*. 2009, IEEE. Waikoloa, Big Island, Hawaii. pp. 1–10. DOI: http://doi.ieeecomputersociety.org/10.1109/HICSS.2009.750.

87. S. Ghemawat, H. Gobioff, and S. -T. Leung. The Google File System. *SIGOPS Operating Systems Review*, 2003. 37(5): 29–43.

88. *Hadoop Distributed File System HDFS*. [accessed December 2009]; Available from: http://hadoop.apache.org/hdfs/.

89. Y. Yu, M. Isard, D. Fetterly, M. Budiu, U. Erlingsson, P. K. Gunda, and J. Currey. DryadLINQ: A System for General-Purpose Distributed Data-Parallel Computing Using a High-Level Language. In *Symposium on Operating System Design and Implementation (OSDI)*. December 8–10, 2008. San Diego, CA.

90. M. Isard, M. Budiu, Y. Yu, A. Birrell, and D. Fetterly. Dryad: Distributed Data-Parallel Programs from Sequential Building Blocks. *SIGOPS Operating Systems Review*, 2007. 41(3): 59–72.

91. J. Ekanayake, T. Gunarathne, J. Qiu, G. Fox, S. Beason, J. Y. Choi, Y. Ruan, S.-H. Bae, and H. Li. *Applicability of DryadLINQ to Scientific Applications.* January 30, 2010, Community Grids Laboratory, Indiana University. [accessed January 19, 2013]; Available from: http://grids.ucs.indiana.edu /ptliupages/publications/DryadReport.pdf.

92. Y. Gu and R. Grossman. *Sector and Sphere: The Design and Implementation of a High Performance Data Cloud.* Crossing boundaries: Computational science, e-Science and global e-Infrastructure I. Selected papers from the UK e-Science All Hands Meeting 2008 *Philosophical Transactions of the Royal Society of London. Series A,* 2009. 367: 2429–2445.

93. J. Gray. Jim Gray on eScience: A Transformed Scientific Method. In *The Fourth Paradigm: Data-Intensive Scientific Discovery.* Tony Hey, Kristin Tolle and Stewart Tansley, Editors. 2009, Microsoft Research. Redmond, Washington. p. xvii–xxxi.

94. NOSQL Movement. *Wikipedia list of Resources.* 2010 [accessed June 5, 2010]; Available from: http://en.wikipedia.org/wiki/NoSQL.

95. NOSQL Link Archive. *LIST OF NOSQL DATABASES.* 2010 [accessed June 5, 2010]; Available from: http://nosql-database.org/.

96. F. Chang, J. Dean, S. Ghemawat, W. C. Hsieh, D. A. Wallach, M. Burrows, T. Chandra, A. Fikes, and R. E. Gruber. *Bigtable: A Distributed Storage System for Structured Data,* in *OSDI'06: Seventh Symposium on Operating System Design and Implementation.* 2006, USENIX. Seattle, WA [accessed January 19, 2013].

97. Amazon. *Welcome to Amazon SimpleDB.* 2010 [accessed 2010 June 5]; Available from: http://docs.amazonwebservices.com/AmazonSimpleDB/latest /DeveloperGuide/index.html.

98. J. Haridas, N. Nilakantan, and B. Calder. *Windows Azure Table.* 2009 May [accessed June 5, 2010]; Available from: http://go.microsoft.com/fwlink /?LinkId = 153401.

99. International Virtual Observatory Alliance. *VOTable Format Definition Version 1.1.* 2004 [accessed June 5, 2010]; Available from: http://www.ivoa .net/Documents/VOTable/20040811/.

100. R. Barga, D. Gannon, N. Araujo, J. Jackson, W. Lu, and J. Ekanayake. Excel DataScope for Data Scientists. In *UK e-Science All Hands Meeting 2010.* September 13–16, 2010. Cardiff, Wales UK [accessed January 19, 2013].

101. Apache. *Hbase implementation of Bigtable on Hadoop File System.* [accessed June 5, 2010]; Available from: http://hbase.apache.org/.

102. Apache. *The CouchDB Document-Oriented Database Project.* [accessed June 5, 2010]; Available from: http://couchdb.apache.org/index.html.

103. M/Gateway Developments Ltd. *M/DB Open Source "Plug-Compatible" Alternative to Amazon's SimpleDB Database.* 2009 [accessed June 5, 2010]; Available from: http://gradvs1.mgateway.com/main/index.html?path = mdb.

104. Z. Hill and M. Humphrey. CSAL: A Cloud Storage Abstraction Layer to Enable Portable Cloud Applications. In *Second International Conference on Cloud Computing Technology and Science (CloudCom).* November

30–December 3, 2010, IEEE. Indianapolis. pp. 504–511. Available from: http://www.cs.virginia.edu/~humphrey/papers/CSAL.pdf. DOI: http://dx .doi.org/10.1109/CloudCom.2010.88.

105. P. Donnelly, P. Bui, and D. Thain. Attaching Cloud Storage to a Campus Grid Using Parrot, Chirp, and Hadoop. In *Second International Conference on Cloud Computing Technology and Science*. November 30–December 3, 2010, IEEE. Indianapolis. pp. 488–495. Available from: http://www.cse .nd.edu/~ccl/research/papers/chirp+parrot+hdfs.pdf. DOI http://dx.doi .org/ 10.1109/CloudCom.2010.42.

106. J. Dean and S. Ghemawat. MapReduce: Simplified Data Processing on Large Clusters. *Communications of the ACM*, 2008. 51(1): 107–113.

107. Microsoft Research. *Daytona Iterative MapReduce on Windows Azure*. [accessed November 6, 2011]; Available from: http://research.microsoft .com/en-us/projects/daytona/.

108. J. Ekanayake, H. Li, B. Zhang, T. Gunarathne, S. Bae, J. Qiu, and G. Fox. Twister: A Runtime for Iterative MapReduce. In *Proceedings of the First International Workshop on MapReduce and its Applications of ACM HPDC 2010 Conference June 20–25, 2010*. 2010. ACM. Chicago, IL. [accessed January 19, 2013]; Available from: http://grids.ucs.indiana.edu/ptliupages /publications/hpdc-camera-ready -submission.pdf.

109. T. Gunarathne, B. Zhang, T.-L. Wu, and J. Qiu. Portable Parallel Programming on Cloud and HPC: Scientific Applications of Twister4Azure. In *IEEE /ACM International Conference on Utility and Cloud Computing UCC 2011*. December 5–7, 2011. Melbourne, Australia.

110. B. Palanisamy. *Purlieus: Locality-Aware Resource Allocation for MapReduce in a Cloud*. 2011 [accessed June 12, 2011]; Available from: http://www.cc.gatech .edu/~pbalaji/purlieus/.

111. Zend PHP Company. *The Simple Cloud API for Storage, Queues and Table*. 2010 [accessed June 1, 2010]; Available from: http://simplecloud.org/.

112. X. Gao, Y. Ma, M. Pierce, M. Lowe, and G. Fox. Building a Distributed Block Storage System for Cloud Infrastructure. In *CloudCom 2010*. November 30–December 3, 2010. IUPUI Conference Center Indianapolis. [accessed January 19, 2013]; Available from: http://grids.ucs.indiana.edu/ptliupages /publications/VBS -Lustre_final_v1.pdf.

113. S. Vazhkudai, S. Tuecke, and I. Foster. Replica Selection in the Globus Data Grid. In *Proceedings of the 1st International Symposium on Cluster Computing and the Grid CCGrid*. IEEE Computer Society. Brisbane, Australia. p. 106.

I

Data-Intensive Grand Challenge Science Problems

CHAPTER 3

Large-Scale Microscopy Imaging Analytics for In Silico Biomedicine

Joel Saltz, Fusheng Wang, George Teodoro,
Lee Cooper, Patrick Widener, Jun Kong,
David Gutman, Tony Pan, Sharath Cholleti,
Ashish Sharma, Daniel Brat, and Tahsin Kurc

CONTENTS

3.1 INTRODUCTION

The ability to quantitatively characterize biological structure in detail through in silico experiments has great potential to reveal new insights into disease mechanisms and enable the development of novel preventive approaches and targeted treatments. The term in silico experiment broadly refers to an experiment performed on a computer by analyzing, mining, and integrating biomedical datasets and/or by simulations. This chapter presents the data and computational challenges of using large volumes of digitized microscopic tissue data in in silico experiments and describes some of the data-intensive computing approaches we have used to address these challenges.

Since the first application of digital imaging technology to microscopic data [1] (i.e., to tissue data at cellular and subcellular level), the ability to acquire high-resolution images from whole tissue slides and tissue micro-arrays has become more affordable, faster, and practical [2–10]. Image scanning times have decreased from 6 to 8 hours per whole tissue slide about a decade ago to a few minutes with advanced scanners and improvements in autofocusing and slide holders have facilitated high-throughput image generation from batches of slides with minimal manual intervention. It is rapidly becoming feasible for a research study or health-care operation to routinely generate hundreds to thousands of whole slide images per day. High-resolution whole slide tissue images not only reduce dependence on physical slides, they can also enable more effective ways of screening for disease, classifying disease state, understanding its progression, and evaluating the efficacy of therapeutic strategies. This potential is also fueling the emergence of what we refer to as microscopy imaging analytics, the process of management, analysis, and quantitative characterization and correlation of microanatomic features on high-resolution image data using computer algorithms.

Despite the wider availability of advanced imaging instruments and the demonstrated potential of whole slide tissue images in biomedical research, to date, microscopic imaging has been underutilized in biomedicine, compared to other imaging modalities such as magnetic resonance imaging, which has seen widespread adoption. This is primarily because even moderate numbers of digitized tissue specimens quickly lead to formidable information synthesis and management problems. Software for extraction and interpretation of information from thousands of tissue images has to deal with hundreds of terabytes of data, meet expensive data processing requirements, and manage and query trillions of microscopic objects and their features.

In the rest of the chapter, we discuss these requirements in greater detail. The requirements in large-scale studies easily overwhelm the memory and computation capacity of high-end workstations. We present data-intensive computing techniques and middleware support to address the challenges on high-performance computing machines.

3.2 APPLICATION OF MICROSCOPY IMAGING IN IN SILICO BIOMEDICAL RESEARCH

A basic use of the microscopy imaging technology is telepathology, in which a pathologist can remotely render diagnoses for patient care in the absence of glass slides and a microscope [11,12]. In this form of use, a whole slide imaging system should support the implementation of a "virtual microscope" [11,13–26], which emulates the core operation of a microscope, enabling browsing through a slide to locate an area of interest, local browsing in a region of interest to observe the region surrounding the current view, and changing magnification level and focal plane.

Although a virtual microscope facilitates remote viewing of images, systematic investigations of tissue at microanatomic level (i.e., at the level of blood vessels, cells, and nuclei) can lead to a much more sophisticated understanding of disease types, the function and role of cellular-level processes, and the relationship between cellular-level morphological characteristics, genomic expression data, and clinical outcome in disease progression and response to treatment. Studies of brain tumors conducted using tissue slide images and genomic data at the In Silico Brain Tumor Research Center (ISBTRC) [27], for example, have produced results that reveal morphological subtypes of glioblastoma not previously recognized by pathologists and that demonstrate significantly correlated genes through correlation of the extent of necrosis and angiogenesis with gene expression data [28–30]. In these studies, in silico experiments analyzing 462 tissue slide images, where each image is approximately 50k × 50k pixels, or about 10 GB uncompressed, from 162 patients discovered that the morphological signatures in glioblastoma self-aggregate into four distinct clusters. The survival characteristics of this morphology-driven stratification are significant when compared to the survival of molecular subtypes, suggesting that morphology is a good predictor of prognosis.

These types of in silico studies involve the extraction, management, and quantitative characterization of microanatomic structures, such as cells, nuclei, macrophages, and tissue blood vessels, from imaged tissue specimens. We refer to this process as microscopy imaging analytics. Section 3.3 presents the data management and computation challenges of using microscopy imaging analytics in large-scale studies.

3.3 DATA AND COMPUTATION CHALLENGES IN MICROSCOPY IMAGING ANALYTICS

State-of-the-art scanners can generate a whole slide tissue image at up to 120k × 120k pixel resolutions. An uncompressed, four-channel representation of such an image is about 53 GB. A multilayer image stack that provides a focus capability typically contains tens of such images. Furthermore, a single scanner can generate hundreds of images, and a study may generate or reference thousands of slides. For instance, studies at the ISBTRC have so far collected about 2000 whole slide glioma tissue images and counting. These images have been gathered from "The Cancer Genome Atlas (TCGA) repository," obtained from the collaborators at Henry Ford Hospital and Thomas Jefferson University, and scanned from tissues at Emory University. The ISBTRC will continue to collect approximately 3500 slides from about 700 patients over the next couple of years. The TCGA repository itself contains about 12,000 slide images for various types of cancers including brain cancer.

A typical microscopy imaging analytics workflow consists of a pipeline of data processing steps [31]: (1) Stitching and registration. Some instruments capture a whole slide image as a set of image tiles. These tiles need to be aligned and stitched together to form a full image. (2) Preprocessing. This step performs operations such as color normalization to compensate for variations in image acquisition processes and instruments. (3) Segmentation of objects and regions. Often the entities to be segmented are composed of collections of simple and complex objects and structures and are defined by a complex shape and textural appearance. Examples include cell nuclei, cell membranes, the boundaries of blood vessels, and the extent of necrosis regions. (4) Feature computation. This is the process of calculating informative descriptions of objects or regions, and often precedes classification tasks. It can be applied on the whole image or individual segmented objects to describe characteristics such as shape and texture. (5) Classification. Segmented objects, regions, or whole slides can be classified into meaningful groups based on computed features. Grouping of cell types, nuclei, or entire slides into categories is a common classification theme. A high-level depiction of this typical analysis workflow for segmentation and classification of nuclei in a whole slide tissue image is shown in Figure 3.1. The figure shows the major steps of the analysis. Individual steps are often composed of a series of operations. For instance, the segmentation step may have 20–25 pipelined operations.

Analysis via this workflow of a single 20k × 20k pixel image takes approximately 10 hours on a single processor workstation. The computational requirements are exacerbated in large-scale studies in which detailed characterization of morphology often involves (1) Coordinated use of many interrelated analysis pipelines, (2) performing algorithm sensitivity analyses, and (3) comparison of analysis results from multiple analysis pipelines and analysis runs. This analysis approach stems from the fact that the effectiveness of an analysis pipeline depends on many factors, including the nature of the histological structures being segmented, the classifications being carried out, and on sample preparation and staining. When the number of images is in the hundreds to thousands, it is not feasible to manually inspect each image for every feature and fine-tune the analysis pipelines. Thus, many variations of analysis pipelines may be evaluated by systematic comparison and evaluation of the results obtained from a subset of images to identify a smaller number of priority pipelines. These pipelines are then executed on the larger collection of images.

This approach leads to a difficult data management, querying, and integration problem. Image analysis algorithms segment and classify 10^5 to 10^7 cells in each virtual slide of size $10^5 \times 10^5$ pixels. Brain tumor tissue analyses, for instance, can encompass the identification and quantification of subcellular structures, which is done through processing in cells or

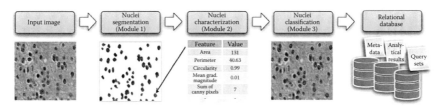

FIGURE 3.1 (**See color insert.**) Nuclear segmentation and classification pipeline. Images are processed through a set of operations for detecting boundaries of nuclei, computing a set of features for each nucleus, and classifying nuclei into categories using machine learning algorithms on these features. The results are stored in a database for further analysis and correlative evaluation.

regions identified as being brain tumor, as well as of angiogenesis regions and pseudopalisades, which requires a synthesis of regional texture analyses. As these analyses may execute multiple interrelated analysis pipelines as described earlier, a systematic analysis of large-scale image data, therefore, can involve classification of roughly 10^9 to 10^{12} microanatomic structures. The process of classifying a given cell is done using roughly 10–100 shape, texture, and (when appropriate) stain quantification features. Hence, a thorough data analysis limited to classifying cells could encompass 10^{10} to 10^{13} features.

It can take an hour or longer to compare results generated from two algorithms for a single image with a database without parallelization. Comparing two result sets from a hundred images can take a week. In addition to comparing results from multiple analyses, scalable mechanisms are needed for producing biologically or computationally meaningful data aggregates (e.g., machine learning-based clustering) from spatial objects and features. Computation of data aggregates on large number of images can take days or weeks.

These data-intensive computational requirements are a major obstacle to widespread use of microscopy imaging data in research and healthcare delivery. Addressing them would facilitate novel studies, new methods of patient diagnosis, and improvements in targeted treatments. With a system capable of synthesizing information from large digitized tissue slide datasets rapidly, a clinician could create morphological signatures of large groups of patients and create a representation of the patient he/she is treating with respect to these groups. In research, data-intensive science solutions would enable scientists to broaden the scale of their research and test new hypotheses by synthesizing information from larger datasets and multiple imaging modalities.

3.4 DATA-INTENSIVE AND HIGH-PERFORMANCE COMPUTING APPROACHES FOR LARGE-SCALE MICROSCOPY IMAGING ANALYTICS

This section presents a set of high-performance computing techniques we have used to address the challenges and requirements described in Section 3.3. These techniques are grouped in three main sections: (1) Support for processing large volumes of image data, (2) support for representing image data and analysis results, and (3) support for metadata and spatial queries on analysis results.

3.4.1 Processing Large Volumes of Image Data

The typical analysis workflow structure described in Section 3.3 (also see Figure 3.1) encapsulates multiple processing patterns. First of all, images can be partitioned into rectangular image tiles, and multiple images or image tiles can be processed concurrently. This leads to a bag-of-tasks style parallelism. Several systems have used this type of parallelism on cluster systems and in grid computing environments. Gurcan et al. [32] reported the successful application of distributed computing in a pilot project to support automated characterization of neuroblastoma using the Shimada classification system. The ImageMiner system used IBM's World Community Grid in July 2006, using more than 100,000 imaged tissue discs [24,33]. Second, the processing of a single image or image tile is expressed as a pipeline of operations—the stitching, preprocessing, segmentation, feature computation, and classification stages form the main steps of the pipeline; a main step can itself be composed of pipelines of data processing operations. This processing pattern is more suitably implemented using a coarse-grain dataflow (or filter-stream) parallelism approach, in which application processing is carried out as a network of components (data processing operations) connected through logical pipes. Each component performs a portion of the application-specific processing, and interactions between the components are realized by flow of data and control information. The out-of-core virtual microscope system [34,35], based on the DataCutter infrastructure [36], supports this type of parallel processing. The runtime system coordinates Input and Output (I/O) operations for reading image tiles from and writing analysis results to one or more discs in the system. The processing operations constituting the analysis workflow are mapped to processors to reduce I/O and communication overheads. Multiple instances of an operation can be instantiated to allow for data parallelism and pipelined processing.

The systems described earlier primarily target distributed and parallel clusters with multicore CPUs. The processing and memory capacity of graphics processing units (GPUs) have improved significantly in recent years. GPUs are increasingly being deployed as complementary accelerators on high-end machines and large-scale clusters. Several research projects have implemented GPU-enabled versions of image processing operations [37–52] to take advantage of low-latency and high-bandwidth GPU memories and massively multithreaded execution models. Teodoro et al. [53] have developed a runtime system to use multicore CPUs and multiple GPUs on a computation node in a coordinated manner. This system implements a performance-aware scheduling strategy, which assigns tasks in an analysis workflow to CPUs and GPUs to optimize the use of aggregate power of CPU cores and GPUs on a computation node. Each task is represented as a (data operation, input data chunk) tuple and scheduled for execution on a CPU core or a GPU based on the relative GPU versus CPU performance gain of the task.

Another possible way to reduce analysis execution times is to trade off analysis accuracy for performance—in some cases, there may not be enough resources available to carry out an analysis at the highest resolution or an exploratory study may not need the highest accuracy to process a large set of potential data points quickly. Kumar et al. [34] have developed a framework that supports adaptive processing and performance-accuracy trade-offs. The framework exploits spatial locality of image features to create dynamic data processing schedules to improve performance while meeting quality of output requirements.

3.4.2 Representing Image Data and Analysis Results

Efficient data repositories anchored on rich and flexible data models play a crucial role in interpretation, reusability, and reproducibility of image analyses as well as efficient storage and query of analysis results. Several projects have developed data models for representation and management of microscopy images and analysis results [54–58], although there are yet no official standard models. The Open Microscopy Environment (OME) [54] project has developed a data model and a database system that can be used to represent, exchange, and manage image data and metadata. The OME provides a data model of common specification for storing details of microscope setup and image acquisition. Cell Centered Database [57,58] provides a data model to capture image analysis output, image data, and information on the specimen preparation and imaging conditions that generated the image data. Digital Imaging and Communications in Medicine (DICOM) Working Group 26 is developing a DICOM-based standard for storing microscopy images [59]. The metadata in this model captures information such as patient, study, and equipment information. The Pathology Analytical Imaging Standards (PAIS) model [55,56] is designed to provide an object-oriented, extensible, and semantically enabled data model to support large-scale analytical imaging and human observations to be storage and performance efficiency oriented, and to support alternative implementations.

These models represent (1) Context relating to patient data, specimen preparation, special stains, and so on; (2) human observations involving pathology classification and characteristics; and (3) algorithm and human-described segmentations (markups), features, and annotations. Markups can be either geometric shapes or image masks; annotations can be calculations, observations, disease inferences, or external annotations. Additional annotations can also be derived from existing annotations. An extensible markup language (XML)-based realization of the model attributes can facilitate exchange and sharing of results in a format more compatible with web standards and tools. However, for very large result sets, the XML representation is not efficient, even with compression of the documents. An alternative approach is to use self-describing structured container technologies such as HDF5. Such container technologies provide more efficient storage than text-based file formats such as XML, while still making available the structure of the data for query purposes. We have observed that for a given set of analysis results, the HDF5 representation of the PAIS model is on average 6–7 times smaller, in compressed form, than compressed XML representation of the model.

3.4.3 Metadata and Spatial Queries on Analysis Results

In silico experiments with microscopy images involve a wide range of queries on analysis results for data mining and correlation purposes. The examples of query types include the following: (1) Retrieval of image data and metadata (e.g., count nuclei where their grades are less than 3); (2) queries to compare results generated from different approaches and validate machine generated results against human observations (e.g., find nuclei that are classified by observer A and by algorithm B and whose feature f is within the range of a and b); (3) queries on assessing relative prevalence of features or classified objects, or assessing spatial coincidence of combinations of features or objects (e.g., which nuclei types preserve nuclei features: distance and shape between two images); (4) queries to support selection of collections of segmented regions, features, and objects for further machine learning or content-based retrieval applications (e.g., find nuclei with an area between 50 pixels and 200 pixels in selected region of interest); and (5) semantic queries based on spatial relationships and annotations and properties drawn from domain ontologies (e.g., search for objects with an observation concept astrocytoma and that are within 100 pixels of each other, but also expand to include all the subclass concepts, gliosarcoma and giant cell glioblastoma, of astrocytoma).

To scale to large volumes of data, parallel database configurations should be considered. We have investigated in a recent work [60] the performance of different database configurations for a common type of query: spatial join and crossmatch [61,62]. The configurations included a parallel database management system with active disc style execution support for some types of database operations, a database system designed for high availability and high throughput (MySQL Cluster), and a distributed collection of database management systems with data replication. Our experimental evaluation has shown that the choice of a database configuration can significantly impact query performance. The configuration with distributed database management systems with replication (i.e., replication of portions of the database) provides a flexible environment, which can be adjusted to the data access patterns and dataset characteristics.

There are some disadvantages to using traditional database systems. Data have to be organized and loaded to the database system for query support. The data load process may be prohibitively expensive when the volume of analysis results is very large. Moreover, it may not be feasible to install database systems on backend nodes of a large cluster system. Finally, the granularity of runtime optimizations is generally limited to partitioning and declustering of data across multiple database instances, since queries are largely executed within the database management system. An alternative approach is to implement MapReduce style processing [63] from the domain of enterprise data analysis.

A recent work by Wang et al. [62] has shown the implementation using Hadoop [63,64] of spatial query processing in the context of microscopy imaging analytics (see Figure 3.2). The implementation provides a declarative query language, based on Hive [64] (Hive provides a SQL like query language and supports major aggregation queries running on MapReduce), and an efficient real-time spatial query engine with dynamically built spatial indexes to support query processing on clusters with multicore CPUs. Processing of a query is accomplished in several steps: (1) Analysis results with spatial boundaries are retrieved by the query engine, and R*-tree indices are built on the fly; (2) initial spatial filtering (spatial join) is done through minimal bounding boundary based on the entries in the R*-tree indices [64]; and (3) computational geometry algorithms for query refinement and spatial measurement are performed to generate the final results. Image tiles form natural units for MapReduce-based execution and are staged on Hadoop Distributed File System (HDFS) [63] (see Figure 3.2a).

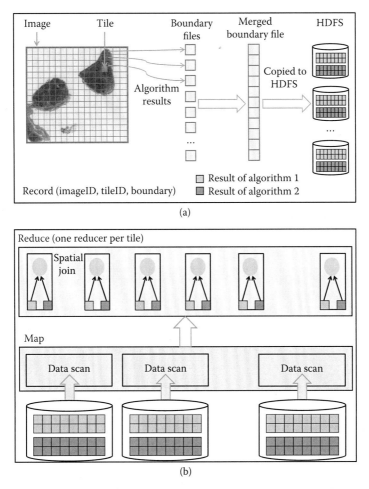

FIGURE 3.2 MapReduce-based query processing for comparison of results from multiple analysis runs. (a) HDFS data staging. (b) MapReduce-based queries.

Steps 1 and 2 are implemented as Map functions, whereas step 3 is executed as the Reduce function (see Figure 3.2b).

Comprehensive query support in a microscopy imaging analytics framework should include semantic query capabilities, in addition to spatial query functionality, because annotations on microanatomic objects may draw from domain ontologies (e.g., cell ontology to describe different cell types). An important requirement in semantic query support is that the query engine should be able to support additional assertions (annotations), which are inferred from initial assertions (also called explicit assertions or annotations) based on the characteristics of the underlying ontology.

On-the-fly computation of assertions for each query may take too long. Precomputation of inferred assertions, also referred to as the materialization process, can reduce the execution time of queries. Combined use of semantic stores [65–67] and rule engines [68] can offer a system capable of evaluating queries with spatial and semantic predicates. In such a system, rules can be used to express spatial relationships in association with ontology concepts—a rule for example may be used to state that type A cells that are within 30 pixels of type B cells should also be annotated as type C cells. During the materialization process, the rule engine and the semantic store/inference engine interact to compute inferred assertions based on the ontology in the system, the set of rules, and the initial set of explicit assertions (annotations) [69,70]. To scale to large volumes of data, one high-performance computing approach is to use data parallelism by partitioning the space in which microanatomic objects are embedded. Another approach is to partition the ontology axioms and rules, distributing the computation of axioms and rules to processors. This parallelization strategy attempts to leverage axiom-level parallelism and will benefit applications where the ontology contains many axioms with few dependencies. A third possible approach is to combine the first two approaches with task parallelism. In this strategy, N copies of the semantic store engine and M copies of the rule engine are instantiated on the parallel machine. The system coordinates the exchange of information and the partitioning of workload between the semantic store engine instances and the rule engine instances. The values of N and M depend on the cost of inference execution and the partitioning of the workload based on spatial domain and/or ontology axioms.

3.5 DISCUSSION AND CONCLUSIONS

High-resolution, high-throughput instruments are being used routinely not only in medical sciences, but also in health-care delivery settings at an accelerating rate. As this decade progresses, significant advances in medical information technologies will transform very large volumes of multiscale, multidimensional data into actionable information to drive the discovery, development, and delivery of new mechanisms of preventing, diagnosing, and healing complex disease. Data produced by advances in digitization and image analysis are outpacing the storage and computation capacities of workstations and small cluster systems. The "big data" from image analysis has similar high-performance and scalability requirements as enterprise health-care data, but presents unique challenges. In the future, even medium-scale hospitals and research projects will require capabilities to manage thousands of high-resolution images, execute and manage inter-related analysis pipelines, and query trillions of microscopic objects and their features. These applications demand fast loading and query response, as well as declarative query interfaces for high usability.

Computational systems with multiple levels of computing and memory hierarchies, such as high-performance computing systems consisting of multicore CPUs and multiple GPUs and composed of multiple levels of coupled spinning drives and solid-state drives (SSDs) in Redundant Array of Independent Disks (RAID) configurations, are becoming mainstream configurations, replacing more traditional homogeneous computing clusters. These systems offer tremendous computing power and low-latency and high-throughput I/O capabilities. Many challenges, however, remain for the effective use of these new technologies. Novel storage, indexing, data staging, and scheduling techniques and middleware support are needed to manage storage hierarchies in tandem with executing computations on clusters of hybrid CPU–GPU nodes. There have been substantial advances in network switches and networking protocols for intracluster communications. Technologies such as InfiniBand provide low-latency, high-bandwidth communication substrates. However, progress in wide-area networking has been relatively slow. Although multi-gigabit networks are becoming more widely deployed within institutions, access to remote resources is still hindered by slow, high-latency networks. Efficient compression, progressive data transmission, and intelligent data caching and computation reuse methods will continue to play critical roles in enabling digital pathology and scientific collaborations involving large pathology image datasets.

ACKNOWLEDGMENTS

This work was supported in part by SAIC/NCI Contract No. HHSN261200800001E from the National Cancer Institute and Contract No. R24HL085343 from the National Heart Lung and Blood Institute, by Grants 1R01LM011119-01 and R01LM009239 from the National Library of Medicine, by Grant RC4MD005964 from National Institutes of Health, by PHS Grant UL1RR025008 from the Clinical and Translational Science Awards program, and by Grant P20 EB000591 from the Biomedical Information Science and Technology Initiative program.

REFERENCES

1. B. H. Mayall and M. L. Mendelsohn. "Deoxyribonucleic acid cytophotometry of stained human leukocytes. II. The mechanical scanner od CYDAC, the theory of scanning photometry and the magnitude of residual errors." *J Histochem Cytochem*, 18 (6), 383–407, 1970.

2. T. J. Eide, I. Nordrum, and H. Stalsberg. "The validity of frozen section diagnosis based on video-microscopy." *Zentralbl Pathol*, 138 (6), 405–407, 1992.

3. T. J. Eide and I. Nordrum. "Frozen section service via the telenetwork in northern Norway." *Zentralbl Pathol*, 138 (6), 409–412, 1992.

4. K. J. Kaplan, J. R. Burgess, G. D. Sandberg, et al. "Use of robotic telepathology for frozen-section diagnosis: a retrospective trial of a telepathology system for intraoperative consultation." *Mod Pathol*, 15, (11), 1197–1204, 2002.

5. I. Nordrum, B. Engum, E. Rinde, et al. "Remote frozen section service: a telepathology project in northern Norway." *Hum Pathol*, 22 (6), 514–518, 1991.

6. E. G. Fey and S. Penman. "The morphological oncogenic signature. Reorganization of epithelial cytoarchitecture and metabolic regulation by tumor promoters and by transformation." *Dev Biol (N Y 1985)*, 3, 81–100, 1986.

7. R. S. Weinstein, K. J. Bloom, and L. S. Rozek. "Telepathology and the networking of pathology diagnostic services." *Arch Pathol Lab Med*, 111 (7), 646–652, 1987.

8. R. S. Weinstein, A. R. Graham, L. C. Richter, et al. "Overview of telepathology, virtual microscopy, and whole slide imaging: prospects for the future." *Hum Pathol*, 40 (8), 1057–1069, 2009.

9. S. Williams, W. H. Henricks, M. J. Becich, et al. "Telepathology for patient care: what am I getting myself into?" *Adv Anat Pathol*, 17 (2), 130–149, 2010.

10. M. G. Rojo, G. B. Garcia, C. P. Mateos, et al. "Critical comparison of 31 commercially available digital slide systems in pathology." *Int J Surg Pathol*, 14 (4), 285–305, 2006.

11. D. C. Wilbur, K. Madi, R. B. Colvin, et al. "Whole-slide imaging digital pathology as a platform for teleconsultation: A pilot study using paired subspecialist correlations." *Arch Pathol Lab Med*, 133 (12), 1949–1953, 2009.

12. J. R. Gilbertson, J. Ho, L. Anthony, et al. "Primary histologic diagnosis using automated whole slide imaging: a validation study." *BMC Clin Pathol*, 6, 4, 2006.

13. A. Afework, M. D. Beynon, F. Bustamante, et al. "Digital dynamic telepathology—the virtual microscope." *Proceedings of the AMIA Symposium*, pp. 912–916, the American Medical Informatics Association, Bethesda, MD, 1998.

14. U. Catalyurek, M. D. Beynon, C. Chang, et al. "The virtual microscope." *IEEE Trans Inf Technol Biomed*, 7 (4), 230–248, 2003.

15. R. Ferreira, B. Moon, J. Humphries, et al. "The virtual microscope." *Proceedings of the AMIA Annual Fall Symposium*, pp. 449–453, the American Medical Informatics Association, Bethesda, MD, 1997.

16. U. J. Balis. "Telemedicine and telepathology." *Clin Lab Med*, 17 (2), 245–261, 1997.

17. M. Dziegielewski, G. M. Velan, and R. K. Kumar. "Teaching pathology using 'hotspotted' digital images." *Med Educ*, 37 (11), 1047–1048, 2003.
18. C. S. Farah and T. Maybury. "Implementing digital technology to enhance student learning of pathology." *Eur J Dent Educ*, 13 (3), 172–178, 2009.
19. P. N. Furness. "The use of digital images in pathology." *J Pathol*, 183 (3), 253–263, 1997.
20. M. Guzman and A. R. Judkins. "Digital pathology: A tool for 21st century neuropathology." *Brain Pathol*, 19 (2), 305–316, 2009.
21. F. J. Leong and A. S. Leong. "Digital imaging in pathology: Theoretical and practical considerations, and applications." *Pathology*, 36 (3), 234–241, 2004.
22. A. M. Marchevsky, R. Dulbandzhyan, K. Seely, et al. "Storage and distribution of pathology digital images using integrated web-based viewing systems." *Arch Pathol Lab Med*, 126 (5), 533–539, 2002.
23. J. H. Saltz. "Digital pathology—the big picture." *Hum Pathol*, 31 (7), 779–780, 2000.
24. L. Yang, W. Chen, P. Meer, et al. "Virtual microscopy and grid-enabled decision support for large-scale analysis of imaged pathology specimens." *IEEE Trans Inf Technol Biomed*, 13 (4), 636–644, 2009.
25. L. Zheng, A. W. Wetzel, J. Gilbertson, et al. "Design and analysis of a content-based pathology image retrieval system." *IEEE Trans Inf Technol Biomed*, 7 (4), 249–255, 2003.
26. M. Hadida-Hassan, S. J. Young, S. T. Peltier, et al. "Web-based telemicroscopy." *J Struct Biol*, 125 (2–3), 235–245, 1999.
27. J. Saltz, T. Kurc, L. Cooper, et al. "Multi-Scale, integrative study of brain tumor: In silico brain tumor research center." *Proceedings of the Annual Symposium of American Medical Informatics Association 2010 Summit on Translational Bioinformatics (AMIA-TBI 2010)*, the American Medical Informatics Association, San Francisco, CA, 2010.
28. L. A. Cooper, J. Kong, D. A. Gutman, et al. "An integrative approach for in silico glioma research." *IEEE Trans Biomed Eng*, 57 (10), 2617–2621, 2010.
29. L. A. D. Cooper, J. Kong, F. Wang, et al. "Morphological Signatures and Genomic Correlates in Glioblastoma." *International Symposium on Biomedical Imaging*, the Institute of Electrical and Electronics Engineers (IEEE), Chicago, 2011.
30. L. Cooper, J. Kong, D. Gutman, et al. "Integrated morphologic analysis for the identification and characterization of disease subtypes." *J Am Med Inform Assoc*, 19 (2), 317–323, 2012.
31. M. N. Gurcan, L. Boucheron, A. Can, et al. "Histopathological image analysis: a review." *IEEE Rev Biomed Eng*, 2, 147–171, 2009.
32. M. N. Gurcan, J. Kong, O. Sertel, et al. "Computerized pathological image analysis for neuroblastoma prognosis." *AMIA Annu Symp Proc*, 304–308, 2007.
33. L. Yang, W. Chen, P. Meer, et al. "High throughput analysis of breast cancer specimens on the grid." *Med Image Comput Comput Assist Interv Int Conf Med Image Comput Comput Assist Interv*, 10 (Pt 1), 617–625, 2007.

34. V. Kumar, T. Kurc, M. Hall, et al. "An integrated framework for parameter-based optimization of scientific workflows." *The 18th International Symposium on High Performance and Distributed Computing (HPDC 2009)*, Germany, Association for Computing Machinery (ACM), New York, 2009.

35. V. Kumar, B. Rutt, T. Kurc, et al. "Large-scale biomedical image analysis in grid environments." *IEEE Trans Inf Technol Biomed*, 12 (2), 154–161, 2008.

36. M. Beynon, T. Kurc, U. Catalyurek, et al. "Distributed processing of very large datasets with datacutter." *Parallel Comput*, 27 (11), 1457–2478, 2001.

37. D. Meilander, M. Schellmann, S. Gorlatch, et al. "Parallel medical image reconstruction: from graphics processing units (GPU) to Grids." *J Supercomput*, 57 (2), 151–160, 2011.

38. Z. G. Ying, C. Yong, J. K. Udupa, et al. "Parallel fuzzy connected image segmentation on GPU." *Med Phys*, 38 (7), 4365–4371, 2011.

39. M. C. Huang, F. Liu, and E. H. Wu. "A GPU-based matting Laplacian solver for high resolution image matting." *Vis Comput*, 26 (6–8), 943–950, 2010.

40. R. Shams, P. Sadeghi, R. A. Kennedy, et al. "A survey of medical image registration on multicore and the GPU." *IEEE Signal Process Mag*, 27 (2), 50–60, 2010.

41. A. Abramov, T. Kulvicius, F. Worgotter, et al. "Real-time image segmentation on a GPU." In *Facing the Multicore-Challenge: Aspects of New Paradigms and Technologies in Parallel Computing*, R. Keller, D. Kramer, J-P. Weiss, eds., Vol. 6310, pp. 131–142, Springer-Verlag Berlin, Heidelberg, Germany, 2010.

42. N. Singhal, I. K. Park, and S. Cho. "Implementation and optimization of image processing algorithms on handheld GPU." *IEEE International Conference on Image Processing*, pp. 4481–4484, 2010.

43. N. Zhang, J. L. Wang, and Y. S. Chen. "Image parallel processing based on GPU." *The 2nd IEEE International Conference on Advanced Computer Control*, Vol. 3, pp. 367–370, 2010.

44. A. Herout, R. Josth, P. Zemcik, et al. "GP-GPU implementation of the 'local rank differences' image feature." *Comput Vis Graph*, 5337, 380–390, 2009.

45. Y. Allusse, P. Horain, A. Agarwal, et al. "GpuCV: A GPU-accelerated framework for image processing and computer vision." *Advances in Visual Computing, Part II, Proceedings*, Vol. 5359, pp. 430–439, Springer-Verlag Berlin, Heidelberg, Germany, 2008.

46. Z. P. Xu and W. B. Xu. "GPU in texture image processing." *Proceedings of International Symposium on Distributed Computing and Applications to Business, Engineering and Science (DCABES)*, Vols. 1 and 2, pp. 380–383, 2006.

47. M. Schmeisser, B. C. Heisen, M. Luettich, et al. "Parallel, distributed and GPU computing technologies in single-particle electron microscopy." *Acta Crystallogr D Biol*, 65, 659–671, 2009.

48. D. Crookes, P. Miller, H. Gribben, et al. "GPU implementation of Map-Mrf for microscopy imagery segmentation." *IEEE International Symposium on Biomedical Imaging: From Nano to Macro*, Vols. 1 and 2, pp. 526–529, 2009.

49. G. M. Tan, Z. Y. Guo, M. Y. Chen, et al. "Single-particle 3D reconstruction from cryo-electron microscopy images on GPU." *ICS'09: Proceedings of the 2009 ACM SIGARCH International Conference on Supercomputing*, pp. 380–389, Association for Computing Machinery (ACM), New York, 2009.

50. T. D. R. Hartley, A. R. Fasih, C. A. Berdanier, et al. "Investigating the use of GPU-accelerated nodes for SAR image formation." *IEEE International Conference on Cluster Computing and Workshops*, pp. 663–670, 2009.

51. A. Ruiz, O. Sertel, M. Ujaldon, et al. "Pathological image analysis using the GPU: Stroma classification for neuroblastoma." *IEEE International Conference on Bioinformatics and Biomedicine*, pp. 78–85, 2007.

52. G. Teodoro, R. Sachetto, O. Sertel, et al. "Coordinating the use of GPU and CPU for improving performance of compute intensive applications." *IEEE International Conference on Cluster Computing and Workshops*, pp. 437–446, 2009.

53. G. Teodoro, T. Kurc, T. Pan, et al. "Accelerating large scale image analyses on parallel, CPU-GPU equipped systems." *The 26th IEEE International Parallel and Distributed Processing Symposium (IPDPS)*, 2012.

54. I. G. Goldberg, C. Allan, J. M. Burel, et al. "The open microscopy environment (OME) data model and XML file: Open tools for informatics and quantitative analysis in biological imaging." *Genome Biol*, 6 (5), R47, 2005.

55. D. J. Foran, L. Yang, W. Chen, et al. "ImageMiner: A software system for comparative analysis of tissue microarrays using content-based image retrieval, high-performance computing, and grid technology." *J Am Med Inform Assoc*, 18(4), 403–415, 2011.

56. F. Wang, J. Kong, L. Cooper, et al. "A data model and database for high-resolution pathology analytical image informatics." *J Pathol Inform*, 2, 32, 2011.

57. M. E. Martone, A. Gupta, M. Wong, et al. "A cell-centered database for electron tomographic data." *J Struct Biol*, 138 (1–2), 145–155, 2002.

58. M. E. Martone, S. Zhang, A. Gupta, et al. "The cell-centered database: a database for multiscale structural and protein localization data from light and electron microscopy." *Neuroinformatics*, 1 (4), 379–395, 2003.

59. DICOM. "Digital imaging and communications in medicine." May 2011; http://medical.nema.org/.

60. V. Kumar, T. Kurc, J. Saltz, et al. "Architectural implications for spatial object association algorithms." *The 23rd IEEE International Parallel and Distributed Processing Symposium (IPDPS 09), Rome, Italy*, the Institute of Electrical and Electronics Engineers (IEEE), Washington, DC, 2009.

61. J. Gray, M. Nieto-Santisteban, and A. Szalay. "The zones algorithm for finding points-near-a-point or cross-matching spatial datasets." *The ACM Computing Research Repository (CoRR), abs/cs/0701171*, pp. 18., 2007.

62. J. Becla, K.-T. Lim, S. Monkewitz, et al. "Organizing the extremely large LSST database for real-time astronomical processing." *The 17th Annual Astronomical Data Analysis Software and Systems Conference (ADASS 2007), London, England*, vol. 1, pp. 4, 2007.

63. J. Dean and S. Ghemawat. "MapReduce: Simplified data processing on large clusters." *USENIX Association Proceedings of the Sixth Symposium on Operating Systems Design and Implementation*, pp. 137–149, 2004.

64. N. Beckmann, H. Kriegel, R. Schneider, et al. "The R* tree: An effcient and robust access method for points and rectangles." *Proceedings of the ACM SIGMOD International Conference on Management of Data*, pp. 322–331, ACM Press, New York, 1990.

65. K. Wilkinson, C. Sayers, H. A. Kuno, et al. "Efficient RDF storage and retrieval in Jena2." *Proceedings of VLDB Workshop on Semantic Web and Databases*, pp. 131–150, Berlin, Germany, 2003.

66. J. Broekstra, A. Kampman, and F. van Harmelen. "Sesame: A generic architecture for storing and querying RDF and RDF schema." *International Semantic Web Conference, Lecture Notes in Computer Science*, no. 2342, pp. 54–68, 2002.

67. A. Kiryakov, D. Ognyanov, and D. Manov. "OWLIM—A pragmatic semantic repository for OWL." *WISE Workshops, volume 3807 of Lecture Notes in Computer Science*, pp. 182–192, 2005.

68. E. F. Hill. *Jess in Action: Java Rule-Based Systems.* Manning Publications, Greenwich, CT, 2003.

69. S. Narayanan, U. Catalyurek, T. Kurc, et al. "Parallel materialization of large Aboxes." *The 24th Annual ACM Symposium on Applied Computing (SAC 2009), Hawaii, March* 2009.

70. V. Kumar, S. Narayanan, T. Kurc, et al. "Analysis and semantic querying in large biomedical image datasets." *IEEE Computer Magazine, special issue on Data-Intensive Computing*, 41 (4), 52–59, 2008.

CHAPTER 4

Answering Fundamental Questions about the Universe

Eric S. Myra and F. Douglas Swesty

CONTENTS

4.1 INTRODUCTION: CHARACTERIZING PROBLEMS IN ASTROPHYSICS

Over the past few decades, computational modeling of astrophysical phenomena has emerged as the premiere tool for the theoretical astrophysicist to gain a better understanding of the universe. The state of the art in computational astrophysics has rapidly advanced, enabled by the rapid growth in floating-point computational power as described by Moore's law.

Scientific discovery, enabled by computational modeling, has allowed us to gain a better understanding of the conditions present at the very birth of the universe as well as to better understand the birth and death of stars and the ultimate fate of the universe itself.

However, from its nascent years, limits on the field of computational astrophysics have been set by, among other things, the ability to manage large amounts of data from numerical simulations. The modifier "large" needs some qualification since what qualifies as large depends strongly on the time frame being considered. In the 1970s, a few hundred kilobytes of data may have been considered large. In contrast, at the time of writing, in 2013, we have entered an era when dataset sizes of petabytes have become a reality, while dataset sizes of terabytes are almost ubiquitous within the field.

Problems in computational astrophysics and closely related applied fields, such as the study of high-temperature shock-wave and explosive phenomena, have always been considered compute-intensive and have played a key role in driving the development of supercomputing technology. In more recent years, although the characterization of these applications as compute-intensive remains as accurate as ever, they have become increasingly data-intensive as well. It is this data-intensive attribute that forms the subject of this chapter.

As a field of research, astrophysics seeks to apply the laws of physics, which can act over an extremely wide range of distance and timescales, to explain phenomena in the universe. These astrophysical phenomena almost always occur at vastly larger scales than terrestrial events. They also involve an elaborate interplay among diverse branches of physics. Given this nature, problems in computational astrophysics rapidly become complex and can almost always be characterized as follows:

- *Multiscale*: Comprised of physical processes ranging from the microscopic to the macroscopic that are coupled in some way.

- *Multiphysics*: Comprised of diverse branches of physics; a simulation in computational nuclear astrophysics can combine, for example, both the strong and weak nuclear interactions, electromagnetism, radiation, hydrodynamics, statistical mechanics, and gravitational physics.

- *Multiregime*: Having components that can exhibit, for example, classical behavior in one region or scale of the problem and quantum mechanical behavior in another; or exhibiting transport phenomena that cover a wide range in the magnitude of coefficients and, hence, transport behavior.

- *Interdisciplinary*: Having solution and analysis techniques that combine diverse branches of physics with one or more of mathematics, computational science, computer science, statistics, and perhaps other fields as well.

As a *multiscale* environment, the spatial and temporal scale of problems in astrophysics is immense. Microscopic processes of interest may cover subnuclear, nuclear, atomic, and molecular distance scales, which range from $<10^{-15}$ to 10^{-10} m and beyond. The scales of macroscopic systems being modeled can range from planetary and stellar (10^6–10^{11} m) to stellar cluster (~10^{15} m), to galactic (~10^{21} m), to clusters (~10^{23} m) and to superclusters (~10^{24} m) of galaxies, and beyond (see Figure 4.1). Interesting temporal scales are equally wide ranging, with microscopic reactions occurring on a timescale of as little as 10^{-15} s or less, while the upper end is bounded only by the age of the universe itself (~10^{17} s).

Although any given astrophysics problem may present a wide range of possible scales, many of the important processes (and those that drive the evolution of a system under study) are interesting on scales intermediate (in space, time, energy, etc.) to the microscopic and to the largest macroscales in a simulation. This might include such multiscale processes as shock waves, development and evolution of instabilities, turbulence, and mixing. Complicating effects might include the presence of gravitational fields, electromagnetic fields, radiation, reactive flow, and lack of thermodynamic equilibrium in some key processes.

Although the steady advances in compute- and data-intensive science have enabled significant new achievements, no foreseeable technological advance will ever enable adequate modeling of all processes at all scales. (However, see later in this section for how some significant inroads into this issue are being envisioned.) As Rosner (2008) has put it, much computational astrophysics is in a class of problems whose computing and data requirements can be labeled as "voracious." Solutions of such problems can consume every potentially available computer cycle in a boundless quest for greater accuracy, higher precision, and increased fidelity of the input physics. A small sample of voracious multiscale astrophysics problems includes modeling shock waves, neutrino transport, and resulting instabilities in core-collapse supernovae (Bruenn et al. 2005; Keil et al. 1996;

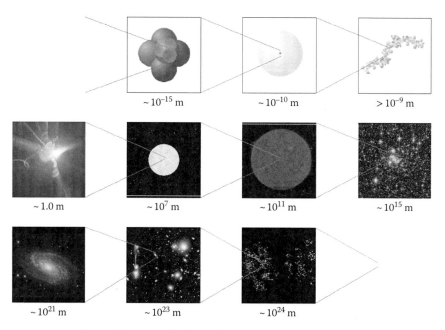

FIGURE 4.1 **(See color insert.)** The multiple scales of computational astrophysics. Processes that are the primary drivers of the evolution of astrophysical systems are frequently microscopic in origin, but the phenomena of interest are usually macroscopic process on one or more macroscopic length scales. The top row shows typical length scales for microscopic processes, ranging from the quark and nuclear scales of 10^{-15} m (themselves, scales that likely originate from smaller constituent scales) through to atomic and molecular scales (which can have numerous scales, depending on the molecule of interest). The macroscales on which simulations, experiments, and observations take place are on the bottom two rows: on the middle row (from the left), the laser-driven laboratory-astrophysics scale, planetary and white-dwarf scale, stellar giant scale, and star-cluster scale. On the bottom row, one reaches scales that are increasingly cosmological in size: (from the left) galaxy, galaxy cluster, and galaxy supercluster. (Image credits: middle row left: Carolyn Kuranz, University of Michigan; center right and far right: NASA; bottom row: NASA.)

Swesty and Myra 2005); modeling flame physics in thermonuclear supernovae (Townsley 2009); modeling turbulence, as might occur in interstellar molecular clouds during the process of star formation (Kritsuk and Norman 2004); investigating the role of supermassive black holes and accretion in cosmological structure formation (Sutter and Ricker 2010).

Nearly all large-scale problems in astrophysics are *multiphysics* problems in the way outlined in the preceding. Almost invariably, a problem in astrophysics consists of modeling collections of discrete objects and/or fluids whose time-evolutionary behavior is sought. These items are subject to one or more interactions that drive their subsequent evolution: nuclear reactions, chemical reactions, electromagnetic fields, radiation, and so on. Each interaction introduces additional physics modules into the model, along with coupling terms that govern how their presence affects the other components of the model. Such coupling can be complex, with strengths that vary spatially and temporally in various ways. The ability to solve multiphysics problems and the ability to perform data-intensive science are closely correlated, since each additional physics module adds to the number of physical quantities whose evolution is followed across the problem domain.

Computational astrophysics is *multiregime*, and even more strongly so, if we include the emerging field of laboratory astrophysics (or high-energy-density physics). In laboratory-astrophysics experiments, terrestrial facilities irradiate test materials using high-power lasers to create conditions similar to that found in some key astrophysical environments (Remington et al. 2006). This capability gives experimental insight into the astrophysical processes and the high-temperature environments in which they occur. This capability also provides validation data for astrophysics codes. When combined, laboratory and conventional astrophysical codes may model diverse conditions, including solids, liquids, gases, and plasmas, with materials perhaps transitioning between these states as part of the model; they can include pure elements, molecules, and mixtures of materials; materials having anisotropies and/or inhomogeneities; radiation, perhaps consisting of photons, light, and/or heavy particles. Each of these possible constituents may find itself in several physical regimes during the evolution of a problem of interest and, thus, may require several different physical models to describe its state or evolution.

Advances in computing and computational technology have enabled data-intensive astrophysics by permitting the relaxation of two longstanding, limiting constraints: *model dimensionality* and *model resolution*.

From the early days of electronic computing, all but the simplest problems in computational astrophysics have been artificially constrained with respect to model dimensionality. Although simple observation tells us that most interesting astrophysical processes are laden with multidimensional phenomena, limits in computational power have required dimensional reduction through the imposition of symmetry to the problem, for example, spherical symmetry for models of stellar evolution and axial symmetry for models incorporating accretion disks.

Increasingly, one of the enabling benefits being realized is that compute- and data-intensive science are being relieved of dimensional constraints. It is now common for astrophysical simulations to be performed in three spatial dimensions. This freedom unleashes the full consequences of any inherently multidimensional terms in the modeling equations. Thus, problems whose equations contain cross products or tensors, such as rotational mechanics, fluid mechanics, electromagnetism, anisotropic media, and general relativity, are affected. Furthermore, this is the only way these problems can really be modeled correctly.

When we look at treating radiation in astrophysical problems, dimensionality is even more important. Full solution of a problem in radiation transport requires a complete description of six-dimensional phase space (three spatial and three momentum dimensions) evolved in time. Although data-intensive science seeks to expand the realm of computationally tractable problems, as of this writing, the routine execution of radiation hydrodynamics problems that are fully six-dimensional still presents a challenge for today's computer architectures. High-resolution simulations, with a full complement of physics, are presently restricted to occasional "hero runs" on one of today's large capability systems.

The issue of model resolution enters computational astrophysics in several ways. First, and most straightforwardly, one often desires higher resolution without necessarily increasing the fidelity of any included physics modules. Alone, this higher resolution can allow better resolution of shock waves, more detailed study of instabilities, more accurate tracking of features, and better characterization of such quantities as peak temperatures and interface locations. Increasing resolution also contributes great value to the process of code verification: It can be used to show evidence that implemented algorithms show satisfactory convergence properties in space and time.

A second way resolution benefits a simulation is by permitting better refinement of microscopic or subgrid models of important physical processes. So, in addition to the better grid resolution just discussed, that finer grid can additionally contain information is more faithful to the underlying physics. This improved physics is often only of value when a model's grid resolution is sufficiently fine that important features and processes can be identified. Improved microphysics can enter in several ways, including better microscopic models for local quantities (e.g., thermodynamic properties of materials; coupling coefficients between multiphysics components, such as matter or radiation; and improved reaction networks), better coefficients for nonlocal effects (e.g., transport properties, including multiphysics coupling between matter and radiation), or improved models for tracking interfaces or surface effects.

Improved resolution can further benefit data-intensive astrophysics through a third, less direct way, namely, through the implementation of tiered subgrid multiscale models. Although a multiscale approach has long been used implicitly in essentially every computational astrophysics simulation, it has only been relatively recently that its fuller potential has started to be demonstrated. As a first example, we consider models of nuclear burning and flame front propagation in thermonuclear (type Ia) supernovae (Townsley 2009), as shown in Figure 4.2. The problem contains many length scales that are much finer than any feasible computational grid containing the complete problem. (The smallest length scale can be

FIGURE 4.2 **(See color insert.)** Image showing fluid-mixing instabilities as part of a thermonuclear flame model in a subgrid simulation of a type Ia supernova using the FLASH code. The figure shows the developing flame surface influenced by Rayleigh–Taylor instabilities. (Figure adapted from Calder, A. C. et al., *Astrophysical Journal Supplement Series*, 143:201–229, 2002, and provided courtesy of the Flash Center for Computational Science, University of Chicago.)

as much as a factor of 10^{-10} smaller than the macroscale grid spacing.) An approach to this issue is to compute the flame physics on a tractable scale for a sufficient number of representative portions of the complete physical domain. This microscale solution is then used to provide subgrid-based input for the macroscale model.

As a second example, we divert to the field of material science, where Abraham et al. (1998) demonstrate the propagation of cracks in materials by introducing intermediate-scale models between the largest macroscopic continuum scale of ultimate interest (>1 mm) and the smallest *ab initio* quantum mechanical approach at the level of the atomic constituents (at a scale of ~1 nm). Two mesoscales are used to link these extremes—one on a larger scale, which models dislocations, and the other on a smaller scale, which models atomic interactions by a semiclassical mass-on-springs approach. An appropriate modification of this four-tiered approach may find astrophysical application modeling the possible formation of cracks in the crusts of neutron stars. Additionally, a broader version of this approach may be useful in developing more detailed response models of materials used in laboratory-astrophysics experiments.

Besides the increased accuracy afforded by this class of techniques (as compared against purely macroscopic approximations), there is also opportunity to account for subgrid microscopic geometric effects such as lattices, molecular chains, filaments, and voids. By using mesoscale models, the need for assumptions such as local thermodynamic equilibrium can be relaxed. Macroscopic state variables, such as energies, pressures, and material strengths, which are needed by the hydrodynamic equations, can be computed directly, rather than assuming a possibly inaccurate equilibrium model. Enriched with these techniques, there is some hope of being able to model rapid timescale, multiple-distance-scale processes more realistically. This approach actually represents a hierarchy of data-intensive science, since each scale can represent a data-intensive problem in itself.

When faced with what are clearly voracious computing problems, we might ask how much resolution and how much multiscale modeling are ultimately enough? What additional science is delivered as resolution increases? Are there thresholds in computational and data-intensive capability that lead to the delivery of new science? Can we assume that "what's past is prologue" guides predictions of future thresholds?

We have seen that, in the recent past, additional dimensionality and the addition of multiphysics modules have clearly been important thresholds in many areas of astrophysical discovery. As these thresholds have been crossed, major qualitative improvements in science have resulted. Anticipating future thresholds, there are a number of unexplored areas of astrophysics where added dimensionality and additional physics will allow future study. As mentioned previously, the possible roles of magnetic fields or the effects of radiation are areas that benefit. These are obvious future science thresholds that we can predict, based on past experience. In contrast, and further to our preceding discussion, a less obvious threshold might be the ability to run entire multiscale models, using a "sufficient" number of tiered scales, all executing concurrently. Can we anticipate some future advances, analogous to the development of adaptive mesh refinement (AMR), that will easily permit adaptive computation at multiple scales within a single computational domain?

Data-intensive science has enabled support for more serious verification and validation (V&V) studies. Large multiphysics codes present special problems for systematic verification (Keyes et al. 2013). This is because standard multiphysics test problems are few and, in a large code, the number of possible regime and coupling combinations are many. Although it is difficult to see how this will not always be a challenge to manage from a logistical perspective, data-intensive capability at least gives us the system resources to make inroads that can lead to better approaches. Additionally, we now have an improving ability to perform detailed code-to-code comparisons, which contribute to increased confidence in all codes involved in a study. Finally, a data-intensive environment has given us ability to make more thorough analysis of existing validation data. The detail now provided by simulations can also be used to design experiments that themselves yield additional validation data. Although this entire area remains challenging, without a data-intensive computing environment, it would be prohibitive to approach V&V studies in any serious way. Similarly, a data-intensive paradigm supports reasonable approaches to Uncertainty Quantification (UQ) and Quantification of Margins and Uncertainties (QMU), which are of growing importance to the astrophysics community.

The creation/gathering, processing, analysis, application, and archiving of astrophysical data are large subjects and their details can be explored in this chapter only briefly. So that we can have some chance of discussing important issues, we find it necessary to limit our discussion here to the field of *computational* astrophysics. Unfortunately, this restriction does not allow us to present any detail on the advances in observational branches of astronomy and space science that result from emerging data-rich environments. In passing, we mention two of the most well-known observational repositories: (1) the Multimission Archive at Space Telescope Science Institute (http://archive.stsci.edu/index.html), which contains data from numerous NASA astronomy missions, including data from the Hubble Space Telescope, and (2) the Sloan Digital Sky Survey (http://www.sdss.org). Both repositories contain multiple databases, the largest among them individually contain several tens of terabytes. Although these databases are no longer considered massive by current leading-edge standards, they have nevertheless been groundbreaking, both in the scientific discoveries they have enabled and in their approaches to management of digital astronomical data. We also mention the planned ground-based Large Synoptic Survey Telescope (LSST, http://www.lsst.org/lsst), which is anticipated to collect more than 1 petabyte of data per year and requires in excess of 100 teraflops of computing power when fully operational.

Even by restricting the scope of our discussion in this chapter to computational astrophysics, we leave ourselves much ground to cover, since the field represents a rich mixture of problems and possible solutions representing the birth, life, and death of all astronomical objects from the basic subatomic particles up through the elements, planets, stars, galaxies, and ultimately the universe itself. Although the dominant physics and the computational approaches vary widely among these problems, and among the people who study them, they do share many common features and issues. It is these characteristics that we address in the sections that follow. To do this effectively, we will describe a case study taken primarily from our own study of core-collapse supernovae, a branch computational nuclear astrophysics. Throughout, we make frequent reference to V2D, a two-dimensional radiation-hydrodynamics code for applications in computational nuclear astrophysics and high-energy-density physics, which we have developed (Myra et al. 2009; Swesty and Myra 2009). It has been designed to take best advantage of current-generation and foreseeable high-performance computing (HPC) system architectures. It is developed with advanced algorithms and state-of-the-art code-development practices. Throughout, we distinguish design and practice used in the V2D code and workflows that derive from it, from the designs and practices of other computational astrophysics codes with which we are familiar. We also point out the commonalities and differences that exist within the computational astrophysics community with respect to algorithmic, coding, and workflow practices.

In the DOE Office of Science Data-Management Challenge (Mount 2004, p. 19), classification of data is made into three primary types: (1)

simulation driven, (2) observation/experimentally driven, and (3) information intensive. Each of these is applicable to astrophysics. Relevant to discussion in this chapter, category (1) is presently the most significant source of data. Data in category (2) are primary restricted to observational astronomy and include the data in astronomical databases mentioned earlier in this section. However, validation data from astronomical observations and laboratory-based astrophysics experiments also contribute to this category. Finally, data in category (3) were almost completely absent from computational astrophysics more than about a decade ago. However, information-intensive data are increasing both in importance and in dataset size, and it will possibly make up a significant portion of computational astrophysics data in the future.

The remainder of the chapter is organized as follows: in Section 4.2, we describe data types that are encountered in computational astrophysics. In Section 4.3, we outline the scope of dataset sizes we encounter. Section 4.4 contains a description of how these datasets are created and organized by applications that produce them. Data nonlocality and network issues are the subjects of Section 4.5; how nonlocality affects data analysis is the subject of Section 4.6. In Section 4.7, we discuss data archiving, and in Section 4.8, we discuss data and workflow. Scientific visualization of large datasets is the subject of Section 4.9. We conclude by giving our outlook in Section 4.10.

4.2 DATA TYPES

Most computer simulations in astrophysics involve the numerical solution of partial differential equations that describe either the static structure or the time-evolution of matter, radiation, and the gravitational field. Such simulations can, in general, be complicated by the inclusion of a variety of microphysical process for the chemical evolution of matter and the interaction of radiation and matter. Furthermore, adding another layer of complexity, problems may involve widely disparate scales in time and space. The most common approach to solving these sets of partial differential equations (PDEs) involves the use of finite-difference or finite-volume techniques that rely on the discretization of the underlying PDEs on structured meshes. The computational algorithm may also make use of AMR techniques (Plewa et al. 2005) to address multiple spatial scales present in a problem.

This particular set of simulation methodologies gives rise to data in the form of arrays representing discretized values of data on a (perhaps hierarchically) structured mesh. Because of the relatively simple problem geometries found in astrophysical applications, unstructured meshes, which are widely utilized in engineering applications, are less likely to be employed. The use of structured meshes has three main advantages: the meshes are easier to partition for parallel simulations, parallel input/output (I/O) of the mesh variables may be more easily handled, and the code will exhibit regular patterns of memory access allowing codes to be highly optimized for cache efficiency.

Data on these meshes is typically stored in the form of multidimensional arrays of floating-point data representing the physical quantities being modeled, that is, matter density, temperature, gravitational potential, and so on. These quantities can be scalar, vector, or tensor in nature. The number of variables required is problem-dependent. In a two-dimensional fluid dynamic simulation of the evolution of an ideal gas, this may involve as few as four variables at each spatial point: the gas density, the gas temperature (or equivalently gas or total energy), and the two components of the fluid velocity. In more complex simulations involving the modeling of the evolution of a radiation spectrum, with reactive-flow chemistry present, there could be hundreds or thousands of variables required to fully represent the system at each point in space.

4.3 DATASET DIMENSIONALITY AND SIZES

Two decades ago, the limitations of computer architectures placed severe restrictions on the types and sizes of astrophysical simulations. In the present day, the capability of architectures has grown to allow simulations that were undreamt of just a quarter century ago. Nevertheless, even at the present time, there are astrophysical problems that are intractable at the petascale and even, in the future, at the exascale. The problem size is set by the dimensionality of the problem and the need for resolution in each dimension.

In the first few decades of computational astrophysics, problems that were considered computationally tractable were those that were primarily limited to a single spatial dimension. Typically, such computational models of that era were stellar evolution models, which solved the partial differential equations of stellar equilibrium, or time-dependent hydrodynamic models, which addressed spherically symmetric problems. It was not until the late 1980s that HPC architectures achieved sufficient capacity to address complex multidimensional problems. The issue of dataset size is easily illustrated by the idea of a basic three-dimensional fluid dynamic simulation with an ideal gas. If we suppose that our three-dimensional computational domain is discretized into 100 zones in each dimension, we have a total of $100 \times 100 \times 100 = 10^6$ spatial zones. For a simple fluid dynamic simulation, five variables must be stored in each zone: the fluid density, the fluid internal energy, and the three components of the velocity vector (Mihalas and Mihalas 1999). The concept of dimensionality encompasses not only spatial dimensions but also those of phase space. In many astrophysical applications, one seeks to model the transport of radiation through the fluid of an object such as a star. Ideally, one would like to do this by solving the Boltzmann transport equation, which describes the temporal evolution of the radiation distribution problem in a six-dimensional phase space (three spatial and three momentum dimensions). If we were to discretize each dimension of phase space with 100 zones, we would obtain a total of 10^{12} zones in the problem. Such a problem is clearly in the petascale domain. At 1000 zones for each dimension, the problem is well into the exascale. At the present time, astrophysicists often assume that radiation flow is diffusive in nature, that is, that the distribution of radiation is nearly isotropic, in which case the problem reduces to a lower dimensionality, for example, three spatial dimensions and one spectral (energy) dimension. This could still render a very large dataset that is well into the terabyte range even for a modestly resolved problem.

Modeling fluid flows with complex chemistry also increases the size of problems. This is especially true of nuclear-astrophysical phenomena, where many different isotopes of the various elements involved may be synthesized or subsumed in complex chains of nuclear reactions. Ideally, one may wish to follow the temporal evolution of hundreds or thousands of nuclear species. For each species, one must solve a continuity equation that describes the temporal evolution of each species due to both fluid advection and nuclear chemical evolution. It is quite possible that a problem may have more nuclear species than it has spatial zones in each dimension.

A final aspect of dataset size that should be mentioned relates to the organization of the data. As mentioned, in virtually all astrophysical modeling applications at this time, structured grids are utilized. The use of structured grids allows simulation codes and data processing codes to be highly optimized with regard to both data access speeds and floating-point performance. Nevertheless, an additional complexity enters many of today's simulations (Plewa et al. 2005), where use of patch-based adaptive-mesh refinement introduces another layer of complexity in dataset management for large-scale datasets.

4.4 CREATION AND ORGANIZATION OF SIMULATION DATA: PARALLEL COMPUTING AND I/O

Although the term "data-intensive" usually conjures up associations of vast amounts of output data sitting somewhere on disk or tape, this image provides an incomplete picture of the data in simulation-based science. Most data-intensive applications in astrophysics are data-intensive in their use of nearly every available system resource. The process of running a simulation can stress system memory, I/O bandwidth, and network resources, before its being transferred to storage, where a "final" dataset eventually resides. Furthermore, concurrent with or following the storage process, datasets are subjected to various kinds of analysis, filtering, visualization, reordering, and rewriting—data-intensive processes that we consider in this and the next sections.

In typical astrophysical simulations, output data that are either directly read in or written out by the simulation code can typically be classified by one or more of the following functions:

1. *Input data, starting conditions, code options.* These are data contained in files required to set up initial conditions and specify a preselected set of switches for the code to use during a simulation. Traditionally, these files have been small but, increasingly, data-intensive capability has enabled use of real-world data from observations or detailed output from other simulations as initial conditions. Although such files are no longer necessarily small, it is unusual for I/O processing on them to occur more than once per run. (In V2D, a common usage practice involves using the output of another code [RH1D] as initialization data for a V2D run [see Figure 4.4]. The same practice is frequently used for simulations involving the CRASH code at the University of Michigan [Drake et al. 2011].)

2. *Checkpoint data.* These are data, stored periodically, to enable recovery of an application in mid-run, should the application fail due to some computer system problem, expiration of allotted wall-clock time, an unfortunate choice of input option when the application was started, or perhaps a software bug. Files containing these data save the state of the simulation at predefined points within a run and contain all the information needed to restart the application from the point where the checkpoint file was written. This includes grid

information, timestep information at the point the file was written, and key state variables of the system across the complete computational domain.

3. *Diagnostic data.* These data contain information that may be employed by the user (both mid-run and post-run) to determine whether a particular run has produced valid results by some set of criteria. Files in this category might be as basic as a sequential log file or as sophisticated as complex visualization data. It has been our experience that suitably chosen visualizations of diagnostic data can be invaluable as a debugging tool by revealing possible problems in a simulation code.

4. *Output snapshot data.* These are files that contain the raw data of a run—typically snapshots of all relevant quantities in the problem made periodically throughout a run. Since each file typically involves a large amount of data, they tend to be produced sparingly, or else the time to write out this data may come to dominate the wall-clock time required to complete a simulation. In V2D and related codes, some I/O economy is achieved by having a single class of files serve as both checkpoint and output data.

5. *Output time-series data.* These are files containing data for tracking time evolution of a particular quantity of interest, where standard output data dumps are too infrequent to obtain sufficient time resolution. Files in this class include only a small, specialized portion of a complete dataset.

The issue of checkpoint data requires more discussion. There has been much recent speculation as to whether it even makes sense to perform checkpointing on future exascale systems (Cappello et al. 2009). This is because it is anticipated that the time needed to write a checkpoint file on an exascale system will be comparable to the mean time to interrupt of some key system component directly related to a running application. Irrespective of this prediction, checkpointing remains very much a going concern at the present time, and it is important to have an effective strategy for carrying it out for as long as it is feasible. In V2D, we employ user-initiated checkpointing, that is, the source code that governs the writing of checkpoint files is contained in the V2D codebase. Thus, the application directly controls the frequency of writing and the format of the files. As mentioned above in this section, we have chosen to have checkpoint files that also serve as output-snapshot-data files. This helps amortize the cost of I/O. In designing this I/O utility, we were faced with the choice between writing out every quantity in the simulation or writing out a "minimal dataset,"—a dataset that is just large enough such that any unsaved quantities can be recalculated as functions of those that have been saved. Since I/O operations are expensive timewise, V2D uses the approach of saving only a minimal dataset. Although this produces more efficient I/O performance during the run of a simulation, it comes at a price. In any post-run analysis, a minimal dataset needs to be "reconstituted" to full size at the expense of both more computing and I/O. The justification for this approach is that this additional processing can generally take place on smaller, more available, and less expensive systems than where the simulation was originally run. If data reconstitution is done locally (see Section 4.5), rather than on a remote HPC system, file transfer across the network is more efficient with minimal datasets. Furthermore, a certain amount of data reconstitution and expansion needs to take place in any event, since there are physical quantities (uncalculated during the run) that do not play any direct role in a simulation, but which lend understanding toward analyzing and interpreting a run. Finally, and perhaps of greatest importance, a minimal dataset approach mitigates the issue of stalling the progress of a simulation at points where it is blocked waiting for I/O operations to complete.

In a typical simulation, we desire periodic output at a frequency sufficient to characterize the specific model being run. In typical large-scale runs of V2D, this results in thousands of checkpoint/output snapshots per simulation. The I/O architecture of V2D is designed to write each snapshot as a single parallel-HDF5-formatted file, regardless of the number of processes involved in the job run. A single file per snapshot simplifies file management and eliminates the need for a post-processing step to coalesce multiple files. In making this choice, the costs to weigh have been (1) the need to develop and maintain a more complex I/O module in the code; (2) adoption of a community-standard file format, which probably brings net benefits, but runs the risk of not being supported in the future; and (3) the risk that this parallel I/O strategy may not perform well on future system architectures. We address these concerns in turn: Any required additional code complexity referred to in item (1) can be classified as code that is "written once; used forever." In fact, our I/O interface to HDF5 is used not just by V2D, but all radiation hydrodynamics codes and analysis packages we currently support. By using this model, we gain data portability. We write each file once on the system where a simulation takes place and subsequently read these output files many times on other systems of diverse sizes, architectures, and capabilities—literally from laptops to supercomputers. As far as item (2) is concerned, our adoption of a community standard follows a strategy that is increasingly employed elsewhere in computational astrophysics. As an example, the FLASH code (Fryxell et al. 2000) also uses HDF5 (Ross et al. 2001). As further demonstration of the benefit of this standard, FLASH's I/O module has recently been ported at the University of Michigan for use in the CRASH code. The adoption of a standard file format is a double-edged sword. Its benefit is immediate portability to many computer system architectures and potentially many applications and analysis tools. However, there is always the risk that the investment may not pay off indefinitely if the community at large does not continue its support by way of required government or industrial involvement. Finally, regarding item (3) concerning future architectures, the history of I/O performance on massively parallel systems has generally lagged somewhat relative to advances in raw computing horsepower. The same is likely to be true on exascale systems. It is our calculated assumption that useful, standardized, I/O formats will outlive any specific architecture or implementation. Although, underlying software may change, we expect interfaces and formats to be longer lived. Furthermore, the frequently used alternative of writing one file per process is fraught with even greater risk in the future. It simply ceases to be a viable method once high-performance applications feature the million-way parallelism destined for the exascale era.

The overarching issues of I/O scalability and the adequate provisioning of I/O on capability and large capacity systems are ongoing. There are certainly persistent technical issues in trying to achieve a parallelism of I/O that is comparable to the parallelism achievable in pure computation. This issue becomes especially acute at the exascale and beyond, where scalability of applications may become severely limited by lack of scalability in the I/O. (For a code to achieve, e.g., $\sim10^6$ times parallel speedup, Amdahl's law requires that $<10^{-6}$ of code execute sequentially—including the I/O!) As parallel I/O technology has advanced, systems implementations have become better at hiding some problematic I/O operations from an application. This includes such solutions as the implementation of automatic read-and-broadcast routines for situations where all processes read the same (small) file and tiered coalescing of I/O to avoid many-to-one problems. It will be of great benefit to the community if research and development on analogous issues continue.

In addition to technical challenges, there is also what might be best termed a political issue with I/O performance, since maximizing floating-point operations is almost always seen as a more publicity-worthy expenditure of limited financial resources than achieving the optimal balance of compute-to-I/O capacity. The application community needs to find a suitable forum for this concern to be addressed.

4.5 GEOGRAPHIC NONLOCALITY OF DATA: NETWORKING ISSUES

In the world of data-intensive science, reality dictates that some portion of a scientific dataset is always nonlocal to the scientist. Even if one lives and works at a HPC installation, there are local network issues to consider, since it is the rare scientist who can do his or her work sitting at the console of an HPC system with all the data sitting on the local disk! Instead, a more realistic situation presently exists where a researcher can often be 1000 km or more removed from the system where a simulation is performed. In this section, we discuss the issues of network communication. Arguments as to where data should reside and models for how it can be effectively analyzed are topics for the Section 4.5.

Although it is perhaps perpetually stylish to moan about network performance, there is no question that network capacity and bandwidth improvements of the past decades have followed a performance curve similar to that of the microprocessor. Since data *are* frequently nonlocal, network improvement has been a *sine qua non* upon which the progress of data-intensive science has relied.

At least one issue remains troublesome: While bandwidth is constantly improving, and this improvement is likely to continue, network latency is ultimately limited. (The speed of light is finite!) Logistical networking (Beck and Moore 2005), which was developed as part of U.S. Department of Energy's SciDAC program, demonstrated one method by which issues inherent in long-range file transfer could be circumvented through the introduction of intermediate storage depots. This led to much improved file-transfer speeds over the coast-to-coast file transfers taking place as part of a number of SciDAC projects. Its capability was heavily utilized by the present authors. However, no technology has much to offer in the way of improving interactive latency for tools based on a graphical user interface (GUI) when used remotely.

The so-called last-mile problem, where the effective performance of an otherwise high-speed network is impaired by slow network hardware at the destination site, continues to plague researchers and institutions. When this issue occurs, it is not always due to inferior network technology at one end or the other. It may also be due to legitimate network security concerns, the result of which is an introduction of overhead and a resultant throttling of data transfer.

4.6 EXTRACTING SCIENTIFIC MEANING: POST-PROCESSING AND ANALYSIS OF DATA

In furthering achievement of better answers for the fundamental questions of astrophysics, we need to be prepared for a vast increase in data volume. Although computational astrophysics simulations have always had the ability to produce more output data than a human being could easily assimilate, the situation has changed qualitatively over the careers of the present authors. In the present age of petabyte datasets, assimilating the output of even a single simulation cannot be made, unassisted, by a single human being over the course of an entire career. Hence, we are presented with the requirement of effective tools for large-scale data analysis.

This requirement, in turn, vastly increases post-processing computing demands. Indeed, in many branches of computational astrophysics, performance specifications for computer systems dedicated to post-processing analysis are becoming increasingly comparable to the systems on which the original simulations take place.

In Section 4.4, we discussed relying on post-run processing to reconstitute a minimal-output snapshot written by a simulation. In doing this, we have access to the physical quantities of interest needed to analyze a run. In addition to enabling routine visualizations of a simulation, post-run analysis is becoming more and more sophisticated as a means to recognize and track the evolution of transient and evolving features (e.g., shocks, vortices, and interfaces) and to perform statistical analysis of such features. Figure 4.3 shows the stratification of convective vortices observed during a simulation of an early phase of proto-neutron-star formation (Myra and Swesty 2009). The ability of automatically identifying and tracking the development and evolution of such features is a major benefit.

Post-run analysis of output data may also function as an early stage in an analysis pipeline or tree, which prepares datasets for subsequent use by other software (either commercial or homegrown). Such software may perform visualization (Section 4.9), UQ (e.g., sensitivity or regression

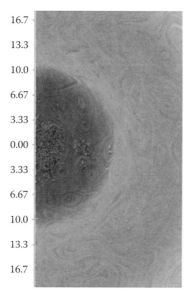

16.7

13.3

10.0

6.67

3.33

0.00

3.33

6.67

10.0

13.3

16.7

FIGURE 4.3   (**See color insert.**) Proto-neutron star convection, visualized with LEA. In this close-up of the core of a proto-neutron star, we observe stratified flow in the form of small-scale vortices. (The numerical scale shown is in kilometers.) This is an example of a class of features whose creation and evolution is of interest—one that can benefit from feature detection and tracking technology. (Visualization courtesy of Polly Baker and Ed Bachta, Indiana University; figure adapted from Myra, E. S. et al., *Computing in Science and Engineering*, 11(2):34–44, 2009.)

analysis), or data mining (e.g., extracting feature sets common to multiple runs and finding the correlation among them), to name a few possible functions. In some cases, subsequent use of analysis software may require intermediate reformatting of data or some other type of data reorganization or conversion (e.g., averaging, unit conversion, type conversion, and reordering). Although it is possible to minimize data conversion with careful design of homegrown analysis tools, this step seems unavoidable when one wants to use available commercial or other third-party software.

A question arises as to the location where it makes sense to do post-processing and analysis—remotely, where the data are created, or local to the scientist, where network response is fastest. Latency can make remote processing problematic, especially whenever visualization or a GUI is a part of the process. Latency is less of an issue for automated nonvisually based post-processing. Regardless of the choice, the network comes into play one way or another. If *all* analysis is done remotely, one can be left with the awkward problem of GUIs powered by a remote system being displayed and used on local systems. For the transcontinental distances we have faced in analyzing runs of V2D, the never-to-be-made-smaller latency (the light-travel time between local systems in New York and remote systems in California) of 10 ms, when combined with other latency overheads, makes remote GUIs frustrating to use. In practice, the typical round-trip transit time for packets between New York and California is closer to 80 ms. The optimal way to use the remote analysis model is in a client–server mode where the GUI runs on a local workstation and the analysis software run remotely. Although this can be technically sound and an effective usage model, we do not find ourselves able to use it frequently. Logistical, software, networking, and security issues often stand in the way.

An alternative model is to retrieve all the raw data back across the network and do post-processing and analysis locally. The downside to this is that data transfer times can be long and disk space requirements can often press up against the limit of what is available locally. If these issues are not prohibitive, this model can often be the most productive, since analysis tools can be used locally. In practice, we have found ourselves using this model most frequently, but it may not work for all sites or for all kinds of desired analysis.

It is also possible to do a hybrid model, where some of the data is analyzed remotely and then, as needed, is transferred to the local site. In the case of V2D, where post-processing data reconstitution expands the amount of data (potentially dramatically), use of this model requires even a greater consumption of network resources. In practice, we rarely operate this way, although it can be effective especially if, by doing this, one can somehow avoid the need to transfer all the data to the local site.

Finally, sometimes, the optimal post-processing and analysis model is dictated purely by software considerations. Licensing, availability of software for a target architecture, and multisite collaborations can lead to analysis following a combination of the preceding methods.

4.7 DATA ARCHIVING AND THE RE-ACCESS OF DATA

It has often been lightheartedly suggested that almost all data in computational astrophysics can be classified as "write once; read never." Although, at times, there has been more truth in this adage than any researcher would care to admit, the situation is changing. Frequently, raw simulation-based data are now being used as, or converted to, information-intensive data. Much of this change is the result of the increasingly important role UQ is playing in astrophysics research. Because of this, data archiving deserves attention. We briefly concern ourselves here with two broad issues in archiving: (1) how much, what form, where, and on what medium to store the data and (2) the metadata and technology needed to access this data in a "smart" way for use by analysis tools.

Although the situation may change with the introduction of exascale technology, we have found that current tape-based archiving systems, such as HPSS (http://www.hpss-collaboration.org), have proven adequate for long-term storage for a number of data-intensive projects in which we have been involved. Given the high capacity of such systems, we generally err on the side of saving too much data, rather than too little. As a result, it is common for us to save an entire raw dataset as well as a certain amount of post-processed data. Given that archival storage systems are a part of HPC facilities, archival data is almost always stored there.

However, storage of raw data can present problems when it is destined for re-access. It is our observation that, within the computational astrophysics community, there is presently little software in use to organize archival data. As data volume and the amount of re-access continues, we can foresee need for some kind of database management system (DBMS) that can keep track of data location as well as associated metadata to guide retrieval. At the University of Michigan, as part of the CRASH project, we make the practice of saving, in a data repository, key files associated with important runs. These files include input files, a full source tree for the version of the code with which a run was made, any source files that may differ from the tagged version, location information for archived output data, and annotation regarding the simulation. This is certainly a reasonable start to organizing archival data. However, with this scheme, even tasks such as simple searches remain difficult in the absence of an organized DBMS.

4.8 THE DATA PIPELINE AS PART OF WORKFLOW

Figure 4.4 shows a typical workflow of a simulation run of V2D together with some post-run processing and analysis steps. The RH1D step (shown in the shaded box at the top left of the figure) is one option for starting the workflow: this one-dimensional radiation hydrodynamics code can be used to produce input for the two-dimensional radiation-hydrodynamics initialization program V2D_IDAT. There are other options as well (not shown), including using data input from other codes or using a module within V2D_IDAT to calculate initial conditions directly. The function of V2D_IDAT is to set up the initial input data for the full simulation code

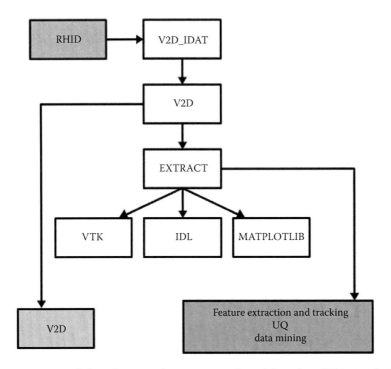

FIGURE 4.4   Workflow diagram showing typical workflow for a V2D simulation run and analysis. Also shown are some optional or occasionally performed steps. The output of the RH1D application is one possible way of creating input data for V2D_IDAT, the application that creates initial data for the simulation code, V2D. EXTRACT takes output from V2D and processes it for use by visualization packages (e.g., VTK, IDL, and MATPLOTLIB) or for other analysis software. The output from V2D may also be used to create another instance of a V2D run.

V2D. V2D_IDAT completes its task by writing a checkpoint file, which is read by V2D, when it starts execution. The program EXTRACT is our post-processing step, which uses output from V2D to generate all the quantities of interest for analysis. In Figure 4.4, we show primarily visualization packages as tools for further analysis although, as shown, such a toolset could also include feature extraction software, UQ analysis software, data mining software, and other packages. We also note that a run of V2D can trigger another run of V2D. In a subsequent run, V2D may perhaps be run with a different set of options to solve a problem whose specifications have been modified (perhaps significantly) from the previous run.

Although Figure 4.4 could be interpreted as implying that each stage of the V2D workflow is an atomic operation, that is, that a given step cannot commence until the previous step has completed, this need not be the case.

Achieving concurrency by overlapping independent steps in the pipeline is key to good wall-clock performance. For example, while an instance of V2D is running, it is possible to execute an EXTRACT post-processing step on any already-existing snapshot or time-series data. It is also possible to create and analyze diagnostics, move any available data across the networks, and archive that data. We emphasize that post-processing on the fly is necessary to enable real-time diagnosis of runs. The ability to see and respond to diagnostic data while a run is in progress is vital, especially when using capability HPC systems. Time on such systems is a precious resource that precludes waiting until the end of a run to determine whether a simulation has evolved in a valid and scientifically useful way. (For a more generic picture of workflow processes, we refer the reader to Mount [2004], where sample workflow and control flow figures are shown on pages 20 and 21.)

To attain high efficiency in workflow processes, automation of tasks is essential. Among the tasks we have had success automating are job monitoring; migration of files to tertiary (long-term) storage and migration of files to local sites for analysis; zeroth-pass analysis to confirm model validity and the value of continuing a simulation run; and first-pass analysis to do "standard" analysis to identify areas where human intervention will pay off. Many of these tasks can be accomplished with rudimentary shell scripting. However, we have been pleased with the outcome of demonstration projects using workflow-management software from the Kepler Project (Ludäscher et al. 2009).

The advent of large-scale UQ studies in computational astrophysics makes automation of tasks an even greater necessity. In addition to managing workflow of an individual job, there is the task of managing collections of these workflows as part of a UQ study. Aside from the obvious operational complexities involved in doing this effectively, there are other issues at play. One of the most important involves the selected range of input parameters within a study. Some choices can result in the failure of a run as the execution explores unvalidated areas of the simulation code. Additionally, in QMU studies, there will be runs having qualitatively different outcomes (some of which may also lead to code failure). An automated way of treating and deconvolving these issues is clearly desirable.

4.9 SCIENTIFIC VISUALIZATION OF DATA

In the early days of computational astrophysics, when data was output by line printers on green and white paper, the line plot served as the main way in which scientific data was presented so that humans could accomplish scientific discovery. With the advent of multidimensional astrophysical simulations, the demands grew, and the 1990s and the first decade of the 21st century saw the development of the field of scientific visualization as a means of analyzing the large datasets resulting from computational science and engineering efforts. This has resulted in new visualization tools such as VisIt (http://visit.llnl.gov) that are designed to handle terascale datasets. Many of the tools have resulted from the development of object-oriented classes of rendering software, like the Visualization Tool Kit (Schroeder et al. 2006), which have allowed software builder to create such new tools with relative ease. However, the increasing size of datasets adds challenges to the process of scientific discovery that are not necessarily solved by the ability to manipulate large datasets, using parallel architectures, with standard techniques such as isosurface rendering, streamline tracing, and so on. The increase in dataset sizes, as a result of the expansion of the field of computational science to multidimensional and multiphysics models, has made it more difficult to enable scientific discovery through conventional visualization techniques. The level of fine-scale structure enabled by petascale computational models means that traditional visualization techniques are insufficient to enable discovery. An excellent example of this is seen in Figure 4.3, which shows the velocity field from a two-dimensional proto-neutron star convection simulation depicted using the technique of Lagrangian–Eulerian Advection (LEA). This technique reveals a very fine-scale structure in the fluid flow that would not be readily seen using a traditional quiver plot or using arrow glyphs to depict the direction of fluid flow throughout the domain. This is an example of an innovative technique in visualization that has allowed users to visually process information in a manner that was heretofore impossible. Unfortunately, this particular technique does not easily generalize to three-dimensional problems.

The multiscale nature of modern astrophysical simulations poses ever more challenges for scientific visualization. This is also true for the increasing physical realism of models. New visualization techniques, not just new tools, are needed to enable scientific discovery at the petascale and the exascale. Unfortunately, there is no obvious road map to the creation of innovative techniques, and progress in this area is likely to be unsteady.

4.10 CONCLUSIONS

What challenges does the future hold? What opportunities lie in store? As we head inexorably toward the era of exascale computing, there are numerous obstacles that data-intensive computational astrophysics modelers will face. Among these challenges is the ability to carry out simulations and data analysis in a fault-tolerant fashion. In an era of 100K or million-processor simulations or visualizations, there will be increasingly frequent hardware failures to contend with. This is true for both processing hardware and storage hardware. As growth in datasets continues to match growth in memory capacity, how will we accommodate the inevitable bit errors that are encountered in storage and memory? What adjustments are required of us to carry out computational endeavors in this context? In addition, there is the issue of data movement into and out of computational platforms. The clear trend in HPC is toward architectures with a deeper hierarchy of latencies for data access. How will we achieve parallel I/O to checkpoint our simulations and move our results to disk where we can begin analysis? And how will we actually accomplish that analysis? With the continual growth of computational power allowing for larger problem sizes, and a wider range of spatial scales in simulations—as well as increasingly complex physics—how will we be able to make physical sense of our simulation results at the exascale and beyond?

One might be tempted to think that the answer to these questions lies in the continual growth of computing power and network throughput speeds. However, the speed of light is finite and unchanging. This ultimately determines the physical sizes of computing systems, although it is perfectly conceivable that advances in technology will continue to increase the computing density in the near-term future and perhaps beyond. The speed of light also limits response-time latencies across continental and intercontinental networks. This limit has essentially been reached as of this writing, and it is difficult to conceive of ways of effectively circumventing it.

It may very well be that the future of computational astrophysics, and computational science in general, lies not with innovation in hardware, as much as it lies with the development of new paradigms for how we address the problems that we are faced with in constructing and executing a computational model. Significant progress might compel advancements in computational techniques on a scale not seen since the advent of electronic computers.

As we conclude this chapter on computational astrophysics, it is perhaps helpful to perform the exercise of temporarily banishing some of these overwhelming issues that are expected to continue afflicting the field. This allows us to create a vision of what the field can transform into in the absence of technological limitations. In doing this, we cannot, of course, ignore the laws of physics, but it is possible to conceive that there are workarounds by which their grip might be loosened.

In much of our foregoing discussion on the current state of the field, the most vexing challenges are largely a result of data nonlocality. This includes both the large-scale geographical distribution of data (an issue that we have spent much time discussing) and the smaller system-scale nonlocality issues (e.g., cache hierarchies) faced by current and foreseeable computer architectures. It is worth asking what a research environment would look like and what capabilities could be unlocked if these constraints could somehow be relaxed. Although the scope of such problems is daunting, advances in usable remote technology (to say nothing of more generous travel budgets) could do much to relieve the large-scale nonlocality problem. Furthermore, the small-scale system-level constraints on data locality are (at least presently) better classified as issues of economics connected with creating commercially viable computer designs rather than fundamental physical constraints.

So, what could be enabled? One of the holy grails of simulation science is the ability to steer applications as they are executing. When we eliminate data-locality issues and the operational issue of rationed system resources, we find that real-time steering becomes a real possibility. Of course, to enable full steering in any useful sense, we also require full on-the-fly use of our favorite suite of analysis tools. At some level, this capability has been demonstrated for certain, more modest applications, but the ability to do this on capability systems, *at scale*, for true multiphysics applications would represent a major accomplishment. The derived benefit of enabling improved physical insight and faster research progress is difficult to overstate. Additionally, since visualization tools have demonstrated value in software debugging, such a capability also enables faster application development and code that is more likely to be correct.

To achieve true application steering, one would require an immersive virtual-reality environment of the sort available in a CAVE (Cruz-Neira et al. 1992) or similar device. Once equipped with this capability, the multidimensional, multiphysics nature of astrophysics simulations then presents some challenges: first, how to distill the crushing deluge of output data and use it to give guidance to the user steering the application, and second, how to train the user to "drive" the application. The analogy with driving an automobile is appropriate since the ability to brake, back up, and choose the speed of forward progress for an application would provide the greatest benefit—a benefit that would be further enhanced by the ability to "turn," by resetting parameters of a stopped run before resuming execution.

Sitting on top of application steering is the ability to perform UQ on the fly and steer that analysis as well. Presently, post-run analysis forms the backbone of standard UQ procedures; a real-time capability could bring benefit by enabling dynamic analysis—exploring the effects of input parameters whose relative importance may change throughout the course of a run. It is likely that this capability would require running an ensemble of applications concurrently. If a single run can be thought of as creating a single thread through solution space, then this form of UQ analysis creates a solution tube. Depending on the problem being explored, the shape and size of this tube can provide information regarding uncertainties in a calculation or perhaps be used to rule out solutions on physical grounds.

In some sense, the story of scientific simulation in the computer age is one of making the act of simulating more of a real-time event. Advances throughout the decades have enabled an ever-more-powerful collection of calculations to move from the batch-processing environment of the computer center to the interactive environment of the desktop. In what we have just described, the goal is no different. Our quest is to make simulation a truly interactive real-time pursuit.

REFERENCES

Abraham, F., Broughton, J., Bernstein, N., et al. 1998. Spanning the length scales in dynamic simulation. *Computers in Physics*, 12:538–546.

Beck, M., and Moore, T. 2005. Logistical networking: a global storage network. *Journal of Physics: Conference Series*, 16:531–535.

Bruenn, S. W., Mezzacappa, A., and Dineva, T. 2005. Dynamic and diffusive instabilities in core collapse supernovae. *Physics Reports*, 256:69–94.

Calder, A. C., Fryxell, B., Plewa, T., et al. 2002. On validating an astrophysical simulation code. *Astrophysical Journal Supplement Series*, 143:201–229.

Cappello, F., Geist, A., Gropp, B., et al. 2009. Toward exascale resilience. *Technical Report of the INRIA-Illinois Joint Laboratory on PetaScale Computing IJHPCA*, 23(4):374–388.

Cruz-Neira, C., Sandin, D. J., DeFanti, T. A., et al. 1992. The CAVE: Audio visual experience automatic virtual environment, *Communications of the ACM*, 35(6):64–72. DOI:10.1145/129888.129892.

Drake, R. P., Doss, F. W., McClarren, R. G., et al. 2011. Radiative effects in radiative shocks in shock tubes, *High Energy Density Physics*, 7:130–140.

Fryxell, B., Olson, K., Ricker, P., et al. 2000. FLASH: An adaptive mesh hydrodynamics code for modeling astrophysical thermonuclear flashes. *Astrophysical Journal Supplement Series*, 131:273–334.

Keil, W., Janka, H.-T., and Müller, E. 1996. Ledoux convection in proto-neutron stars—A clue to supernova nucleosynthesis? *Astrophysical Journal*, 473: L111–L114.

Keyes, D., McInnes, L. C., Woodward, C., et al. 2013. Multiphysics simulations: Challenges and opportunities. *The International Journal of High Performance Computing Applications*, 27:4–83.

Kritsuk, A. G., and Norman, M. L. 2004. Scaling relations for turbulence in the multiphase interstellar medium. *Astrophysical Journal*, 601:L55–L58.

Ludäscher, B., Altintas, I., Bowers, S., et al. 2009. Scientific process automation and workflow management. In *Scientific Data Management: Challenges, Technology, and Deployment*. ed. A. Shoshani, and D. Rotem, 467. Boca Raton, FL: Chapman & Hall.

Mihalas, D., and Mihalas, B. W. 1999. *Foundations of Radiation Hydrodynamics*. Mineola: Dover.

Mount, R., ed. 2004. *The Office of Science Data-Management Challenge—Report from the DOE Office of Science Data-Management Workshops*, www.slac.stanford.edu/cgi-wrap/getdoc/slac-r-782.pdf, May.

Myra, E. S., Swesty, F. D., and Smolarski, D. C. 2009. Stellar core collapse: A case study in the design of numerical algorithms for scalable radiation hydrodynamics. *Computing in Science and Engineering*, 11(2):34–44.

Plewa, T., Linde, T., and Weirs, V. G., eds. 2005. *Adaptive Mesh Refinement—Theory and Applications, Lecture Notes in Computational Science and Engineering, Vol. 41, Proceedings of the Chicago Workshop on Adaptive Mesh Refinement Methods, Chicago, Sept. 3–5, 2003*. Berlin: Springer.

Remington, B. A., Drake, R. P., and Ryutov, D. D. 2006. Experimental astrophysics with high power lasers and Z pinches. *Reviews of Modern Physics*, 78:755–807.

Rosner, R. 2008. Where are we, and where might we be going? *Presentation at the Workshop on Scientific Challenges for Understanding the Quantum Universe and the Role of Computing at the Extreme Scale*. Palo Alto, CA: Stanford University.

Ross, R., Nurmi, D., Cheng, A., et al. 2001. A Case Study in Application I/O on Linux Clusters, Presented at SC2001 November 2001, Denver, CO, www.sc2001.org/papers/pap.pap166.pdf.

Schroeder, W., Martin, K., and Lorensen, B. 2006. *The Visualization Toolkit: An Object-Oriented Approach to 3D Graphics*, 4th Edition. New York, NY: Kitware, Inc.

Sutter, P.M., and Ricker, P.M. 2010. Examining subgrid models of supermassive black holes in cosmological simulation. *Astrophysical Journal*, 723:1308–1318.

Swesty, F. D., and Myra, E. S. 2005. Multigroup models of the convective epoch in core collapse supernovae. *Journal of Physics: Conference Series*, 16:380–389.

Swesty, F. D., and Myra, E. S. 2009. A numerical algorithm for modeling multigroup neutrino-radiation hydrodynamics in two spatial dimensions. *Astrophysical Journal Supplement Series*, 181:1–52.

Townsley, D. M. 2009. Treating unresolvable flame physics in simulations of thermonuclear supernovae. *Computing in Science and Engineering*, 11(2):18–23.

CHAPTER 5

Materials of the Future

From Business Suits to Space Suits

Mark F. Horstemeyer

CONTENTS

Materials science is quickly changing the face of what is possible. Looking to the future, it is easy to envision a number of areas where materials engineering and design can impact our daily lives: cars that are lighter, more efficient, and yet handle crashes better; batteries that can last longer and can be recharged more effectively; lightweight insulations that virtually eliminate heat transfer; paper that uses digital signals instead of ink to display messages, and can be reused indefinitely; and a suit that monitors your health as you wear it.

When Richard Feynman, the Nobel Laureate in physics, in his famous presentation in 1959 at Caltech (document in 1960) stated that "there is plenty of room at the bottom," he meant the creation of new materials. He also stated that there was not just room at the bottom but plenty of room. He discussed how if we understood the nanoscale information that we could design new products related to miniaturized evaporation methods, nanolubrication methods, miniature computers, novel biological methods, better electron microscopes, and small robots all by rearranging the atomic structure. He was casting a vision for the future, and indeed he did. Although we have made some progress toward his vision, we still have a ways to go.

How do we get where we are today to this new world of materials envisioned in the first paragraph? We need to consider Richard Feynman's perspective of how do we rearrange atoms to create the particular structures that can be engineered for the new application. It is more than just science; it is groundbreaking engineering. In Theodore von Karman's terms: "a scientist studies that which exists; an engineer creates that which never was." What is the process for creating new structures? What are the associated manufacturing processes by which to create these structures? How do we rearrange the atoms in the manufacturing state so that they are stable for the structural integrity to exist in their life cycle under various conditions (temperatures, stresses, humidities, etc.)?

Let us start with the end in mind. Imagine that an engineer is asked by his or her boss to design a new bulletproof shirt that needs to be as light and flexible as the current cotton shirts, does not overheat the individual, withstands high rate impacts, and must be procured in a mass production manner via its material processing. With these requirements in mind, now imagine that the engineer has only to log onto a cyberinfrastructure for materials informatics, codes, and data that already have tied together the myriads of options starting with each element in the periodic table and associated links for the composite-based shirt. The designer puts the requirements into the cybersystem and four options are suggested—all that meet the thermal, mechanical, and materials processing requirements. Now, the designer must consider the life cycle costs associated with the manufacturing process and potential market (colors, sizes, etc.) and then make decisions about which path to pursue.

Although this dream is still off in the future, it is closer than what we think and much closer than when Feynman made his comments about "plenty of room at the bottom" in 1959. The confluence of large-scale computing with greater precision experimental capabilities has offered the theoretical community new realms of accuracy and precision that have not been realizable before. If the cause-and-effect relationships from materials processing-to-structure-to-properties-to-performance (Olson 1997; 2000; Horstemeyer and Wang, 2003), as shown in Table 5.1 and the cause-and-effect relationships from multiple scales (the electronics scale up to the structural scale) have been quantified (see Table 5.2), then one can start with the end in mind and go backward to optimize the material and structural design (mathematicians call this an inverse problem). The process is illustrated in general in Figure 5.1 and for an example of creep in Figure 5.2. However, the key is quantifying the cause-and-effect relationships.

TABLE 5.1 Different Materials Processing Methods That Lead to the Various Internal Structures within the Materials That Finally Lead to the Mechanical Properties and Life Cycle Performance

Materials Processing	Structures	Properties/Performance
Casting	Porosity	Fatigue
Rolling	Precipitates	Ageing
Extrusion	Inclusions	Fracture toughness
Forging	Second phases	Impact resistance
Stamping	Intermetallics	Energy absorption
Heat treatment	Grains	Plasticity
Annealing	Vacancies	Corrosion
Pultrusion	Dislocations	Stiffness
Machining/cutting	Texture	Creep/stress Relaxation
Blow molding	Cross-links	Yield
Fiber spinning		Ductility
Filament winding		Hardness
Injection molding		Ultimate strength

TABLE 5.2 Different Entities at the Different Length Scales for Metals, Polymers, and Ceramics

Size Scale	Metals	Polymers	Ceramics
Highest	Structures	Structures	Structures
	Continuum	Continuum	Continuum
	Grains	Fibers	
	Particles/inclusions	Hard phases	Second Phases
	Precipitates	Cross-links/entanglements	
	Dislocations/ vacancies	Molecules/chains	
	Atoms	Atoms	Atoms
Lowest	Electrons	Electrons	Electrons

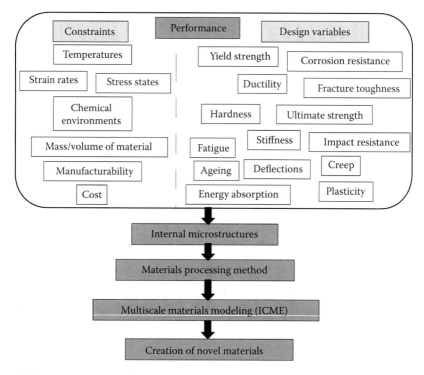

FIGURE 5.1 Schematic illustrating the solving of the inverse problem where the performance requirements are examined first and then the creation of new materials is backed out at the end.

The multiscale physics basis can be described today using a multiscale modeling philosophy (see Horstemeyer 2009 for a review) as shown in Figure 5.3 for a metallic system; experiments can be conducted at each length scale to validate the computational analyses at each length scale, as shown schematically in Figure 5.4. Once one can garner the correct physics in the multiscale modeling and then experimentally validate it, then the tools for the manufacturing and performance analysis are ready for design optimization. The goal is to use this simulation-based tool earlier in design to achieve more optimized components and systems (see Figure 5.5). For a continuum-level model and simulation, the method to accomplish this bridge of science and engineering is by using thermodynamically constrained internal state variables that are physically based on microstructure-property relations. When the microstructure-property relations are included in the internal state variable rate equations, history effects can be captured. Hence, the cradle-to-grave notion arises. As such,

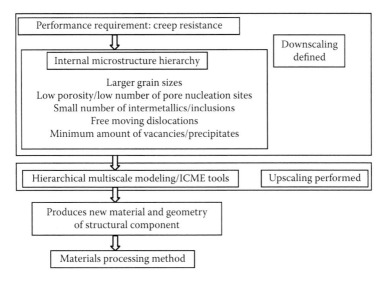

FIGURE 5.2 Schematic illustrating the solving of the inverse problem when creep is considered as the main performance criterion. When others are considered, then one must consider a multiobjective design optimization scheme.

scientifically oriented research occurs in the multiscale methodology, and the engineering design practice uses the cradle-to-grave internal state variable model. In Figure 5.6, an example of cradle-to-grave simulation-based design is shown for a stamped automotive product used in a crash scenario and incorporates the multiscale methodology for wrought 6022 aluminum alloy.

The typical design practice focuses on the system or component designer, who determines the materials; evaluates the design space; understands the static, dynamic, and thermal constraints; lays out a test matrix; and works with the material scientist and finite element analyst as a team leader. For the most part, this component/system designer does not have knowledge of all of these areas. Design teams may certainly differ from the straw man design team setup here, but the essential members and tasks are similar. The designer, material scientist, and finite element analyst function as a team with clear, independent tasks that come together as the designer systemizes the information. In the past century, with the advent of the automobile, aircraft, and space flight, successes have been achieved with this type of design team. For the most part, the design teams have recognized the need for research to make breakthroughs in their next-generation designs. However, current industrial trends require that the next-generation designer must not only be a designer but a material scientist

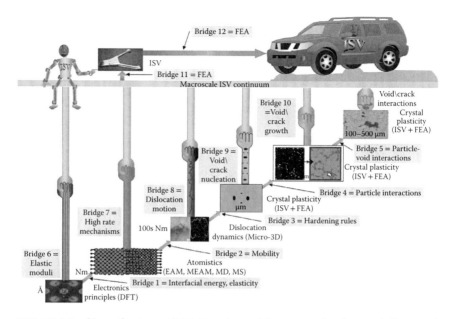

FIGURE 5.3 **(See color insert.)** Multiscale modeling example of a metal alloy used for design in an automotive component. The hierarchical methodology illustrates the different length scale analyses used and various bridges needed. ISV, internal state variable; FEA, finite element analysis; EAM, embedded atom method; MEAM, modified embedded atom method; MD, molecular dynamics; MS, molecular statics; DFT, density functional theory. (From Horstemeyer, M.F., *Practical Aspects of Computational Chemistry*, Springer Science+Business Media, 2009).

and finite element analyst as well. This requires a paradigm shift, because the manufacturing process and design scenarios are being pushed to one person who is to integrate the design, materials science information, and finite element analysis results. Therefore, advanced computational tools are necessary to help this next-generation designer; one such design tool is presented in this chapter. This next-generation designer/material scientist/finite element analyst requires a tool or suite of tools incorporated into a methodology that comprises and synergizes information from the perspective of design optimization, materials data, and applied mechanics. This is the idea of the new terminology of Integrated Computational Materials Engineering (ICME) (2008), which essentially is the modernized terminology for converging "simulation-based design," "cradle-to-grave modeling," and "multiscale modeling" into one paradigm.

"Simulation-based design" means more than just using finite element analysis late in the design process to help analyze the design of a component

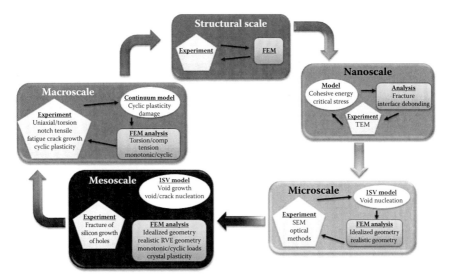

FIGURE 5.4 **(See color insert.)** Schematic illustrating the corroboration of multiscale modeling and multiscale experiments at each length scale and the pertinent effects being pushed up to the next higher length scale. (From Horstemeyer, M.F., *Practical Aspects of Computational Chemistry*, Springer Science+Business Media, 2009).

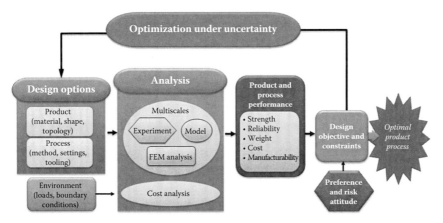

FIGURE 5.5 **(See color insert.)** Design optimization under uncertainty using the multiscale modeling and multiscale experimental methodologies to optimize processing and products.

or system. Here, it means to integrate the finite element analysis with optimization methods and tools during the early design phase. The "cradle-to-grave" term means to capture the history effects. The "cradle-to-grave" concept literally considers the birth of the material until its death (or last prime usage) by monitoring changes in the stress and strain states and the

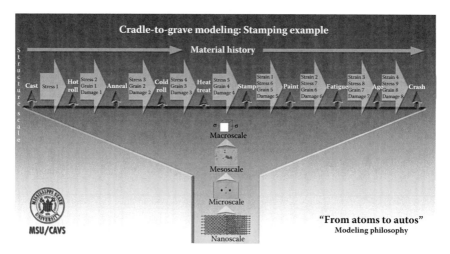

FIGURE 5.6 **(See color insert.)** To capture the cradle-to-grave history, robust models must be able to capture various manufacturing and in-service design scenarios.

microstructure/defect features. As described more in Horstemeyer (2009), "multiscale modeling" captures the bridges from the different length scales to solve the boundary value problem in a continuum manner but admitting lower length scale structures.

Figure 5.5 shows the ICME modeling paradigm with an optimization step included in the decision-making process. The schematic in Figure 5.5 shows that computations, manufacturing costs, and design constraints are all important in the design process as well as information from a materials database. All of these features are critical for an integrated toolkit for the next-generation designer. This figure represents a notion that is in direct contrast with current practice. The standard practice is to make a finite element mesh from the solid model when a part is geometrically designed. The stress analysis is performed and the highest stresses are used to determine the hot spots. This standard practice has only brought so far and has limited our designs because of the lack of accuracy due to the lack of integration of all of the different objectives and constraints. The new design methodology shown in Figure 5.5 optimizes the part based on the damage metric that arises from the internal material structures that in turn arise from the materials processing method. The damage state is used to determine the new geometry and hence the new material processing history that changes the microstructural features and defects/inclusions. And so goes the cycle.

This new design paradigm extends some current, limited practice of hierarchical material modeling by incorporating quantum theory simulations into product design optimization for the entire life cycle (e.g., for an optimization of a structural scale automotive component). Recent successes in materials design using this paradigm show that the methodology is obtainable. Horstemeyer and colleagues (Horstemeyer and Wang 2003; Horstemeyer 2009), Olson and colleagues (Kuehmann et al. 2001; 2007), and Dulikravich and colleagues (Egorov-Yegorov and Dulikravich 2005) have used methods described here to create modern engineered materials and structures. Although these examples are for aluminum and steel alloys, this type of ICME paradigm can be applied to titanium alloys, magnesium alloys, and other structural materials such as polymers, ceramics, biomaterials, and geomaterials as hoped by Feynman (1960).

The cyberinfrastructure (Haupt 2011; Haupt et al. 2011a,b) shown in Figure 5.7, which can be found at http://icme.hpc.msstate.edu, shows the different length scales of models, materials, experimental data, and simulation codes available to address the multiscale, history modeling approach.

There has been great progress in connecting materials processing to structure to property/performance in the past two decades. Much of this work has focused on failure analysis so that better designs can

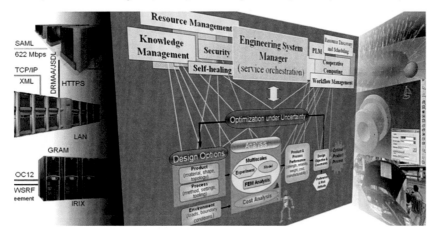

FIGURE 5.7 **(See color insert.)** Cyberinfrastructure that includes materials informatics for the different models and codes at different length scales along with the experimental data for metals, polymers (biological materials included), ceramics, and geo-materials. The website is http://icme.hpc.msstate.edu.

result. Figures 5.8 and 5.9 show the connections with the materials processing methods and the associated microstructures that affect fracture and fatigue.

Figure 5.8 shows the materials processing method for metals related to the associated structures that cause fracture (mechanical property). For wrought materials (rolling, extrusion, forging, and stamping), the main fracture causing entity is pore/void crack nucleation from particles. For cast materials, there are different amounts of porosity based on the casting method; hence, sometimes one method, which allows more porosity, will have a porosity-induced fracture mode from pore growth and coalescence, whereas another method with minimal porosity will have fracture arising from the voids/pores nucleating from second phase particles. For powder metals that are compacted and sintered, porosity is the key structure and pore coalescence is the main mechanism leading to final fracture.

Figure 5.9 shows the materials processing method for metals related to the associated structures that cause fatigue failure (mechanical property) and the rough estimates of the number of cycles for incubation (INC), microstructurally small crack (MSC) regime, and long cracks (LC). Note

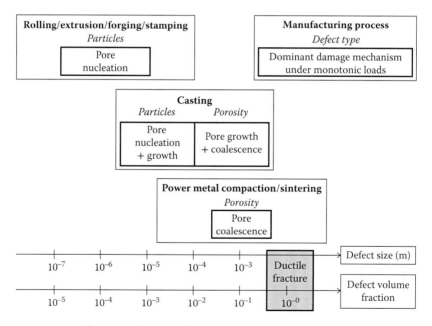

FIGURE 5.8   Schematic showing the relationships of the manufacturing process with the associated structures and their contribution to the fracture mode of the material.

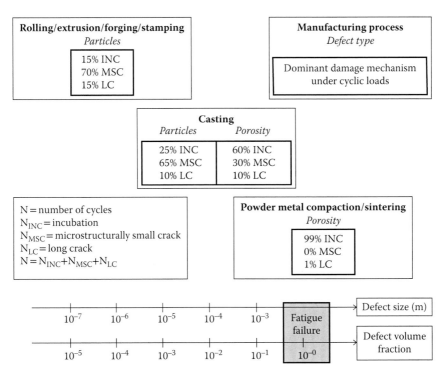

FIGURE 5.9   Schematic showing the relationships of the manufacturing process with the associated structures and their contribution to the fatigue of the material.

that each materials processing method admits a different number of cycles for each regime. For wrought materials (rolling, extrusion, forging, and stamping), the MSC regime is greatest with some incubation and long crack growth. For cast materials, the incubation and MSC regimes predominate with only a small amount of long crack growth. For powder metals, the incubation regime is dominant, because once a crack appears, the driving force is so strong since they incubated from a large pore, and fatigue failure occurs almost immediately.

Now, let us consider the different scale structures and the associated environments of most concern where ICME can have a paradigm shift. Each type of structure is mentioned first below with the environment of consideration in parentheses.

5.1 ENERGY

Nuclear reactors (ageing, creep, corrosion)

Nuclear bombs (high rate phenomena, impact, fracture)

Solar cells (thermal expansion, creep, adhesion behavior, strength)

Engines/motors (thermal expansion, creep, corrosion, strength, fatigue, wear, fatigue)

Hydrogen cells (thermal expansion, creep, corrosion, strength, fatigue, wear, fatigue)

5.2 INFRASTRUCTURE

Bridges (wear, corrosion, creep, fatigue, fracture, crashworthiness, vibrations/resonant frequencies)

Roads (wear, creep, corrosion)

Embankments (creep, strength)

Railroads (wear, creep, corrosion, fatigue, fracture)

5.3 TRANSPORTATION

Automobiles (creep, fatigue, fracture, crashworthiness, vibrations/resonant frequencies, noise)

Trucks (creep, fatigue, fracture, crashworthiness, vibrations/resonant frequencies, noise)

Tanks (corrosion, fatigue, fracture, vibrations/resonant frequencies)

Airplanes (creep, corrosion, fatigue, fracture, vibrations/resonant frequencies)

Helicopters (creep, corrosion, fatigue, fracture, vibrations/resonant frequencies)

Spacecraft (creep, fatigue, fracture, crashworthiness, vibrations/resonant frequencies, noise)

Trains (creep, fatigue, fracture, vibrations/resonant frequencies)

Ships/boats (creep, corrosion, fatigue, fracture, vibrations/resonant frequencies)

Submarines (creep, corrosion, fatigue, fracture, vibrations/resonant frequencies)

5.4 NANO- AND MICROSTRUCTURES/SMALL DEVICES

The modern term "nanotechnology" or "microelectromechanical systems" (MEMS) or "nanoelectromechanical systems" (NEMS) is a critical area both now and in the future where ICME can play a very important role and can also be traced back to Feynman's 1959 presentation on "plenty of room at the bottom." Since the surface area to volume ratio is large in these technologies, different structure–property relations come into play when compared to larger structures such as automobiles. In particular, wetting, friction, and electrostatics are more relevant as well as the geometry, size, and volume of the defects internal to the small volume of material. In general, the smaller the component or device, the more important the "structure" is in the structure–property relationship. Recall how important the heterogeneous microstructures were in the control arm fatigue, and fracture problem as shown in Chapters 8 and 9. The point here is that the "nanostructure" is even more important in these smaller components and devices.

MEMS are devices or components with geometric attributes that are often less than a micrometer in size. MEMS devices can be complicated structures or simple in geometry. The main key is that they have some mechanical functionality to be a MEMS device. In the context of research, they can be categorized as microactuators, microsensors, microelectronics, or microstructures with a focus to convert the energy of one form to another (where the mechanical energy is a key energy type).

One example of a MEMS device is the LIGA (Lithographie, Galvanoformung, Abformung) nickel spur gear used in nuclear weapons. LIGA is an electroplating processing method in which gears on the order of micrometers can be made. Son et al. (2005) studied the tension and fatigue properties of LIGA components that were on the order of 10 μm. They experimentally showed a size scale effect related to the tensile strength and ductility, both of which can be related to the volume-per-surface area's relationship with dislocation nucleation as proposed by Horstemeyer et al. (2001) for nickel.

Another example of a MEMS device is silicon gears developed at Sandia National Laboratories (www.mems.sandia.gov). The MEMS gears have been used for sensors, microfluidic devices, actuators, and micro-optics to name a few applications. Some have been used for weapons, whereas others have been used for examining new energy methods.

Whether one has nanodevices or microdevices, the processing is a key. Two current physical deposition processing exist: chemical vapor deposition (CVD) and physical vapor deposition (PVD). In CVD, a source gas is used to stream onto a substrate that induces a reaction to grow the material of choice for the MEMS structure. In PVD, a material is removed from a target and then deposited onto the material of choice like sputtering. The patterning of the MEMS structure has been realized through different lithographic technologies: photolithography, electron beam lithography, ion beam lithography, and x-ray lithography, all of which have their benefits and downsides. Etching and micromachining become very important and problematic for these NEMS and MEMS devices as well. Clearly, there are opportunities for using ICME tools to help grow the nano- and micro-level of applications, materials, components, and structures.

5.5 CONCLUSIONS

Although materials science and engineering have recently provided the capacity for creating novel materials/structures using computational tools, the full impact of what Richard Feynman's "plenty of room at the bottom" vision has not been realized. In terms of materials, computational tools for metals are probably further along particularly when thinking about the "integrated" aspects of ICME; however, ceramics and polymers are not too far behind, multimaterial structures (composites) are even further behind.

Another key to realizing Feynman's vision is the education of the next-generation materials engineers and scientists. Instead of stovepiping the undergraduates and graduates as either theoreticians or experimentalists, the challenge is to educate students with tools that integrate theory, experimental methods, and computational tools. Granted, this is a major paradigm shift in U.S. education (graduate education in particular) but one that is required if the Feynman's vision is to be realized.

Finally, materials design cannot just start at the bottom when creating a new material or structure. If a computational physicist develops a new material based on electronics structures calculations, say of magnesium, but does not know at the top level what the requirements are for the magnesium alloy, a great science paper might ensue but not a magnesium alloy for practical use. The downscaling requirements for the final application should define the upscaling calculations required. This bridging paradigm allows electronics structures calculations for magnesium to be directed to the final application. Then and only then can Feynman's vision be realized—when we start with the end in mind, which in our case is the creation of novel materials.

ACKNOWLEDGMENTS

MFH acknowledges the support from the Center for Advanced Vehicular Systems at Mississippi State University for this work. Also, he thanks Dr. Mark Tschopp for reviewing this chapter.

REFERENCES

Egorov-Yegorov, I.N., and Dulikravich, G.S. "Chemical composition design of superalloys for maximum stress, temperature, and time-to-rupture using self adapting response surface optimization." *Materials and Manufacturing Processes*, 20, 569–590, 2005.

Feynman, R.P. "There's plenty of room at the bottom." *APS Conference, CalTech Engineering and Science*, 23, 5, 1960.

Haupt, T. "Cyberinfrastructure support for integrated materials engineering." In *The Proceedings of the 1st World Congress on Integrated Computational Materials Engineering (ICME)*, J. Allison, P. Collins, and G. Spanos (eds.), Seven Springs, PA, Wiley, pp. 229–234, 2011.

Haupt, T., Sukhija, N., and Horstemeyer, M.F. "Cyberinfrastructure support for engineering virtual organization for cyberdesign," *Parallel Processing and Applied Mathematics, Lecture Notes in Computer Science*, 7204(2012), 161–170, 2012.

Haupt, T., Sukhija, N., and Zhuk, I. "Autonomic execution of computational workflows," federated conference on computer science and information systems (FedCSIS'11), Szczecin, Poland, September 18–12, 2011b.

Horstemeyer, M.F., Baskes, M.I., and Plimpton, S.J. "Computational nanoscale plasticity simulations using embedded atom potentials." *Theoretical and Applied Fracture Mechanics*, 37(1–3), 49–98, 2001.

Horstemeyer, M.F. and Wang, P.T. "Cradle-to-grave simulation-based design incorporating multiscale microstructure-property Modeling: reinvigorating design with science." *Journal of Computer-Aided Materials Design*, 10(1), 13–34, 2003.

Horstemeyer, M.F. "Multiscale modeling: a review," In *Practical Aspects of Computational Chemistry*, J. Leszczynski and M.K. Shukla (eds.), Dordrecht, the Netherlands, Springer Science+Business Media, pp. 87–135, 2009.

ICME, *Integrated Computational Materials Engineering: A Transformational Discipline for Improved Competitiveness and National Security*, Committee on Integrated Computational Materials Engineering, National Research Council, Washington, DC, National Academy of Engineering. ISBN: 0-309-12000-4, 2008.

Kuehmann, C., Olson, G.B., Wise, J.P., and Campbell, C. "Advanced Case Carburizing Secondary Hardening Steels," U.S. Patent No. 6,176,946 B1, January 23, 2001.

Kuehmann, C., Olson, G.B., and Jou, H.J. "Nanocarbide Precipitation Strengthened Ultrahigh-Strength, Corrosion Resistant, Structural Steels," U.S. Patent No. 7,160,399, January 9, 2007.

Olson, G.B. "Computational design of hierarchically structured materials." *Science*, 277(5330), 1237–1242, 1997.

Olson, G.B. "Pathways of discovery: Designing a new materials world." *Science*, 288, 993–998, 2000.

Son, D., Kim, J.J., Kim, J.Y., and Kwon, D. "Tensile properties and fatigue crack growth in LIGA nickel MEMS structures." *Materials Science and Engineering A*, 406(1–2), 274–278, 2005.

II

Case Studies

Earth System Grid Federation: Infrastructure to Support Climate Science Analysis as an International Collaboration

A Data-Driven Activity for Extreme-Scale Climate Science

Dean N. Williams, Ian T. Foster, Bryan Lawrence, and Michael Lautenschlager

CONTENTS

This chapter will describe the purpose and design of the Earth System Grid Federation for climate science, and how housing, managing, searching, and disseminating climate data on massive scales are critical to national and international climate assessments. The chapter also shows how cooperative data activities organized at the international level play a central role in advancing earth system science. We look at components comprising the data federation and discuss data preparations, secure access protocols and quality assurance, the movement of large-scale data, and network issues. Finally, we discuss data replication for greater data access and product services to meet the needs of a diverse community by supplying visualization and analysis capabilities.

6.1 INTRODUCTION TO CLIMATE SCIENCE

Climate science is the study of long-term trends in average meteorological conditions with an emphasis on climate change. These meteorological climate change conditions are studied over the periods of decades and even centuries. The principal tools climate scientists use are numerical climate models and statistical methods for studying the output of these models. The models are based on the laws of physics and mathematics and other sciences such as biology, chemistry, and geochemistry to predict natural occurring and human-induced climate changes.

6.1.1 Model Intercomparison Projects

Coupled global climate models (GCMs) composed of four major climate subsystems (atmosphere, ocean, land, and sea ice) are used to project and predict future temperatures and climate changes under various scenario conditions. These scenarios, like those put forth by the Intergovernmental Panel on Climate Change (IPCC), are carefully constructed experiments that represent ever-increasing natural and anthropogenic emissions of greenhouse gases such as carbon dioxide, methane, water vapor, ozone, and nitrous oxide. Scenarios help climate scientists identify real climate-change trends and their potential impact on all life species on earth. From these experiments, climate scientists propose methods of adapting to or mitigating the effects of climate change. Adaptation includes things humans can do to cope with the changes, such as constructing seawalls to protect low-lying areas from rising sea levels. Mitigation, on the other hand, consists of activities humans can do to avoid, minimize, slow, or even reverse the climate change—dispersing aerosols in the atmosphere to reflect solar radiation and cooling the planet, for example. Modeling the earth's climate is an ongoing evolution using high-capacity, high-performance computing (HPC) to simulate ever-changing conditions and even greater complex scenarios. Today's HPC machines can paint a more accurate and detailed picture of long-term climate evolution using GCMs representing the sophisticated physical processes of the climate subsystems (Figure 6.1). GCMs are the scientists' primary tool for modern climate modeling, and more than two dozen such high-fidelity systems are available.

Under the governance of the United Nations Environment Program and the World Meteorological Organization, the IPCC was founded as an international body to assess climate change. Succession reports from the first IPCC assessment in 1990 to the most recent Nobel Peace Prize-winning fourth assessment (AR4) in 2007 have increasingly shown evidence edging toward anthropogenic causes for global warming. Figure 6.2 shows AR4 computations based on observed condition (pre-2001) and the average ensemble of two dozen climate models—assuming a continued high rate of greenhouse gas emissions.

The analysis of the data produced the most compelling statement of any IPCC assessment report: "Most of the observed increase in globally averaged temperatures since the mid-20th century is *very likely* due to the observed increase in anthropogenic greenhouse gas concentrations." With

Modeling the climate system

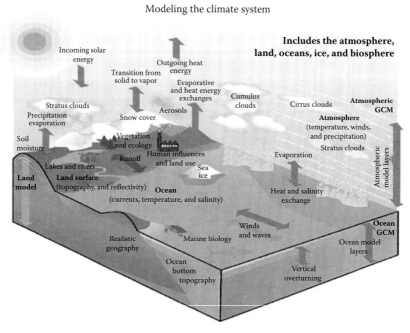

FIGURE 6.1 High-performance computers run coupled global climate models (GCMs) representing four major climate subsystems: atmosphere, ocean, land, and sea ice in an attempt to simulate an accurate depiction of earth's future climate.

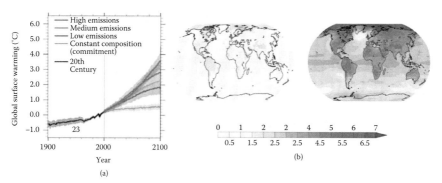

FIGURE 6.2 (a) Two centuries of global surface temperature are shown. Historical observations from 1900 to 2000 show the average plotted in black. In color, four simulated scenarios (plotted within a confidence interval) represent different rates of greenhouse gas emissions. (b) Climate models predict that the greatest rise in temperature will occur in the high latitudes. Shown is the average increase (K) in surface temperature as predicted using the CMIP3 archive, relative to the preindustrial period. The left image depicts early 21st century and the right depicts the late twenty-first century.

each succession report, higher resolution grids, additional components (such as aerosols, biogeochemical, hydrology cycle, etc.), and ensembles of simulations have drastically increased the number and size of files comprising individual data sets.

With that said, over the years, the IPCC has seen a significant rise in requests to hold more data. The IPCC AR4 data archive required 35 terabytes (TB or 10^{12} bytes). In contrast, the upcoming 2013 IPCC Fifth Assessment Report (AR5) archive will require 3.5 petabytes (PB or 10^{15} bytes), with total federated data holdings expected to reach 5–10 PB. To complement the IPCC AR5 data archive, other high-profile data sets are scheduled to coexist within the federated infrastructure.

6.1.1.1 Earth System Grid Federation and Coupled Model Intercomparison Project

Climate scientists collaborate worldwide to inter-compare knowledge gained by examining disparate data. To facilitate this collaboration, the international Earth System Grid Federation (ESGF) developed software for universal climate data access in which data curation, preservation, dissemination, analysis, and visualization can be appropriately, reliably, and readily managed for a diverse user community. Hardware and software investments have enhanced science nationally and internationally by allowing free flowing uniform data for cross model and project intercomparison. Though in some cases the desired stages of development are just now being realized, they ensure future progress for the continued successful dissemination of global climate information. These efforts provide convenient access and maximize scientific productivity. Managing a distributed-data system of this magnitude presents several significant challenges not only to system architectures and application development but also to the existing wide-area networking infrastructures. At the heart of ESGF software development is connecting people, organizations, bureaucracies, government agencies, and academia for the betterment of climate research.

Although ESGF software is universal and opening up to host other forms of data such as observational and reanalysis data, its primary focus is to host the fifth phase of the coupled model intercomparison project (CMIP 5) data archive to be used for the upcoming IPCC AR5. CMIP5 represents an international coordinated effort of climate change experiments and simulations and is expected to provide information to all three IPCC Working Groups (scientists, impacts and adaptation, and mitigation). A CMIP document describes three primary experiments:

1. Near-term decadal prediction simulations (10- to 30-years) initialized in some way with observed ocean state and sea ice

2. Long-term (century time-scale) simulations initialized from the end of freely evolving atmospheric/ocean GCM simulations of the historical period

3. Atmosphere-only (for computationally demanding and numerical weather prediction models)

In all, there are over 50 planned simulations to inform the three experiments. The ESGF data-delivery system can be viewed as two major components: data nodes, where the data actually reside, and gateways, which support portal services and serve as interfaces to end-users who can search, discover, and request data and data products. The ESGF consortium has installed over 20 data nodes to CMIP5 modeling centers that are devoted to serving data to the community.

Figure 6.3 shows the ESGF infrastructure strategy with the national and international collaborations. Users can access ESGF data using web browsers, scripts, and in the near future, UV-CDAT (ultrascale visualization—climate data analysis tools) clients. ESGF infrastructure is separated into gateways and data nodes. Gateways handle user registration and management and allow users to search, discover, and request data. Data nodes are located where the data resides, allowing data to be published (or exposed) on disk or through tertiary mass storage (e.g., tape archive) to any gateway. They also handle UV-CDAT data reduction, analysis, and visualization. ESGF currently comprises eight national and international gateways, four of which hold special status in housing CMIP5/AR5 replication data sets: Lawrence Livermore National Laboratory (LLNL)/Program for Climate Model Diagnosis and Intercomparison (PCMDI), British Atmospheric Data Centre (BADC), Deutsche Klimarechenzentrum (DKRZ—the German Climate Computing Centre), and the Australian National University (ANU) National Computational Infrastructure (NCI). Users have access to all data from the federation, regardless of which gateway is used.

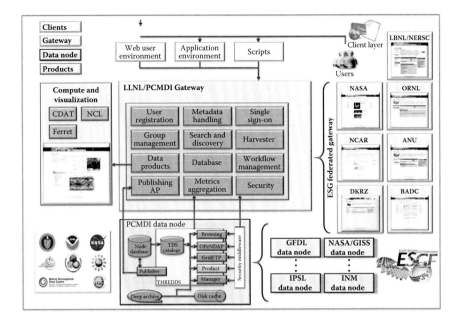

FIGURE 6.3   A combination of policies and standards, international collaborations, and the overall vision and strategy of a comprehensive infrastructure and framework for data planning, management, and access will support CMIP5 and through it, the next IPCC Assessment Report, AR5. ESGF infrastructure is separated into gateways and data nodes.

6.2   FEDERATION

The ESGF comprises a set of distributed and federated *nodes*, which interact with each other based on a peer-to-peer paradigm (see Figures 6.3 and 6.4). In brief, this peer-to-peer network provides for federated identity (users registered with any ESGF site can access data at any other site),

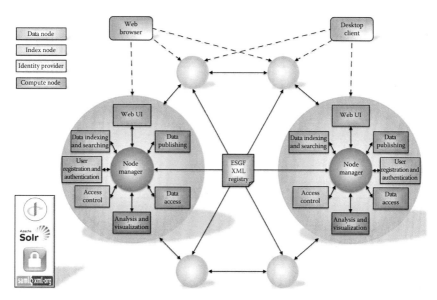

FIGURE 6.4   ESGF peer-to-peer architecture.

federated search (users can search all ESGF data at any of several locations), and data replication (many data are copied to two or more ESGF sites).

ESGF's federated architecture has both reliability and performance benefits. The fact that data are replicated over multiple sites guards against both data loss due to catastrophic failure of a single site and interruption of service due to transient failures. In addition, data replication provides greater total aggregate bandwidth and also reduces demands on scarce trans-oceanic network capacity. ESGF today has more than 23,000 registered users; we expect this number to continue to grow, as will data volumes. Even if all data could be moved to a single location, the bandwidth demands on that site would likely be untenable.

ESGF's federated architecture also has organizational benefits. ESGF can be thought of as a club of data providers. Each participating provider is an independent entity, committed to making some collection of data (e.g., its contributions to CMIP5) available to the international climate research community. A federated architecture allows individual sites to contribute to the community without compromising their ability to meet their own mission.

Based on community protocols, each ESGF node is equipped with the ESGF software stack, which at the time of installation can be configured with one or more of four possible "flavors":

1. *Data Node:* the application servers for data publishing and access

2. *Index Node:* services for publishing and discovering data

3. *Identity Provider:* facilities for registering and authenticating users and establishing access control

4. *Compute Node:* web application and analysis engines to produce high-level data products and visualizations

Each node advertises its capabilities to all of its peers and, in turn, collects the same information from all other nodes. As a result, each node constantly produces a document (the ESGF registry) that contains the most up-to-date information about the state of the whole federation. Specifically, the ESGF registry includes the URLs for all services for all nodes, their public certificates, their descriptive metadata, and other information. If a new node joins the federation, or another node becomes unavailable, the ESGF registry at each node is quickly updated to reflect the changes.

Interoperability between nodes is made possible by the adoption of a set of core technologies from the open source community designed to work in a federated environment. Specific technologies include the following:

- OpenID and OpenSSL are used to enable single sign on throughout the federation. Users can register at any node. Any time users are asked to authenticate in the system, they are always redirected to the node where they provide their username and password. Once authenticated, their credentials (a web cookie for OpenID or an X509 certificate for OpenSSL) are honored throughout the federation.

- SAML (the security assertion markup language) is used for encoding and signing authorization assertions about the ability of users to perform operations on a given resource (e.g., downloading a data file). Because of mutual trust agreements, SAML assertions issued by a node are honored throughout the federation.

- The native Solr protocol for distributed searches is currently used to allow users to start from a node and find data across the system. This protocol might be replaced or supplemented in the future by a more implementation-neutral protocol based on a popular standard such as open search.

A net result of the adoption of these federation protocols and technologies and the advertisement of services through the ESGF registry is that users and desktop clients can start the data-discovery process at any node and seamlessly access data throughout the system.

6.3 DATA PREPARATION

The ESGF data-preparation process encompasses data set and metadata generation, publication, quality control, and metadata preparation steps.

6.3.1 Data Format and Metadata

Because metadata is absolutely critical to the discovery, interpretation, and analysis of data, the ESGF team, in collaboration with the providers of CMIP5 model output and observational data sets, have defined a set of conventions for data format and metadata standards for all data that are ingested into the system (both models and observations).

Data sets must be formatted as NetCDF single-variable time series, possibly split across multiple files depending on the time extent. Having only one variable per file guarantees that users who intend to study that physical field do not have to download large amounts of unwanted data. Additionally, files must follow the Climate and Forecast 1.5 convention (CF-1.5), which establishes constructs for expressing geographic coordinates, grids, and standard names for the measured or simulated quantities. The use of standard names is particularly important because it allows users to compare the same physical parameter across different models and across models and observations. Data set level metadata is encoded within each NetCDF file as global file attributes, with standardized names and values that are harvested by the publishing infrastructure (discussed later) and made available to users and clients in the discovery process.

Before data can be ingested into the system, it is checked for compliance to the above format and metadata specifications by post-processing it through the climate model output rewriter (CMOR) program. CMOR also reorganizes the data on disk according to a conventional data structure (the DRS or data reference syntax) intended to facilitate browsing and access of the data holdings throughout the federation, including easy modification of bulk download scripts.

6.3.2 Publisher

Data providers put data into the system by running the ESGF publisher application, an extensible python-based package designed for scripted as well as interactive publishing of the data. The publisher parses the files in the specified directory tree and extracts all relevant metadata information from the files or the directory structure itself. Metadata is stored in a Postgres relational database at the data node where the data reside and then aggregated to build metadata catalogs in the THREDDS XML format, which are used to configure the main ESGF application servers—the THREDDS data server (TDS) and the live access server (LAS). The final step of the publishing process consists of notifying the ESGF publishing web application of the newly created catalogs, so they can be parsed into metadata (name, value) pairs that are ingested into a Solr index to drive data discovery. The publisher is also used to publish replicated data on other data nodes.

6.3.3 MetaFor Questionnaire and CIM

In the case of CMIP5 model output, an additional source of metadata is constituted by the online questionnaire developed by the European project MetaFor. Modeling groups participating in the CMIP5 activity are required to fill in very detailed metadata about their runs online, from which the information is captured into XML documents conforming to the CIM (Common information model) schema and in turn stored into an XML database. CIM documents are propagated to the ESGF gateways as an atom feed and provide a wealth of information to climate researchers who need to know the details of a model run or inter-compare model runs with each other.

6.3.4 Quality Control Levels and DOI

Under the leadership of the World Data Center for Climate (WDC Climate), the ESGF has established a rigorous quality control (QC) pipeline that is used to verify the technical and scientific integrity of all model output distributed through the system (similar procedures are also being put in place for observations). The pipeline is composed of three successive stages, each of which results in a QC flag being assigned to the data:

1. *QC1* is defined as automatic software checks on data and metadata. It is performed by each center before the data can be published and the CMOR and Publisher software, which verify general compliance with the required data.

2. *QC2* is defined as subjective quality control on data and metadata. For a data set to pass QC2, the author must manually inspect its

content to verify its integrity, and at the same time, the metadata from the MetaFor questionnaire must be fully populated. After this stage, data can be replicated among the three main CMIP5 archives: PCMDI, BADC, and DKRZ.

3. QC3 is defined as double and cross checks on data and metadata. This stage is composed of a scientific quality assurance process and a technical quality assurance process. Once a data set has passed QC3, it is assigned a digital object identifier (DOI), which can be used to reference the data in formal publications. For model output, a DOI is assigned to each model and experiment combination and spans multiple parameters produced by the same model runs and multiple ensemble members.

6.3.5 Federated Metadata

Once the data is harvested at each single node, metadata is used to drive federation-wide searches for models and observations through a mechanism known as "distributed search." When a client sends a search query to an ESGF node, that node propagates the query to all other nodes and then takes care of reassembling all the responses into a single summary response document, which is returned to the client. The response document contains all the data set access points at the distributed nodes, so the client has all information necessary to download the data from anywhere in the federation.

In particular, metadata for the following high-level collections is propagated through the system so that the enclosing data sets can be found and downloaded throughout:

- *CMIP*: The coupled model intercomparison project, phase 5. This collection comprises the output of more than 50 models, run by approximately 25 modeling groups in 17 countries. The overall size of the collection between 5 and 10 PB, with a core set of approximately 3.5 PB, replicated across the main CMIP5 archives.

- *NASA observation data*: This collection consists of selected satellite measurements specially prepared by NASA for comparison with the CMIP5 model output. It includes about 15 variables (air temperature, specific humidity, etc.), chosen from corresponding NASA flagship missions, that have been uniformly gridded and structured to follow the same data format and metadata conventions as the CMIP5 models.

- *North American Regional Climate Change Assessment Program*: This is an international program that serves the high resolution climate scenario needs of the United States, Canada, and northern Mexico, using regional climate model, coupled GCM, and time-slice experiments.

- The community climate system model (CCSM) and the community earth system model (CESM) operated at the National Center for Atmospheric Research (NCAR) is widespread in the university community and at some U.S. national laboratories. NCAR and its U.S. Department of Energy partners developed CCSM and CESM as coupled models, including the atmosphere, land surface, ocean, and sea ice.

- Select data sets from the Atmospheric Radiation Measurement Climate Research Facility and the Carbon Dioxide Information Analysis Center including the AmeriFlux observational data hosted by Oak Ridge National Laboratory (ORNL) for the atmospheric science research and terrestrial carbon and ecosystem research programs.

- The model data from the carbon-land model intercomparison project (C-LAMP) and the advanced very high-resolution climate model simulations resulting from earth system modeling and regional and global modeling programs at ORNL.

6.4 SECURITY

ESGF data repositories contain climate-model data widely shared across the climate community and discoverable via metadata. While access to metadata is unrestricted, access to the data is restricted in some cases and tracked in all cases. Thus, every user of the ESGF system must have an identity, potential roles, and access permissions for specific resources.

6.4.1 Metadata

Anyone can browse, search, and access ESGF metadata and thus discover the data endpoints at which particular data sets are located.

The metadata itself is harvested during the publication process from the data at modeling centers. The metadata harvesting process scans the local data, generates the metadata for the required data, and pushes that information to the hosted metadata catalog. To ensure metadata integrity and to ensure publication by only the owner of the metadata, security processes are in place to identify and authorize the publisher. User in the system who publish data need special roles assigned to them, and they login to the system before they publish metadata. Thus, we can establish user identity, check the user's role in the federation, and allow the metadata to be published and made available. Modifying the metadata also requires similar privileges, and with all interactions requiring secure connection (HTTPS), similar security can be enforced. (Data access is tracked and sometimes restricted from commercial use.)

As mentioned earlier, most of the data in the system are open for access. But all data owners require that access to the data be logged—that is, metrics collected on who accesses the data and when. This allows collection of usage reports, for example, to report to funding agencies but more importantly to notify users of data changes. To allow such metrics collection, the system requires that all users login and provide their identity before they download any data.

All data services require that users authenticate to download data, and the system requires the use of X.509 credentials for authentication. For improved usability, an online credential generation and translation system allows the user to provide their username/password and obtain an X.509 credential for use in the download process. This latter technology is built into the download solution, so the user need not deal with certificates and related files. These mechanisms allow ESGF data services to identify the user downloading the data and to log that information for metrics and notification.

Once the user is identified, the data services call out to the ESGF authorization service that protects the data to determine download eligibility. Permissions to download are based on the user's group membership in the federation, and the authorization services look up the membership information to return a decision. The federation uses standard interfaces to communicate this information and relies on SAML for the data format and protocol.

6.4.2 Data Integrity

It is imperative that data published in the system can be verified to ensure that it has not been modified, maliciously or accidentally. This is achieved in the federation by use of checksums at the time of publish, with options for all data-transfer commands to validate checksums at the destination.

6.5 DATA MOVEMENT

Convenient, reliable, and high-performance data-movement mechanisms are fundamental to ESGF's operations. These mechanisms are used extensively by ESGF users to download data from ESGF data nodes and by

ESGF sites for site-to-site data movement for replication purposes. ESGF employs a variety of tools for this purpose. Wget, Bulk Data Mover (BDM), and DataMover-Lite (DML) are older tools that have been in use for some time; the Globus Online hosted data-movement service is a newer option that provides alternative usability and performance.

6.5.1 Wget

One of the most direct ways of downloading large numbers of files through the ESG infrastructure is through automatically generated *wget* scripts (*wget* is a ubiquitous tool for requesting web resources by URL and is installed by default on most systems). Once users have located the files of interest and added them to the gateway data cart, they have the option of requesting a script that contains a sequence of wget requests to the file URLs. The wget command is encoded with the SSL options for transmitting the user X509 certificate as part of the request, so that the data node security filters for authentication and authorization can use it. Consequently, each wget script can be used multiple times, provided the user certificate (located in a standard location) is renewed once it expires. Additionally, because of the standard layout of each data archive mandated by the DRS syntax, a wget script can be manually edited to replace one or more fields with a new value chosen from the DRS-controlled vocabulary and can be used again to download a completely new collection of files.

6.5.2 Bulk Data Mover

BDM [1] is a high-level data transfer management component that manages file transfers using adaptive data transfer algorithms [2] and existing transfer services and tools. BDM specializes in handling large variance in file sizes and many small files, as is common in climate data. BDM supports several underlying transfer protocols and services and provides optimized management of transfer requests adaptively based on the data set characteristics. It achieves high performance using various techniques, including multithreaded concurrent transfer connections, data channel caching, load balancing over multiple transfer servers, and storage input/output (I/O) prefetching. It can accept a request composed of multiple files, multiple directories, or an entire directory. BDM supports transferring multiple files concurrently as well as using parallel TCP streams using GridFTP as the underlying transfer protocol. The optimal level of concurrency or parallel streams depends on the bandwidth capacity of the storage systems at both ends of the transfer as well as achievable bandwidth on the wide-area network. Setting up the optimal level of concurrency is important, especially for applications with file-size variance. Concurrency that is too high becomes ineffective (high overheads and increased congestion), and concurrency that is too low does not take advantage of available bandwidth. A similar phenomenon is observed when selecting the level of parallel streams.

Transfer queue management and concurrent connection management contribute to higher transfer throughput, including both network and storage. When the data set has a large variance in the file sizes, continuous data flow from the storage into the network can be achieved by prefetching data from storage onto the transfer queue of each concurrent transfer connection. This overlapping of storage I/O with the network I/O helps improve the transfer performance. As shown in Figure 6.5, BDM manages a database queue of the concurrent transfer connections and also manages the transfer queues for the concurrent file transfers. Each transfer queue checks the configurable threshold for the queued total file size and gets more files to transfer from the database queue when the queued total file

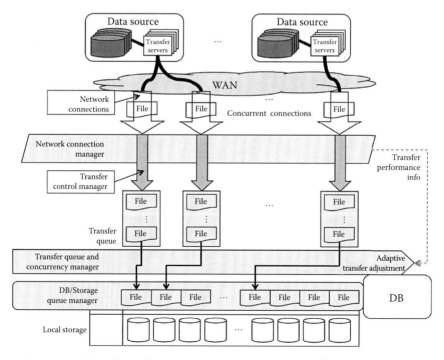

FIGURE 6.5   Transfer and concurrency management in Bulk Data Mover, showing adaptive transfer adjustment.

size goes below the configured threshold. Storage I/O prefetching includes inode creation for writing files at the destination. In many file system cases, many inode creations at the same time cause a significant overhead in file system performance, and this overhead affects transfer performance. By creating inodes at the destination paths when files are being staged on the transfer queue, BDM achieves faster storage I/O during the transfers.

6.5.3 DataMover-Lite

DML [3] is an ESG-specific file download tool with an intuitive graphical user interface that works with ESG authentication and authorization from the gateways and data nodes. DML is a versatile tool that works with multiple transfer protocols such as HTTP, HTTPS, and GridFTP supported by the ESG data servers. DML also supports concurrent file transfers with file-transfer monitoring, fault tolerance capability in case of transient failures, and management of ESGF credentials during the file transfers using OpenID login. DML includes ESG data set catalog browsing and search capability, which allows one-stop downloading convenience to users without having to use multiple components such as a web browser and different downloading tools for specific transfer protocols. Figure 6.6 shows the DML data transfers with ESG data set catalog browsing and search window.

FIGURE 6.6 Earth System Grid data catalog browsing and search capability and data downloading interface.

DML also includes HTTP parallel streaming (http-ps) capability, similar to GridFTP parallel streaming, which makes HTTP data downloads faster than usual HTTP downloads. With the http-ps feature, DML downloads a single file by splitting it into multiple blocks and streaming through parallel HTTP connections to an ESG data server. Each HTTP connection streams partial blocks of files to compose the full file at the target. This capability permits DML to perform HTTP parallel streaming (http-ps) from multiple files from several ESG data nodes when data replicas are available on the nodes. Figure 6.7 shows an example of how DML makes parallel HTTP connections to transfer blocks from a source file for streaming using http-ps capability. The large source file is perceived as multiple small manageable blocks, and the available number of HTTP connections is opened with ESG data transfer server to access the source file. When the data transfer for one block such as block 1 (B1) is completed earlier than other blocks, DML initiates the next block transfer for block 6 (B6). Similarly, the next transfer shifts to B7 after B4 completes. However, the download progress for B7 shows more progress than B6 because of changes in the network conditions. The block size in http-ps can be as small as 1 MB. After all blocks for the source file are streamed to the target site, the final downloaded file is created through a fast merging process.

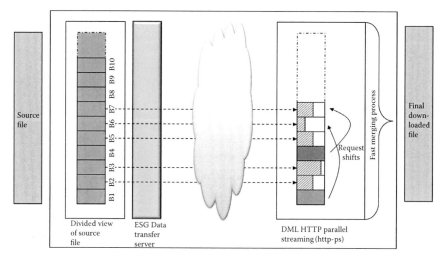

FIGURE 6.7   HTTP parallel streaming (http-ps) feature in DML.

6.5.4 Globus Online

Globus Online [4,5] is a fast, reliable file transfer service that simplifies the process of secure data movement for climate scientists. Provided by Argonne National Laboratory and the University of Chicago, Globus Online is a hosted, high-performance data movement service that builds on GridFTP-based [6] data transfer capabilities. There is no client software to download and install, and users can choose among three interfaces to submit a transfer and monitor progress: the website (see Figure 6.8), hosted SSH client, and REST API. Launched in November 2010, Globus Online already has more than 1200 registered users and more than 100 registered data-transfer endpoints.

With Globus Online, users submit a request to move data from an ESG server to another server or their own computers. Users provide their ESG usernames and passwords; Globus Online then handles all required

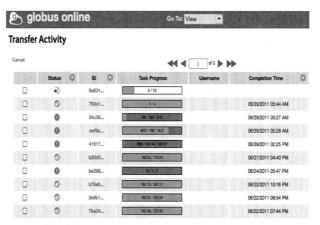

Ensure that the data was transferred intact.

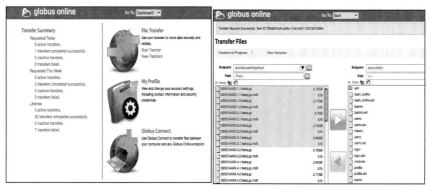

FIGURE 6.8   Globus Online dashboard, transfer, and monitor screens.

security credentials and transfer processes on their behalf. Globus Online applies appropriate performance-tuning parameters based on the file sizes used for the transfer, recovers from faults, and retries transfers until a specified deadline. The user is notified of any relevant status, including successful data transfer, credential expiration, and the like. Data-integrity checks are supported by the service to ensure that the data were transferred intact.

To facilitate download of data from a server to a user's laptop, Globus Online provides a component called Globus connect, a simple one-click download and installation, that allows users to move data to and from any local machine. Globus connect requires only an outbound connection, thereby allowing data transfers even when the machine is behind a firewall. The install does not require administrative privilege on the machine and provides a simple graphical install interface, as shown in Figure 6.9.

Globus Online works with the existing ESG security and data infrastructure, requiring no additional deployments at these sites. ESG administrators can configure their GridFTP servers as "endpoints" in Globus Online for users to discover and use as sources for their data transfers. In addition, Globus Online provides mechanisms to provide logical names for a given GridFTP endpoint, making it simpler for end users to identify and use the endpoint. Users can add their own destination machines as endpoints for transfers.

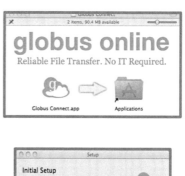

FIGURE 6.9   Globus connect screenshots.

As shown in Figure 6.10, Globus Online has been integrated with the ESG gateway to make it easier for ESG users to access the service as part of the search/download interface. The integration leverages the REST API to the Globus Online transfer capabilities, to submit a data-transfer request on behalf of the user from the ESG portal. This approach combines the Globus Online download capability with the ESG metadata search capability. In addition to the reliability and simplicity of transfer this option provides to the Gateway, it allows users to transfer data to both their laptops and to analysis machines of choice, thus not forcing users to download to laptops and then upload to analysis machines.

Globus Online also supports the administrative replication task in ESG. The hosted command-line interface provides a convenient mechanism for ESG site administrators to setup a repetitive synchronization task to pull data sets that need to be replicated from other ESG sites. The mechanism provides reliable and fault tolerant transfers, with high performance provided by the GridFTP protocol. The synchronization mechanism ensures that only changed files are transferred each time, and multiple levels of synchronization, from file timestamp to checksums, are

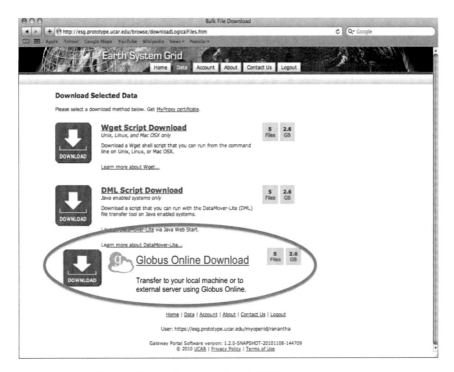

FIGURE 6.10   Globus Online integrated with ESG gateway.

provided. The ESG replication tool interfaces with Globus Online, producing the list of replication endpoints that can be used to drive Globus Online file transfers.

6.5.5 100 Gbps Networks (Climate100)

Climate100 [7] researches, develops, and tests end-to-end capabilities of the next generation networks in collaboration with the ESGF community. Its goal is to ensure that software, services, and applications associated with massive climate data sets can be scaled to the 100-Gbps infrastructure (see Figure 6.11), providing high-performance data movement for distributed networking. ESGF will deploy similar systems on multiple 10- or 100-Gbps connections to move massive data sets between its major data-node sites in the United States. With demonstrated success at the U.S. sites, ESGF will then extend high-performance data transfers to other data nodes in the United States and foreign partner sites. This scaled testing of

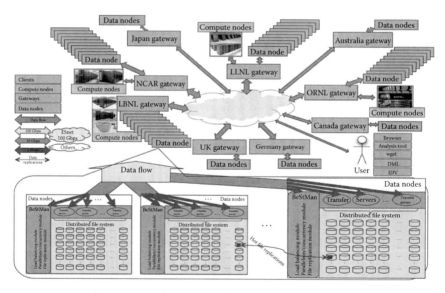

FIGURE 6.11 The envisioned ESGF topology uses 100-Gbps ESnet network connections to provide a network of geographically distributed gateways and data nodes in a globally federated, built-to-share scientific discovery infrastructure. In this way, independent data warehouses can deliver seamless access to vast data archives so that scientists can deploy specialized client applications for extreme analysis. Experts (e.g., model developers, climate researchers) and nonexperts alike will benefit from rich data exploration and manipulation available with such fault-tolerant, end-to-end system integration.

the data transfers and communication connections is important because climate researchers worldwide use many different network service providers. Continued success will enable a production-scale system that will empower scientists to try new and exciting data exchanges that could lead to breakthroughs.

6.6 DATA REPLICATION

The institutions that make up the ESGF have set up data mirrors to host copies of climate-simulation data sets that were originally published at other institutions. ESGF mirror sites include LLNL/PCMDI, the BADC, the DKRZ, and the ANU NCI.

The purpose of data set replication or mirroring is to provide additional copies of climate-data sets closer to end-user scientists and researchers. Because data sets are large, accessing them remotely over transoceanic networks can result in long access latencies, limiting the usefulness of those data sets. Mirroring allows researchers to access data sets of interest faster. The existence of data mirrors also increases the reliability and availability of climate data sets because if one ESGF site should be unavailable, the mirrored data sets can be accessed at other sites.

ESG has developed replication client software that manages the copying and publication of mirrored data sets over wide-area networks. Replication in the ESGF involves three phases. First, the replication client identifies the files that make up a data set by issuing a metadata query to an ESGF gateway. The second phase actually copies those files to the mirror site. In the final phase, a replication site publishes the data set as a replica. These phases are described more fully below.

In the first phase, the replication client queries a federation gateway to obtain metadata associated with a data set to be copied. Federation gateways maintain databases with metadata about data sets. This metadata includes the files in the data set, the data nodes that published the data set, the location of the data set catalog, checksums, and sizes of the files. The metadata catalog is provided as a web service. The replication client queries this web service and uses the response to construct a command file that a data movement agent can use to transport the data from one site to another. The mirror site administrator, who runs the replication client software, selects among several options regarding which data movement agent to use. The default is the BDM, but other choices include the globus-url-copy utility from the Globus toolkit and the Globus Online data transfer service.

In the second phase, the mirror site administrator uses the selected data-movement agent to issue data transfer commands to copy the files in the data set to local storage. This step usually takes the greatest amount of real time, given the large numbers and sizes of files.

Once data sets have been copied successfully, the final phase of replication begins. This step involves publishing the replicated data set at the mirror site gateway.

To publish a replica, the mirror site must first perform a scan of the downloaded files. This scan collects information about the sizes, locations on the disk, file checksums, and other data used to create a local catalog of the data set. With the results of the scan, a new THREDDS metadata catalog is generated. This replica catalog will be used to expose the data set to the local site users.

Before replica publication, the new THREDDS replica catalog is compared to a copy of the original THREDDS catalog retrieved from the original publication site during the metadata query (phase 1). The comparison checks that certain values, such as the names of files, counts of files, file sizes, and checksums, match in each catalog. Although this step is not part of the data quality control for the data content, it does provide assurance that the entire data set has been successfully transferred.

The final step is the publication of the mirrored data set as a replica that can be discovered and accessed by ESGF users. This publication is done using existing ESG publication tools. The mirror site gateway is notified of the new data set and makes it visible to the scientific community.

6.7 PRODUCT SERVICES

Users who do not routinely run numerical climate models themselves are often ill prepared to work with raw ESGF files. The size and complexity of the files can be daunting. ESGF product services are designed to meet the needs of these users by providing scientific visualizations, analysis capabilities, and simplified data subsets that can be obtained with low effort.

The foundation of product services is the ability to view images generated from data sets in a standard web browser. Users can view custom maps, time series plots, vertical profiles and sections, and Hofmöller diagrams. (Graphical representations of model ensemble axes are under development.) Plot characteristics such as contour level, axis scale, and color palette are under the user's control. The system also provides URLs that capture the state of the user interface at the moment a given visualization is generated. A collaborator receiving such a link can re-enter the

ESGF product services at that precise point and pursue further investigation of the data. ESGF visualizations may also be viewed on a virtual 3D globe using Google Earth or as animations. ESGF product services also include routine analysis calculations such as averages, sums, variances, and extrema over the space and time dimensions of the data. Such analyses, because the ESGF server performs them, significantly reduce the volume of data that must be transferred across the Internet.

The ESGF component that provides product services is the LAS, comprising a workflow engine (the "product server"), a user-interface client, and various "backend services." LAS also depends upon an OPeNDAP server, Ferret-TDS situated between the backend services and the data files, to provide analysis capabilities.

The LAS user interface for ESG runs in a standard web browser. It communicates with the product server via Ajax to obtain the names of available data sets, the scientific variables found within each data set, the dimensionality and coordinate limits of each variable, and the operations permissible on each axis or plane of a variable. The user-interface client requests products from the product server via a REST URL. The product server marshals the necessary work to one or more backend services that are invoked through a standardized simple object access protocol (SOAP) interface.

Many distinct product-generating applications can be harnessed behind the SOAP abstraction by creating a "wrapper" in a language convenient to the application. Scientific organizations have made huge investments in specialized analysis and visualization capabilities through custom applications such as UV-CDAT, Ferret, and the NCAR (National Center for Atmospheric Research) Command Language (NCL) and general purpose packages such as Matlab and IDL. Climate scientists rely on these tools day in and day out to do their research. LAS allows an organization to leverage these investments as backend services. Figures 6.12 and 6.13 show examples of LAS products based upon Ferret and NCL.

The ESGF includes LAS servers at many nodes. Each LAS is responsible for the data from its assigned publisher (see Section 6.3.2). LAS servers cooperate with one another by sharing the information from their assigned publishers. Thus, the user-interface client of any LAS is able to present all data sets in the Federation. When a user requests a product from LAS, the REST URL of that product is redirected to the LAS server associated with that data set.

Files are arbitrary units of storage chosen for the convenience of data producers. File boundaries, understandably, present an impediment to

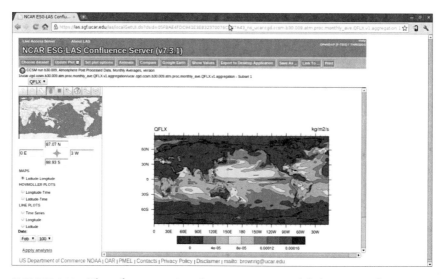

FIGURE 6.12   Plot of community climate system model data using live access server with an NCL backend.

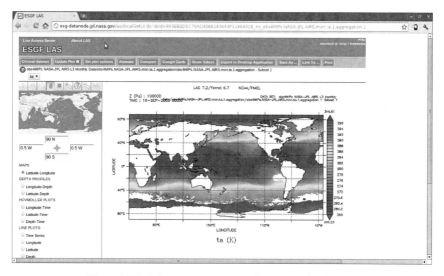

FIGURE 6.13   Plot of AIRS data using LAS with a Ferret backend.

scientific use of the data. Fortunately, OPeNDAP servers can reconstruct logical data sets from files through a process known as aggregation. LAS servers in ESGF are configured to "see" only aggregated data sets, so users are shielded from the impediments of file boundaries.

A key motivation for federating data is the need to intercompare data sets. LAS provides a specialized user interface client (vizGal) for this

FIGURE 6.14 The vizGal user interface client being used to compare vectors and scalar values from two different time steps of a Gulf of Mexico model.

purpose. VizGal allows users to create a custom gallery of related visualizations. VizGal can compute and display the differences between fields, automatically regridding as needed. Figure 6.14 illustrates vizGal in use.

ESGF product services are designed to augment the analysis tools used by scientists at their desktop. LAS supports a seamless segue from the web to desktop analysis by generating command scripts that initialize desktop applications to the user's current data set, variable, and space–time location of interest. The same aggregated data set used by LAS is directly accessible to desktop applications, without the need for file transfers.

6.8 FUTURE DIRECTION

The ESGF effort has begun to look outside the climate community for additional funding opportunities. These efforts intend to broaden the ESGF node development for other scientific domains, such as energy, material, nuclear energy, biology, chemistry, fusion, and others. This by no means indicates that we abandon the climate community; we seek to strengthen the entire scientific community with advances that will be beneficial for all. Besides our usual trademark advances in federation, analysis, and visualization, our efforts will look to expand in the areas of data mining, provenance and metadata, HPC, data movement, and data ontology. These are all challenging areas related to the management, manipulation, storage, access, analysis, and visualization of large-scale scientific data. With that said, we have come to the realization that some communities (such as climate) have a good handle on their data, thanks to efforts such as ESGF. However, satisfying the need for substantial investments in data-driven software and technologies is paramount, as future computing platforms and archives expand and reach extraordinary speeds and capacity. Our goal for these communities is to deliver a comprehensive end-to-end solution for the overall increase in scientific productivity, as has been as our continual effort for climate science.

ESGF is being funded in pieces by smaller, less focused activities with some success. Some of these activities include the climate science for a sustainable energy future to transform the climate model development and testing process; the NASA ACCESS proposal to produce an ESGF node capable of delivering NASA data products to the community; the Climate100, an advanced networking initiative research project to develop and test end-to-end capabilities of the next generation networks in collaboration with the climate community via the ESGF nodes; the large-scale data systems for nuclear energy to provide the nuclear energy community with access to data and information, model codes, analysis tools, and intercomparison capabilities for deeply analyzing nuclear energy simulations; and the NCAR effort to archive and disseminate Arctic data under the Cooperative Arctic Data and Information System. Internationally, small ESGF development efforts are also happening in countries such as the United Kingdom, Australia, Germany, France, Italy, and Japan. In addition, small business innovation research (SBIR) and other commercial efforts are funding their research and development efforts on the access and dissemination of the ESGF infrastructure and data archive.

Many challenges remain and new ones will emerge. Through such efforts mentioned above, we hope to continue and promote the sharing of knowledge, software, and tools among partners that combine model output, observational data, and analysis and visualization tools and the facilities needed to ensure access by all user communities.

ACKNOWLEDGMENTS

The following team members contributed to this chapter: Rachana Ananthakrishnan and Neill Miller from Argonne National Laboratory, Argonne, IL; Mehmet Balman, Junmin Gu, Vijaya Natarajan, Arie Shoshani, and Alex Sim from Lawrence Berkeley National Laboratory, Berkeley, CA; Gavin Bell, Robert Drach, and Michael Ganzberger from Lawrence Livermore National Laboratory, Livermore, CA; Jim Ahrens and Phil Jones from Los Alamos National Laboratory, Los Alamos, NM; Daniel Crichton and Luca Cinquini from Jet Propulsion Laboratory, National Aeronautics and Space Administration, Pasadena, CA; David Brown, Danielle Harper, Nathan Hook, Don E. Middleton, Eric Nienhouse, Gary Strand, Hannah Wilcox, and Nathan Wilhelmi from National Center for Atmospheric Research, Boulder, CO; Steve Hankin and Roland Schweitzer from Pacific

Marine Environmental Laboratory, National Oceanic and Atmospheric Administration, Seattle, WA; John Harney, Ross Miller, Galen Shipman, and Feiyi Wang from Oak Ridge National Laboratory, Oak Ridge, TN; Peter Fox, Patrick West, and Stephan Zednik from Rensselaer Polytechnic Institute, Troy, NY; Ann Chervenak and Craig Ward from Information Sciences Institute, University of Southern California, Marina del Ray, CA.

REFERENCES

1. Sim, A., D. Gunter, V. Natarajan, A. Shoshani, D. Williams, J. Long, J. Hick, J. Lee, and E. Dart. Efficient Bulk Data Replication for the Earth System Grid, *Proceedings of International Symposium on Grid Computing, Data Driven e-Science: Use Cases and Successful Applications of Distributed Computing Infrastructures*, 2010.
2. Sim, A., M. Balman, D. Williams, A. Shoshani, and V. Natarajan. Adaptive Transfer Adjustment in Efficient Bulk Data Transfer Management for Climate Datasets, *Proceedings of the 22nd International Conference on Parallel and Distributed Computing and Systems*, 2010.
3. Scientific Data Management Research Group. DataMover-Lite (DML). http://sdm.lbl.gov/dml/.
4. Allen, B., J. Bresnahan, L. Childers, I. Foster, G. Kandaswamy, R. Kettimuthu, J. Kordas, M. Link, S. Martin, K. Pickett, S. Tuecke. Globus Online: radical simplification of data movement via SaaS. Preprint CI-PP-05-0611, Computation Institute, 2011.
5. Foster, I. Globus Online: Accelerating and democratizing science through cloud-based services. *IEEE Internet Computing* (May/June), 15: 70–73, 2011.
6. Allcock, B., J. Bresnahan, R. Kettimuthu, M. Link, C. Dumitrescu, I. Raicu, I. Foster. The Globus Striped GridFTP Framework and Server. *SC'2005 Proceedings of the 2005 ACM/IEEE Conference on Supercomputing*, 54, 2005. doi 10.1109/SC.2005.72.
7. Scientific Data Management Research Group. Climate 100. http://sdm.lbl .gov/climate100/.

CHAPTER 7

Data-Intensive Production Grids

Bob Jones and Ian Bird

CONTENTS

7.1 INTRODUCTION

Science is fundamental to addressing the problems confronting our planet and our lives, and data are fundamental to science. The science we do now requires ever-increasing datasets that need flexible and powerful computing systems to process them. Computing does not have the power to save our planet from global warming or energy shortages—it does, however, have an underlying role to play in making this happen.

Grids are effective mechanism for bringing together computing and storage resources located in, owned and operated by different organizations. By connecting through the Internet networks, a grid is a means for sharing computer power and data storage capacity and providing secure access. It permits the creation of virtual research communities, making use of computers located all over the globe to become an interwoven computational resource for large-scale compute- and data-intensive grand challenges.

The research community is facing a deluge of data produced by the latest generation of scientific instruments. This increased scale of scientific data is putting pressure on existing information and communications technology (ICT) services that are struggling to store, distribute, process, analyze, and preserve this precious commodity. At the same time, the need to show a return on investment is accelerating the scientific process itself, reducing the time from the acquisition of data to the publication of results. This demands that ICT services be more reliable and dynamic in their ability to serve the research community. As a consequence, the cost of ICT to the research community is continuously increasing and is becoming a significant item in organizations' budgets.

It is necessary to revitalize the ICT models used by the research community to adapt to the changing needs of researchers, embrace new disruptive technologies, and profit from the restructuring of the commercial sector which is introducing new products and services.

CERN, the European Organization for Nuclear Research, is the largest particle physics laboratory in the world and is an international organization with its headquarters in Switzerland. CERN is the home of the Large Hadron Collider (LHC)—the world's largest and most powerful particle accelerator that spans the border between Switzerland and France about 100 m underground. The LHC complex provides research facilities for several thousand high-energy physics researchers from all over the globe. The LHC experiments are designed and constructed by large international collaborations and will collect data over a period of 15–20 years. These experiments run more than one million computing tasks per day and generate more than 15 petabytes (PB) of data per year. The computing capacity required to analyze the data far exceeds the needs of any comparable physics experiments today and relies on the combined resources of some 200 computer centers worldwide. CERN and the particle physics community have chosen grid technology to address the huge data storage and analysis challenge of the LHC. CERN leads the Worldwide LHC Computing Grid project (WLCG), which is a global collaboration linking grid infrastructures and computer centers worldwide. It was launched in 2001 to provide a global computing resource to store, distribute, and analyze the data generated by the LHC. The infrastructure, built by integrating thousands of computers and storage systems in hundreds of data centers worldwide, enables a collaborative computing environment on a scale never seen before. The WLCG serves a community of more than 10,000 physicists around the world with near real-time access to LHC data and the power to process it.

7.2 WHY A GRID?

To understand why a grid infrastructure was chosen, we must consider several factors that had an important impact on the design of the LHC computing environment. Primarily, the volume of data generated by the experiments is estimated at some 15 PB per year. These data are generated at a significant rate: some 300–400 megabyte (MB) per second by the two largest experiments, ATLAS and CMS. During the period when the LHC accelerates heavy ions, this data rate increases to around 2 gigabyte (GB) per second for ALICE, the dedicated heavy ion experiment. The fourth experiment, LHCb, generates data at the lower rate of <100 MB/s. Thus, during most of the LHC running, the total data rate is around 1 GB/s and this rises to close to 3 GB/s during heavy ion running. These data are archived at CERN and a second copy is distributed between 11 regional (Tier 1) centers in real time, together with an equal amount of data resulting from their initial processing. Thus, the distribution system must be capable of supporting these rates continuously.

The Tier 1 centers are large data processing centers capable of providing reliable data archiving for the lifetime of the accelerator (more than 20 years) and they also play a role in distributing the data to a large number of Tier 2 centers (more than 150) once it has been processed in the Tier 1 centers. The analysis of the processed data is subsequently performed at the Tier 2 sites. Thus, another important consideration is the ability to rapidly process and distribute the data to the Tier 2 centers making it available to the thousands of scientists: from a few hundred in each of the LHCb and ALICE experiments, to more than 2000 people in each of ATLAS and CMS.

The required amount of computing capacity—both compute and storage—is driven by the volume of data, and a certain amount of simulation that is required by all experiments. These resources must be accessible by the many scientists using several different workflows from centrally organized and managed data processing for the bulk of the real and simulated data, to more widely varied workflows during physics analysis, where the need for specific datasets is unpredictable and evolves as the analysis progresses.

In considering how to build a system to satisfy all these needs, it was clear from the beginning that putting all of the computing and storage power at CERN would not be feasible for a variety of reasons. First, the infrastructure of CERN's computing facilities would not scale to the required level without significant investments in a power and cooling infrastructure that was many times larger than that which existed at the time. Second, the CERN member states have historically encouraged the provision of a significant fraction of the computing resources for an experiment to be located outside of CERN. The physics institutes and national laboratories participating in LHC experiments have local computing facilities, often of significant scale, and with the expectation of improvements in wide area networking it was reasonable to expect that a distributed system that could couple these resources may achieve the scale needed. There also are sociological advantages in such a distributed system [1]—investment is made locally and benefits the hosting institute in several ways, for example, providing training for students, and the ability to leverage the resources for other local uses.

Figure 7.1 shows the key components of the model for LHC computing. At the time of developing this model (1999), it became clear that grid

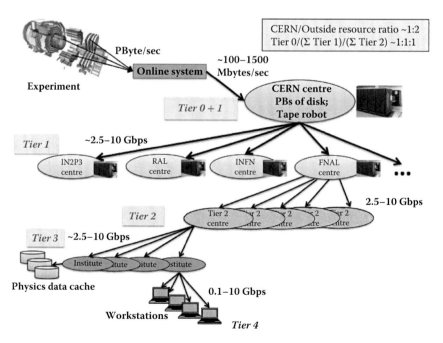

FIGURE 7.1 (**See color insert.**) Original LHC computing model.

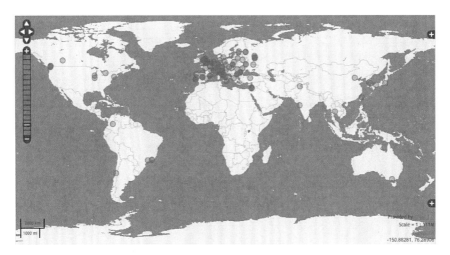

FIGURE 7.2 Worldwide distribution of WLCG grid sites.

computing technology could provide the practical implementation permitting, for example, the integration of computing resources across multiple administrative domains; and the authentication and authorization infrastructure that provides a means for a user to have access to resources that may be distributed around the world.

This model was developed over the following 10 years by the WLCG in conjunction with a series of grid projects funded by the European Commission in Europe [2] and by the Department of Energy and National Science Foundation [3] in the United States. The infrastructure and the tools to manage the exceptional data volumes and rates were developed to provide a production environment that would be usable and manageable by the experiments and the computer centers involved. The Tier 0 provides some 20% of the total computing resources, while Tier 1 and Tier 2 centers each contribute some 40%. In total today, there are approximately 250,000 CPU cores and 150 PB of disk federated by the WLCG.

Figure 7.2 shows the worldwide distribution of the WLCG grid infrastructure.

7.3 PRODUCTION INFRASTRUCTURE

It was essential that the LHC computing environment became an integral part of the production infrastructure of the participating computer centers and that it could be managed by a moderately sized operations team. Thus, in addition to the technical implementation of the grid itself, an entire ecosystem had to be built. This included processes and tools to support daily operation, problem solving, and metrics for reporting. In addition to the service aspects, a set of policies was developed to govern the use of the infrastructure and to address privacy and security concerns. These ranged from acceptable use policies (for users and for virtual organizations) to policies for information retention and publication related to resource usage. A significant achievement was the development of a worldwide network of trust for both authentication (based of X.509 certification authorities) and authorization. This unique trust framework [4] is what enables the many thousands of physicists to seamlessly access compute and storage resources anywhere in the world. When defining these policies it was necessary to strike a balance between defining infrastructure-wide policies and allowing freedom of choice to individual participants.

To bring the infrastructure to production quality with the ability to support all of the technical and management goals of LHC computing, a significant program of testing and challenges was developed and implemented over a period of 6 years. From 2003 onward, starting with the deployment of the initial prototype grid system with only a handful of grid sites, the system has been used for the production of simulated data by the physics experiments, and since 2004 an extensive program of data and service challenges was implemented to show and test all aspects of the infrastructure. These, for example, included achieving the necessary scale and reliability of data transfers and data management, scaling of job workloads, and testing of key support processes including response to simulated security incidents.

In parallel with these organized periods of service challenges, the grid infrastructure was gradually being ramped up in terms of resources in preparation for the start of data taking.

In addition to these tests, from late in 2008 until the LHC started at the end of 2009, significant amounts of cosmic ray data were acquired and processed by the experiments. This allowed for extensive testing under reality conditions of the entire computing models of the experiments from data acquisition to analysis and was important for validating the entire system. By increasing the data acquisition rates, this testing could be done at rates close to the anticipated data rate for future LHC running. The testing program culminated in two major readiness challenges in 2008 and 2009 that showed that the WLCG was prepared for the start of data taking.

7.4 LARGE HADRON COLLIDER ACHIEVEMENTS IN ITS FIRST YEAR

The LHC came on line at the end of 2009, and performance has rapidly improved throughout 2010 and the first half of 2011. During that time, some remarkable achievements of the computing system have been shown. Foremost is the fact that the computing environment has enabled the very rapid delivery of physics results. Processed data are available to physicists for analysis within hours of data taking, and results presented at large international conferences have in some cases been based on data taken only weeks earlier. This rapidity is unprecedented in particle physics.

There are also specific technical achievements that are notable. First, despite the significant testing, the data rates achieved with real data rapidly exceeded those achieved during the test periods. Data rates in excess of 2 GB/s are common and peaks of 5 GB/s have been observed. This ability was made possible by the dedicated and redundant LHC Optical Private Network [5] built between CERN and the Tier 1 sites to enable fast data exchange. Typical data movements worldwide for a large experiment such as ATLAS have been around 10 GB/s.

7.5 LINKING DATA AND PUBLICATION

In parallel with these developments, the scientific process itself is changing, with an increasing dependence on ICT at all stages. Similarly, policy and the public attitude are demanding that research that is publically funded be made publically available. This Open Access movement [6] is particularly relevant for the research community and poses additional demands on the ICT infrastructure. The ability to curate and archive for decades the complete corpus of scientific output, from raw data to final publications, requires novel ICT tools. Such tooling will further accelerate the scientific process, challenging the research communities' culture while permitting a better assessment of the value of scientific output to society. This data continuum will enable multidisciplinary research by connecting information that was previously isolated to create new knowledge, and further justifying the investments in the eyes of the public and funding agencies.

In partnership with other leading High-Energy Physics (HEP) laboratories worldwide, CERN has launched INSPIRE [7], a next-generation digital library system for HEP. Built on the CERN Invenio open source digital library solution, INSPIRE hosts one million records and half a million full-text articles, which are searchable, whose authors are disambiguated, and for which advanced bibliometrics are possible.

7.5.1 An Example from Life Science

The infrastructure on which WLCG is built also serves many other grand challenges from different scientific disciplines. Although the bulk of usage comes from high-energy physics, research communities in astronomy, astrophysics, computational chemistry, earth sciences, fusion, and computer science also successfully use production grid infrastructures. Many communities are using the infrastructure to examine questions they could never have addressed before. In silico drug discovery is one of the most promising strategies to speedup the drug development process [8]. Virtual screening is about selecting, in silico, the best candidate drugs acting on a given target protein. In vitro screening is very expensive as there are now millions of chemicals that can be synthesized. A reliable means of in silico screening could reduce the number of molecules required for in vitro and then in vivo testing from a few million to a few hundred. However, virtual screening represents a computational data challenge requiring intensive computing, in the order of a few teraflops per day to compute one million docking probabilities or the molecular modeling of 1000 compounds on one target protein. Access to very large computing resources is, therefore, needed for successful high-throughput virtual screening.

Large-scale grids for in silico drug discovery open opportunities of particular interest to neglected and emerging diseases. The WISDOM [8] initiative aims to develop new drugs for neglected and emerging diseases with a particular focus on malaria. It relies on emerging information technologies to provide new tools and environments for drug discovery and development. Its main goal is to boost research and development on neglected diseases by fostering the use of open source information technology for drug discovery. The initiative that started in 2005 was initially focused on virtual screening. During the summer of 2005, a first large-scale deployment achieved in silico docking of 42 million compounds in about six weeks against a protein of the parasite responsible for malaria: *Plasmodium falciparum*. The WISDOM initiative subsequently deployed a second data challenge for malaria on several grid infrastructures and, in about 10 weeks, 140 million dockings were achieved, representing an average throughput of almost 80,000 dockings per hour. The workflow of the software involved accessing data in various formats from multiple databases, separately managed around the world, and visualizing the results of computations. This was made possible by the availability of computing resources from several grid infrastructures worldwide. This success led to further challenges targeting avian flu neuraminidase N1. Now results from the WISDOM initiative are being processed and tested in vitro by a number of laboratories around the world.

7.6 EVOLUTION OF THE EUROPEAN GRID INFRASTRUCTURE

The first year of the LHC data taking has been a remarkable success for the WLCG computing infrastructure. The experiments, sites, and the various grid development and operations projects involved have shown the ability to build and run a solid and reliable service.

All sites, large and small, have shown that their resources are accessible and can be used by the experiments. This is important as it shows that the grid really does enable worldwide collaboration and sharing of resources.

This success has been the culmination of close to 10 years of development, testing, and investment by many institutions, people, and contributing projects.

This success was not immediate since it required significant effort, and several rethinks of the way the services should run and be used by the experiments. Now, there is a far better understanding of the actual needs of the physics experiments in terms of the distributed computing infrastructure, than there was 10 years ago. Even in the past two years with the large-scale readiness testing and the beginning of LHC operation, there has been a significant evolution of those requirements. The computing models of the experiments have matured and are tending toward a much more mesh-like topology rather than the strict hierarchy described earlier.

To run the grid service at a very large scale takes a lot of operational effort. Analysis of the failures shows that most are either at the infrastructure level, such as power outages, or related to services that are not redundant across hardware failures. In addition, database service problems engage significant work to restore normal operation including synchronizing data replicas at Tier-1 sites. As a consequence, changes are being considered that will simplify the operation of the grid service and reduce the level of operational effort needed. Technology in this area evolves very rapidly, and there is scope to benefit from recent advances as explained in Section 7.7. All these points will contribute to ensuring the long-term support and sustainability of the infrastructure.

It is still important to recognize that the WLCG uses a grid for a very good reason: to integrate the resources that are distributed worldwide across dozens of different management domains. The collaborative and distributed aspects of the WLCG lie at the heart of the initial design choices and remain valid even as the underlying technologies evolve.

However, it is also apparent that there are many other communities that perceive grids as being hard to use, or not suitable for their needs, and are looking at other solutions such as cloud computing to achieve their objectives [9]. It is necessary to continuously reassess the implementation of the grid infrastructures in light of the changes in the environment noted earlier, but never forgetting the important successes and lessons, and ensuring that any evolution does not disrupt the continued operation for existing communities.

7.6.1 Enabling Grids for e-Science to European Grid Infrastructure Transition

A new organization, called EGI.eu, has been created to coordinate and evolve the European grid infrastructure (EGI) by building on the work of previous European Union–funded projects. This transition was an important step in ensuring the European research community has access to a distributed computing infrastructure. EGI.eu was created on February 8, 2010 to coordinate and maintain a sustainable pan-European computing infrastructure to support European research communities and their international collaborators. The earlier projects nurtured this goal from the initial concept of a scalable, federated, distributed computing system. The conceptual and logistical framework for a permanent organization to oversee the operation and development of the grid on a Europe-wide level was the subject of a dedicated design study (European Grid Initiative Design Study [EGI-DS]), which took place from September 2007 to December 2009. A four-year project called EGI-InSPIRE supports and cofunds EGI.eu operations with a common goal—to create a seamless system ready to serve the demands of present and future scientific endeavors.

7.7 LESSONS LEARNED AND ANTICIPATED EVOLUTION

The successful collection, distribution, and analysis of data from the four LHC experiments represent a major achievement and demonstration of the grid's success. The original concept of a grid—being able to engender collaboration and sharing of resources—is even more relevant today than it was 10 years ago. However, the environment has evolved and today's "grid" infrastructures must be integrated into an overall science e-infrastructure framework, capable of evolving and adapting to the rapidly changing needs of science communities, especially those that are coming to realize the potential of such infrastructures in being able to help them deliver better science.

Networking, grid, and supercomputing have developed independently but are now becoming interwoven and should rather be viewed as building blocks or services within an e-infrastructure ecosystem that a science community can make use of in a coordinated manner. Blurring the boundaries between these elements of e-infrastructure will encourage their interoperation and present a more consistent model to the users. These existing elements must be supplemented by, and integrated with, new components including scientific cloud services, emerging data infrastructures that manage significant slices of the data continuum and other new services. Such an e-infrastructure must also be agile enough to be able to develop the additional services required by new science communities or to help those communities integrate their own specific services, while taking benefit from the overall structure (AAA,* policies, support, etc.).

In such an environment, ease of use and access are vital considerations. The cloud model has been so readily adopted since it is a simple concept from the user point of view. Similar simplicity must be applied to the more traditional services. Enabling the use of existing trusted federated identities would help address some of the initial complexity in being able to access a grid, for example.

It is too simplistic to think that a cloud computing model could, at this point in time, replace the current grid infrastructures. For the foreseeable future, it will remain sociologically and politically attractive for science collaborators to provide resources locally to the benefit of their own local communities, while nevertheless being willing to share these resources and data with their collaborators. Policies and cost models for resource allocation must be developed and must take into account this environment. For example, it will become important that user communities understand the real costs of providing archiving and rapid access to many petabytes of data over many decades. Thus, it is important that those communities contribute in some way to this cost rather than seeing a data infrastructure that is separately funded to manage data. This implies that the costs of these e-infrastructures should be included in the funding of science from the outset as an integral part of what it takes to do

---

* AAA: Authentication, authorization, and accounting; generally used to refer to the underlying security framework.

science. Having a separately funded and governed e-infrastructure potentially hides the true costs from scientists and researches it servers which in turn could lead to inappropriate technical choices and divergence of implementation from need.

REFERENCES

1. I. Bird, B. Jones, and K.F. Kee. 2009, The organization and management of grid infrastructures. *IEEE Computer*, 42(1), ISSN 0018-9162.
2. E. Laure and B. Jones, 2009, Enabling Grids for e-Science: The EGEE Project, in *Grid Computing: Infrastructure, Service, and Applications* Lizhe Wang, Wei Jie, and Jinjun Chen. CRC Press, April, 2009, Boca Raton, FL: Taylor & Francis Group ISBN 9781420067668, http://www.crcnetbase.com/doi /book/10.1201/9781420067682 (last accessed January, 2013).
3. R. Pordes, 2005, The Open Science Grid (OSG), Computing in High Energy Physics and Nuclear Physics 2004, Interlaken, Switzerland, 27 Sep to 1 Oct 2004, p. 963, CERN-2005-002-V-2, http://cdsweb.cern.ch/record/865745 /files/p963.pdf (accessed January, 2013).
4. International Grid Trust Federation, http://www.igtf.net/ (accessed January, 2013).
5. E.-J. Bos, E. Martelli, and P. Moroni, 2005, LHC Tier-0 to Tier-1 High-Level Network Architecture, CERN, Tech. Rep., 2005.
6. The SCOAP3 Working Party, 2007, *Towards Open Access Publishing in High Energy Physics*, ISBN 978-92-9083-292-8, http://cds.cern.ch/record/865745 /files/p963.pdf (accessed January, 2013).
7. A. Holtkamp, S. Mele, T. Simko et al., 2010, INSPIRE: Realizing the Dream of a Global Digital Library in High-Energy Physics, CERN-OPEN-2010-019, 03/06/2010, http://cdsweb.cern.ch/record/1276784/files/CERN-OPEN - 2010-019.pdf.
8. V. Kasam, J. Salzemann, M. Botha et al., 2009, WISDOM-II: screening against multiple targets implicated in malaria using computational grid infrastructures, *Malaria Journal*, 8, p. 88. http://www.malariajournal.com /content/8/1/88.
9. M. Lengert and B. Jones, 2011, Strategic Plan for a Scientific Cloud Computing infrastructure for Europe, CERN-OPEN-2011-036, August 2011, http://cdsweb.cern.ch/record/1374172 (accessed January, 2013).

CHAPTER 8

EUDAT

*Toward a Pan-European Collaborative Data Infrastructure**

D. Lecarpentier, J. Reetz, and P. Wittenburg

CONTENTS

"A fundamental characteristic of our age is the rising tide of data—global, diverse, valuable and complex. In the realm of science, this is both an opportunity and a challenge."[1] This excerpt from a 2010 European report on scientific data captures very well a major trend affecting almost all

---

* See http://www.eudat.eu for more information.

scientific disciplines, and the transformations required if we are to optimally manage and exploit this explosion of data.

Modern information and communication technology (ICT) gives researchers improved access to data-intensive sensors; increasingly comprehensive analysis and simulation facilities; and efficient large-scale communication platforms, enabling, for example, crowd sourcing that engages vast numbers of citizens in data creation and enrichment. As a consequence, politicians speak of ICT as a transformational factor for science,[2] revolutionizing traditions and workflows in almost all research disciplines. As a result, data are now the basis of theorization and proof of results even in disciplines not traditionally data oriented. A completely new paradigm, known as data-intensive science (DIS), is thus being added to the set of methodologies used to discover unexpected phenomena based on statistical operations on large data sets. Data generated by processes in nature, societies, bodies, and minds represent information that can be used in understanding the underlying nature of these processes.

The availability of suitable data is a key pillar of e-Science, enabling teams of investigators to work with, or independently from, the researchers who derived the data, thus reusing and recombining data in new scientific contexts. However, repurposing data—within and especially across disciplinary domains—requires significant integration and interoperability efforts, as well as agreement on standards and sharing practices. In this respect, it is not so much the sheer volume of data, but its increasing complexity and diversity across varying collections that require attention.

In particular, data useful in finding solutions to the grand challenges we face in the areas of climate change, energy production, health, and societies can be seen as highly valuable, although it is impossible, as K. Jeffery points out, to quantify the value of scientific data.[3] Further, data are not just valuable to current generations of researchers, but also to future generations, since they capture information about the state of the world. Thus, proper data management is required to guarantee seamless access to current and future researchers, and special attention is required to ensure proper long-term preservation.

Data reusage in an open data infrastructure scenario also implies that data creators, managers, and users do not necessarily know each other, but nevertheless must rely on each other's quality of work. Thus, new mechanisms are necessary to establish trust between all stakeholders.

Major efforts over a longer period are needed to meet the challenges of integration, interoperability, data life cycle management, and trust-building. Only a systematic and focused approach, encouraging collaboration between the various stakeholders and covering the entire life cycle of data objects, will help us to achieve global reach across disciplines. We call this approach a Collaborative Data Infrastructure (CDI), and anticipate that it can only be established using an organic and agile process at a global level, involving all stakeholders, existing data initiatives and emerging community-specific data solutions. To this end, a top-down holistic construction will not suffice. Thus, the European Collaborative Data Infrastructure (EUDAT) initiative aims to contribute to the development of such a pan-European CDI.

8.1 TOWARD A PAN-EUROPEAN INFRASTRUCTURE FOR SCIENTIFIC DATA

Over the past 5 years, research landscapes and instruments in Europe have undergone significant developments, aiming to facilitate the transition to e-Science and address its related challenges. The Europe 2020 strategy—and its goal of establishing a truly operational European Research Area (ERA) by this date—has triggered cooperation between member states and facilitated the development of cross-national research infrastructures (RIs), most of which rely on digital data for their work. ICT infrastructures (also known as e-infrastructures) have been recognized as "a crucial asset underpinning European research and innovation policies"[4] and significant progress has been made in the deployment of pan-European ICT infrastructures for research supporting multiple disciplines.

8.1.1 Emergence of Pan-European Research Infrastructures

Scientific communities are increasingly organized, bridging with other disciplines and across national borders. The emergence of pan-European RIs in almost all scientific fields reflects this trend. RIs are not new: collaboration between scholars to access shared resources has been around from all time. The great abbey libraries of the Middle Ages, for example, could be considered RIs. Today, modern RIs are more usually instruments and facilities, such as synchrotrons, telescopes, sensor networks, biomedical facilities, and large distributed databases. It is estimated that there are more than 500 RIs of various size in Europe, of which at least 300 have strong international visibility. These RIs represent an aggregate European investment on the order of €100 billion, with an yearly operational and maintenance cost of €10–€15 billion.[5]

Since 2006, through the European Strategy Forum on Research Infrastructures (ESFRI), major progress has been made in terms of agreeing on priorities and securing investments for large-scale RIs. The first ESFRI roadmap for RIs, published in autumn 2006, outlined 35 new large-scale RIs of pan-European interest for construction, each costing from €10 million–€1 billion. The projects service disciplines include social sciences and humanities, environmental science, energy, biological science, medical science, materials science, the physical sciences, and e-infrastructures. This roadmap was updated in 2008 and 2010 and now includes 44 projects with individual construction costs between €2 million and €1.1 billion and operational costs between €2 and €120 million. By the end of 2010, 10 projects had entered their implementation phase, while another 16 are on target for implementation by the end of 2012.[6]

RIs provide valuable platforms for fostering cooperation between researchers across national borders, and are a key driver of construction of the ERA. They also produce and distribute huge volumes of data, and need to rely on a strong and sustainable e-infrastructure to store, share, and protect the growing volume of data they generate.

8.1.2 Move toward Pan-European Horizontal e-Infrastructures

Efficient e-infrastructures are essential facilitators of research and innovation, playing a key role in supporting the deployment of new research facilities on a pan-European level, and addressing their ICT requirements. In recent years, significant investments have been made by the European Commission and European member states to create a pan-European e-infrastructure supporting multiple communities. This e-infrastructure is currently divided into several intertwined domains, together providing a variety of functions and services.

The pan-European research and education network backbone is provided by GÉANT, on behalf of the European consortium of National Research and Education Networks (NRENs). A hierarchical concept is used, in which research organizations connect to each other through their local NRENs, which in turn connect with each other by the pan-European GÉANT backbone network. GÉANT is the world's largest multi-gigabit communication network dedicated to research and education. It already serves around 4000 universities and research centers and connects 36 NRENs. Such a powerful and reliable network is prerequisite to the work of distributed RIs, and to the use of large data storage facilities throughout Europe, independent of the location of the researcher or compute facility.

Researchers also need to rely on a solid computing infrastructure. Until 2010, Europe's largest capacity computing grid for publicly funded research was Enabling Grids for E-sciencE (EGEE). This EGEE infrastructure played a central role in the global distribution and sharing of the huge volume of data generated by the Large Hadron Collider at the European Center for Nuclear Research (CERN), providing the European research community and their international collaborators—representing over 17,000 users across 160 projects—with an e-Research platform for high-throughput data analysis. Since May 2010, a new organizational model has taken over, implemented by the European Grid Initiative (EGI.eu). The objective of EGI.eu (a foundation established under Dutch law) is to create and maintain a pan-European grid infrastructure in collaboration with National Grid Initiatives. Its mission is to enable access to computing resources for European researchers from all fields of science. There have also been significant developments in capability computing because of the Distributed European infrastructure for Supercomputing Applications (DEISA) and Partnership for Advanced Computing in Europe (PRACE) projects. DEISA played a key role in setting up a pan-European high-performance computing (HPC) infrastructure for supercomputing applications and a production-quality environment for EU scientists to use. The PRACE currently represents the top of the European HPC ecosystem, and provides Europe with cutting-edge world-class HPC systems.

A European e-infrastructure ecosystem is thus slowly taking shape, with communication networks, distributed grids, and HPC facilities providing European researchers from all fields with state-of-the-art instruments and services to support collaboration and research at a pan-European level. However, given the accelerated proliferation of data, additional efforts and investments are needed to tackle the specific challenges of data management, and to ensure a coherent approach to scientific data access and preservation. European projects in networking, computing, and grids are moving from one-off projects to sustainable infrastructures and services. EUDAT intends to achieve this goal for the data layer, aiming to develop a high-quality, cost-efficient, and sustainable pan-European data ecosystem, driven by European research needs and user communities.

8.1.3 Data Initiatives: A Fragmented Landscape

There is already a long history of data management projects* and initiatives in Europe, with several existing data infrastructures dealing with established and growing user communities: CERN, the European Space Agency, the European Organisation for Astronomical Research, and the European Bioinformatics Institute are examples of large-scale intergovernmental infrastructures generating and managing massive volumes of data. The numerous RIs being created as part of the ESFRI roadmap[6] can also be seen as data management initiatives, connecting repositories, aggregating and sharing data across national borders, and developing tools to make these data widely available to their communities and beyond.

Although some solid experience exists in dealing with data infrastructures, the current data landscape is still fragmented, with most initiatives addressing the needs of a specific discipline or community. This has resulted in increasing diversity with respect to data architectures, organizations, formats, and semantics. Issues related to interoperability and integration of existing data infrastructures are a growing concern in the context of cross-disciplinary research. Interoperability refers, in particular, to methods that allow us to seamlessly access and process the content of integrated objects. Integration refers to the methods that allow us to bring data objects together by making use of their external properties. Integration, among other things, requires measures to make data visible with the help of metadata of which the schemas and vocabularies are registered and thus openly available and to access the objects based on their persistent identifier (PID) included in the metadata record. Integration also requires that the user can understand what kind of information the objects contain and how it has been created; that is, the metadata records need to cover contextual and provenance information. Long-term sustainability is another issue, with rising costs due to the explosion of data threatening the financial viability of those infrastructures. Thus, the growing volume of data generated by researchers and their instruments—especially as the ESFRI RIs come online—as well as demand for access to those data has

---

* For example the EC-funded projects: DataGrid (2001), http://eu-datagrid.web.cern.ch; DataMiningGrid (2004), http://www.datamininggrid.org; Caspar (2006), http://www.casparpreserves.eu; SeaDataNet (2006), http://www.seadatanet.org; DRIVER (2006), http://www.driver-repository.eu; Planets (2006), http://www.planets-project.eu; Shaman (2008), http://shaman-ip.eu/shaman; PARSE.insight (2008), http://www.parse-insight.eu; Europeana (2008), http://www.europeana.eu; D4Science, http://www.d4science.eu; OpenAIRE (2009), http://www.openaire.eu; Geo-Seas (2009), http://www.geo-seas.eu; PaNdata Europe (2010), http://www.pan-data.eu; ODE (2010), http://www.ode-project.eu; SCAPE (2011), http://www.scape-project.eu.

created a new impetus for the development of a pan-European strategy for data infrastructures.

This view is supported by several European expert groups, including the e-Infrastructure Reflection Group (e-IRG), which, in its 2010 Blue Paper on e-Infrastructures,[7] endorsed by ESFRI, recommended that the European Union "identify and promote common (long term) data related services across different RIs," emphasizing the potential synergies existing between the ICT requirements of individual RIs. The e-IRG reaffirmed the need to create a sustainable European data infrastructure in its 2011 White Paper: "Although well-armed with other infrastructure components from networking, grids and HPC, the European e-infrastructure does not yet have a data infrastructure constituent that would allow the operation and unification of several existing multi-domain data and heterogeneous environments. We do, however, have some prior experience and so the major goal is to unify all existing use cases and provide a global approach for Europe."

In 2009, a High Level Expert Group on Scientific Data was tasked by the Directorate-General for Information Society and Media of European Commission to prepare a Vision 2030 for the evolution of e-infrastructure to scientific data. The group released its final report in 2010, describing long-term scenarios and associated challenges related to scientific data access, curation, and preservation, as well as the strategy and actions necessary to realize their vision. The report contains the following recommendations:

- Develop an international framework for a CDI
- Earmark additional funds for scientific e-infrastructure
- Develop and use new ways to measure data value, and reward those who contribute it
- Train a new generation of data scientists and broaden public understanding
- Create incentives for green technologies in the data infrastructure
- Establish an interministerial group on a global level to plan for data infrastructure

This report also presents the CDI as an important concept guiding future efforts. Indeed, it is likely to shape current and future data initiatives in Europe and is at the heart of the EUDAT initiative.

8.1.4 Collaborative Data Infrastructure

The CDI concept emerged from various initiatives. In its 2009 White Paper,[8] the Partnership for Accessing Data in Europe (PARADE) consortium, which was the origin of EUDAT, presented a two-layer design describing interactions between centers offering community-specific services and centers offering common services. The High Level Expert Group on Scientific Data extended this concept to a three-layer diagram, including data generators and users as part of the CDI ecosystem and resulting in Figure 8.1, which is frequently cited.

The main entities of the CDI are (1) researchers, sensors, and simulation and analysis frameworks that create and consume large data sets with the help of Internet-based tools; (2) data networks and expertise centers that offer community-specific services and have deep knowledge of data content, community traditions, and workflows; (3) networks of data centers that offer cross-disciplinary and thus common data services to the community centers and directly to data consumers and generators.

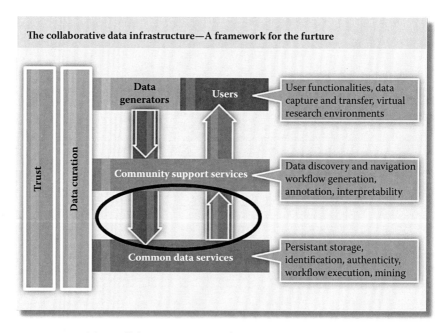

FIGURE 8.1   The Collaborative Data Infrastructure. (From High Level Expert Group on Scientific Data, *Riding the wave. How Europe can gain from the rising tide of scientific data,* Final report submitted to the European Commission, http://cordis.europa.eu/fp7/ict/e-infrastructure/docs/hlg-sdi-report.pdf, 2010.)

This picture can also be seen as an attempt to capture the kind of collaboration required between the different actors. On the top level, we have data generators and users, who rely on community support services specific to their disciplines; in turn, these community support services rely on a set of common data services that can be used by different disciplines. At every layer in the system, there are appropriate provisions to curate data and ensure its trustworthiness.

Establishing the CDI will not be a trivial task. Many communities have recently implemented distributed data organizations and architectures specifically tailored to their needs. This process represents a huge investment, often involving global collaboration, and its results cannot be easily abandoned. Thus for the CDI to succeed, an abstract architecture is required, facilitating integration of preexisting data solutions from participating communities, and data centers that support common data services. The development of such an architecture presents a significant challenge, requiring active collaboration between all actors, and in particular between the communities involved in designing specific services and the data centers willing to provide generic solutions. Within the generic layer, which will be the focus of EUDAT, it is also important to engage with existing initiatives that already provide data services across disciplines.*

Reuse and recombination of data in such open scenarios—covering many different disciplines, each with their own data organizations, domain-specific structures and semantics—is based on suitable integration and interoperability solutions. In this respect, it makes sense to distinguish between the external and internal properties of data objects. Once the definition of a data object is agreed, we can postulate that the external properties of data objects are largely domain independent, while the internal properties are determined by discipline-specific structures and semantics. The CDI thus needs to focus on the external properties of data objects, ignoring their discipline-specific content. Challenges specific to external data properties include: (1) each data object must be identified by a unique PID that should be registered globally and therefore become an official part of the shared data world; (2) the PID records must include

---

* One example of such an initiative is the D4-Science-II project (2009), which aims to develop technology to enable the interoperation of diverse data infrastructures in biodiversity, fisheries, resources management, and high energy physics (http://www.d4science.eu). In the area of libraries and scientific publication, the OpenAIRE project supports the implementation of Open Access in Europe by establishing and operating an e-infrastructure for handling peer-reviewed articles and other scientific publications and data sets from all disciplines; http://www.openaire.eu.

specific information about data objects to ensure, for example, their integrity; (3) each data object—and the various (virtual) collections aggregating them—must be associated with metadata descriptions including PIDs along with contextual and provenance information about the object's content, creation, and history. Kahn[9] describes an abstract data object architecture that comes close to what will be required to establish the CDI as a heterogeneous, competitive, distributed, and organically grown platform for sharing, reusing, repurposing, and preserving data.

A second major principle that will guide the setup of the CDI—allowing establishment of the required trust and proper management of the data object life cycle—is represented by explicit policy rules, as introduced by Moore,[10] designed to govern all operations on officially registered data objects in the emerging complex data domain. Ideally, policy rules can be seen as comparatively simple declarative statements, specified by data managers or creators and addressing all kinds of operations; such policy rules would be necessary for data copying, data transformation, and definition of access rights, for example, and allow users to trace what has happened to objects. Explicit policy rules also form an ideal basis for regular quality—necessary for building trust—with respect to the functioning of data centers and the state of data objects. Sets of consecutively executed policy rules can be seen as workflows, to be interpreted by a workflow execution engine that must operate robustly and also be subject to quality assessment.

8.2 BUILDING THE EUDAT INFRASTRUCTURE

EUDAT is a consortium of 25 European partners from 13 countries. It includes national data centers, technology providers, research communities, and funding agencies. The project is co-funded by the European Commission's 7th Framework Programme starting in October 2011 (grant agreement no. 283304). It is the aim of EUDAT to build a sustainable data infrastructure upon which common services can be deployed for use by diverse communities; a comprehensive approach is required, including several activity strands. First, as mentioned above, active collaboration with all stakeholders will be fundamental to achieving this goal. We must also plan, from the very beginning, the evolution and sustainability of the infrastructure. Among other things, this implies early definition of future partnership and business models for adopting, supporting, and sustaining common services developed for, and partly operated by, the different research communities. To achieve this, we first need to show that our service approach is feasible. Thus, the design and deployment of early services will be critical for the success of the project.

8.2.1 Case for a Multidisciplinary Data Management Approach

Although research communities from different disciplines have different ambitions, particularly with respect to data organization and content, they also share basic service requirements. This commonality makes it possible to establish generic pan-European services designed to support multiple communities. Before the publication of its 2009 White Paper, the PARADE consortium surveyed the needs of nine user communities interested in a common European data infrastructure. They found a high degree of similarity, with unanimous agreement on the importance of long-term data archiving for integrity and authenticity control, and near-unanimous demand for data federation and services enabling discovery, access, data mining, virtual integration, and curation. Complementary surveys, conducted in 2010 by e-IRG[7] and the European e-Infrastructure Forum,[11] also showed strong similarities across RIs in terms of technical requirements and organizational issues. Figure 8.2 illustrates how the requirements of communities from different disciplinary domains can be met by (1) community-specific services such as specific workflow services; (2) services in common with a subset of communities, such as a common meta data registry; (3) domain-specific services, such as discipline-specific

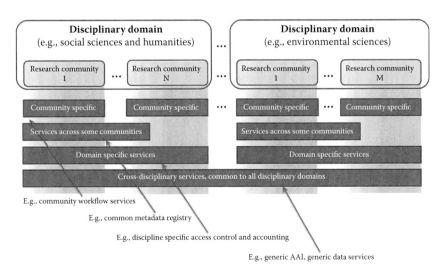

FIGURE 8.2 Different classes of generic services: community-specific, multi-community, domain-specific, and cross-domain services.

mechanisms for access control and accounting; or (4) cross-disciplinary services commonly used by communities from all disciplinary domains.

Perhaps the most challenging requirements relate to Authentication and Authorization Infrastructure (AAI) systems that support fine-grained access control to sensitive data, or that manage access to all e-infrastructure resources via "single identity" and "single sign-on." In recent years, the Authentication and Authorization (AA) landscape has undergone numerous changes. This evolution has been driven by several factors, including the uptake of grid computing, the success of collaborative/community research, and the evolution of the web. Technologies and standards are, however, heterogeneous, and several social, organizational, and administrative issues need to be tackled to improve harmonization before a pan-European solution spanning many organizational and geographical domains can be offered.

There is also widespread recognition among the different communities that better virtual data federation should be achieved, not only for data preservation purposes, but also for access optimization and re-usage of data in new contexts and across the different communities and disciplines. The term "federation" is generally used to refer to the seamless integration of independently managed parts, providing a consistent service that becomes more than the simple aggregation of its constituents. In data infrastructures, we can identify four layers at which federation mechanisms can be applied: storage, file, database, and repository. For the larger e-Science communities and virtual research environments, federated repositories are the infrastructure of choice, since they easily support the creation of virtual collections. Persistent object identifiers are commonly required by several communities. They allow the unique identification of data units across systems, and procedures for resolving object identifiers must be designed to support mechanisms that maintain object integrity and authenticity while fostering a simple, secure, and unique access.

With increasing data volumes and rapid development of new technologies resulting in regular obsolescence of media and data formats, it is a challenge in all disciplines to keep data readable and understandable for future generations of scientists. In general, bit-wise data preservation is required, and this is a primary responsibility of providers of common services, who are generally experienced in maintaining large amounts of data over long periods. With the impending data tsunami, and a growing legacy of digital data stored in large data archives, storage media migrations are high-priority, time-consuming processes that are resource and labor intensive. Although EUDAT's expert teams and services will support the ongoing process of content migration (of object structure and semantics), it will remain a primary task of individual communities to manage to ensure data readability and interpretability.

8.2.2 Designing and Operating Multidisciplinary Data Services

The EUDAT service approach—the cycle of requirements analysis, service and infrastructure design, operation and assessment—follows the principles of agile project management. Three main strands are the basis of EUDAT's service activities.

The first strand involves capturing user requirements. As outlined earlier, some preliminary findings are available; however, this analysis must be deepened and refined in a more systematic way. The generic services to be deployed must be based on user communities' current—and evolving—needs, which can only be achieved through active interaction between service providers and users. To this end, user requirements and corresponding service needs will be typified based on regular surveys. The separation of responsibilities between common and community-specific services will have to be determined in practice. The corresponding boundaries can change over time depending on technological advances. This approach provides sufficient flexibility to react quickly on new developments. Many user communities from various disciplines are already part of the EUDAT initiative, and have been allocated project resources to help specify their requirements and co-design related services. These partners include key representatives from research communities in linguistics (CLARIN), earth sciences (EPOS), climate sciences (ENES), environmental sciences (LIFEWATCH), and biological and medical sciences (VPH).* EUDAT will also gather requirements from the physical sciences represented by STFC (ISIS) and CERN (LHC) as well as others. However, EUDAT does also target disciplines and communities beyond this initial set of communities. For this purpose, User Forums will be held on a regular basis, and be open to all stakeholders interested in adapting their solutions or contributing to the design of the CDI.

The second activity strand concerns the appraisal of technologies and service candidates: this involves identifying, designing, and constructing appropriate services, using existing solutions where possible. Existing

---

* Other communities have accepted to join EUDAT as associated members, representing altogether 15 specific research disciplines across all major fields of science.

technologies will be assessed, along with existing approaches and technological and organizational solutions already favored by relevant communities. Bridging technologies can be usefully deployed when no other appropriate solutions are available. Gaps and market failures will be addressed by a dedicated research activity that will also look at the issues of large and long-term scalability, complexity, and preservation. To this end, a number of core service areas have been identified, and EUDAT will select mature technologies in each area to foster quick start-up where possible:*

- Federated AAI technologies

- Data access and upload services

- Long-term preservation

- PID service

- User workspaces

- Web execution and workflow services

- Monitoring and accounting services

- Network services

- Metadata services

These service areas are essential for the different research communities and the corresponding services can be designed to support multiple communities. The services in the first eight areas are potentially generic, and can be provided without community support. The metadata services are expected to be more community related, and the generalization of metadata services will typically require support from the communities.

Before deployment on the production infrastructure, the candidate service technologies will be tested in a test or pre-production environment.

The third activity strand is operation of the collaborative infrastructure, particularly provisioning of secure and reliable generic services in a production environment, with interfaces for cross-site and cross-community operations. The operation of the infrastructure will provide full life cycle data management services, ensuring the authenticity, integrity, retention,

---

* Data mining was rated highly by the communities, but due to the inherent complexity when working across disciplines, this was not taken up in the first phase.

and preservation of data, especially those marked for long-term archiving. These production-quality services are mostly capable of being shared across different sites and therefore require some coordination. The operational approach comprises:

- The provisioning of network, storage, and compute resources (hardware/system/platform resources) in a scalable way

- The provisioning of services on top of these platform resources, including a registry for all the (shared) services deployed and operated and a service status monitoring

- The coordination of operations (e.g., for change management) by the EUDAT operations team, integration support for new resource/ service providers, community support (including helpdesk, issue tracking, documentation) with interfaces to community or cross-community support teams

- Security, quality control, and compliance, including repository audits and certification

The provisioning of hardware and system resources requires typically specific on-site know-how, whereas the know-how on deploying and running services on top of them can often be shared.

The federated and multifunctional nature of EUDAT requires close collaboration between sites and across departments within a site. The EUDAT operations team will operate a 24/7 service for reporting, tracking, and managing issue, using a trouble ticket system. Detected problems will be classified and promptly reported to the corresponding centers.

Although EUDAT targets all disciplines and communities, it is important to note that the services to be designed and deployed do not need to address all the needs of all the communities. This flexibility allows an incremental approach, where initial and associate communities are taken into account step by step. As indicated in Figure 8.2, we anticipate that some services will support a broader range of communities, while others will only be of interest to a small subset of communities or disciplines.

8.2.3 Expected Benefits of the EUDAT CDI

The benefits associated with creating the EUDAT CDI are many: first, the CDI will facilitate a better and more systematic approach to management of the data life cycle, as well as granting easy access to our scientific records and overcoming fragmentation at the level of data objects and data collections. Increasing the scale of data federations and improving the interoperability of data objects is central to EUDAT's overall approach to the development, deployment, and operation of shared services. EUDAT begins from the principle that individual and community-based data infrastructures should be federated using an architecture that fosters integration without requiring massive changes to existing and proven community-based solutions. We believe that establishing a CDI—using the results of organic discussion, on the one hand, and advocated solutions based on concrete experiments, on the other—is the best way to handle the scale and complexity of data that will be generated over the next 10–20 years.

With the EUDAT CDI in place, better use of synergies can be achieved. The CDI will help to support the infrastructures of existing scientific communities by offering them an infrastructure on which they can rely for their more generic data services. This will allow them to focus a greater part of their effort and investment on services that are discipline specific. The CDI will also provide individual researchers, smaller communities, and projects lacking tailored data management solutions with access to sophisticated shared services, removing the need for large-scale capital investment in infrastructure development. Thus, an overall reduction of costs will be realized through EUDAT's critical mass of infrastructure users and economies of scale.

One of the fundamental goals of EUDAT is the facilitation of cross-disciplinary DIS. By providing opportunity for disciplines from across the spectrum to share data and cross-fertilize ideas, the CDI will encourage progress toward this vision of open and participatory DIS. EUDAT will also facilitate this process through the creation of teams of experts from different disciplines, aiming to develop services cooperatively to meet the needs of several communities.

8.3 GLOBAL COLLABORATION

The challenges embodied by the rising tide of data are global. Equally, the creation of an integrated and interoperable data domain—data as an infrastructure covering several layers—must also be achieved at a global level. In countries such as Australia, Canada, China, Japan, and the United States, the topic is being actively discussed. The visions formulated by the

High Level Expert Group on Scientific Data are thus global and we assume that the following essential principles are shared worldwide:

- Each data object must be identified by a unique and PID officially registered at a registration authority.

- Each data object must be characterized by a metadata description that informs creators, managers, and users of its content, context, and provenance.

- We need an abstract data architecture that specifies ways of integration without hindering scientific dynamics.

- Explicit and declarative policy rules must be the basis for all operations on data objects, facilitating quality assessments.

- A distributed AA system is required to facilitate role-based access in a global trust domain.

- Some basic IT principles, such as registering schemas and element semantics, are required to achieve interoperability.

However, these principles must be turned into concrete specifications, and EUDAT is aware of the need for global collaboration in this respect. Establishing a global Research Data Alliance* (RDA)—organized like the Internet Engineering Task Force and based on concrete suggestions and tested implementations—is seen as a way to create a global interaction framework, pushing harmonization and standardization with respect to the abstract data architecture and all its essential components. Together with OpenAIREplus, EUDAT is calling for international experts from the United States, Japan, Australia and other interested regions to participate in establishing the RDA, defining its scope and organizing international workshops.

8.4 SOCIAL AND ORGANIZATIONAL CHANGE

There is much to achieve before the CDI becomes a reality. Previous chapters have highlighted the main challenges anticipated, and these must be properly addressed if a common infrastructure for scientific data is to be achieved. Delivering high-level multidisciplinary services will not be easy in the current context, which is typified by a diversity of data, disciplines, and practices. The challenges are not only technical, but also social and organizational.

---

* http://rd-alliance.org.

Successful collaboration must be built on trust between service providers and users, but also between the researchers and disciplines themselves. Trust is also essential in the robustness and high availability of the infrastructure, the integrity and authenticity of the data collected and deposited, and the continued existence and persistence of the infrastructure and its components. Although partners in the EUDAT consortium have much experience in collaborating together, these issues will need to be addressed thoroughly to create an open and inclusive infrastructure. This will require the definition of adequate partnership rules, policies, governance structures, control mechanisms, and business and funding models. With respect to future funding models, we can refer to the High Level Expert Group on Scientific Data's document, which states that "an infrastructure for scientific data has a public dimension; it should also have appropriate public funding." Needless to say, the cost of the total data infrastructure ecosystem will be critical.

More generally, we also need to promote a new culture of work practice that encourages the sharing, use, and reuse of data. This is beyond the immediate remit of EUDAT and should be addressed at the policy level, through national and European research funding programs. Additional emphasis must also be put on education and training. Making good use of the exciting opportunities emerging from this new world of data is a complex task. It is crucial that we empower new generations of users with the ability to leverage these possibilities; to this end, we envisage the emergence of a new type of librarian, or data scientist, who will understand the inherent complexity of new technologies and support individual researchers—and all of society—in making the most of the coming opportunities.

REFERENCES

1. High Level Expert Group on Scientific Data. 2010. *Riding the wave. How Europe can gain from the rising tide of scientific data*. Final report submitted to the European Commission. European Union. http://cordis.europa.eu/fp7/ict/e-infrastructure/docs/hlg-sdi-report.pdf.
2. Kroes, N. 2010. Unlocking the full value of scientific data. In *Riding the wave. How Europe can gain from the rising tide of scientific data*. Final report of the High Level Expert Group on Scientific Data submitted to the European Commission.
3. Alliance for Permanent Access. 2008. Keeping the records of science accessible: Can we afford it? Report on the 2008 Annual Conference of the Alliance for Permanent Access, Budapest, November 4, 2008. http://www.alliancepermanentaccess.org/wp-content/uploads/2010/12/documenten_Alliance2008conference_report.pdf.

4. European Commission. 2009. Communication from the Commission to the European Parliament, the Council, the European Economic and Social Committee and the Committee of the Regions—ICT Infrastructures for e-science, 108 final, p. 2. http://eur-lex.europa.eu/LexUriServ/LexUriServ.do?uri=CELEX:52009DC0108:EN:NOT.

5. European Strategy Forum on Research Infrastructures. 2010. Inspiring Excellence. Research Infrastructures and the Europe 2020 Strategy. Luxembourg, Publications Office of the European Union. http://ec.europa.eu/research/infrastructures/pdf/esfri/publications/esfri_inspiring_excellence.pdf.

6. European Strategy Forum on Research Infrastructures. 2011. Strategy Report on Research Infrastructures. Roadmap 2010. Luxembourg, Publications Office of the European Union. http://ec.europa.eu/research/infrastructures/pdf/esfri-strategy_report_and_roadmap.pdf.

7. e-Infrastructure Reflection Group. 2010. e-IRG Blue Paper. http://www.e-irg.eu/images/stories/eirg_bluepaper2010_final.pdf.

8. Koski K., Gheller C., Heinzel S., Kennedy A., Streit A., Wittenburg P. 2009. Strategy for a European Data Infrastructure. White Paper. http://www.csc.fi/english/pages/parade.

9. Kahn R. 2011. A framework for managing the digital object architecture. http://www.euroview2011.com/fileadmin/content/euroview2011/abstracts/abstract_kahn.pdf.

10. Moore R. 2008. Toward a theory of digital preservation. *The International Journal of Digital Curation* 3 (1): 63–75.

11. European E-Infrastructure Forum. 2010. ESFRI project requirements for pan-European e-Infrastructure resources and facilities. http://cordis.europa.eu/fp7/ict/e-infrastructure/docs/eef-report.pdf.

Total domain name registrations
Source: Zooknic, April 2011; Verisign, May 2011

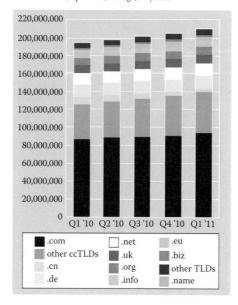

FIGURE 2.1

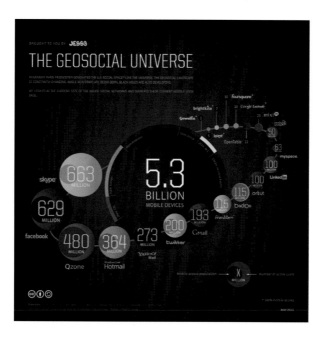

FIGURE 2.2

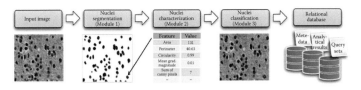

FIGURE 3.1

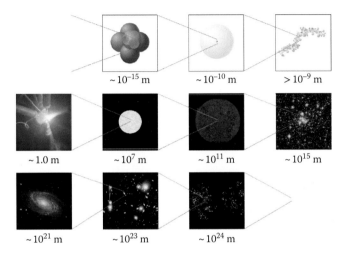

FIGURE 4.1

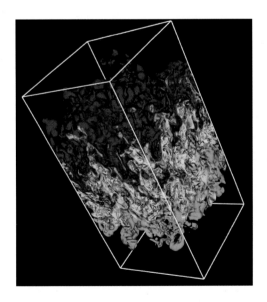

FIGURE 4.2

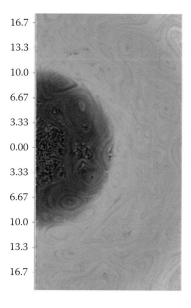

FIGURE 4.3

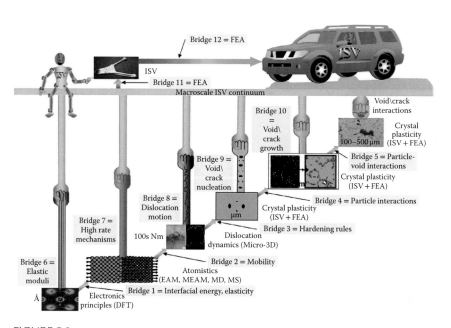

FIGURE 5.3

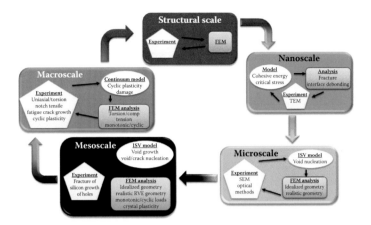

FIGURE 5.4

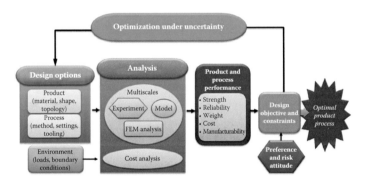

FIGURE 5.5

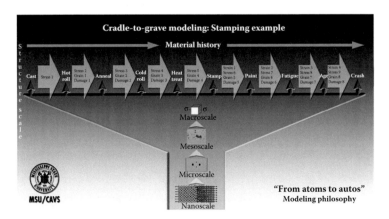

FIGURE 5.6

- IT technologies
- (hidden from the engineer)

- Conceptual design process
- (user-friendly interfaces)

- Engineering tools
- (CAD, CAE, etc.)

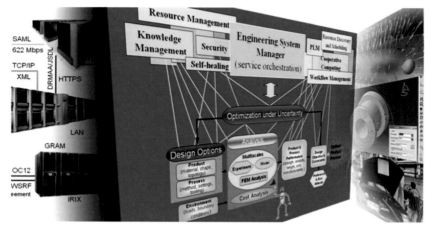

FIGURE 5.7

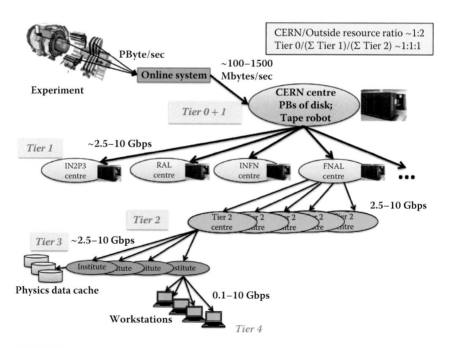

FIGURE 7.1

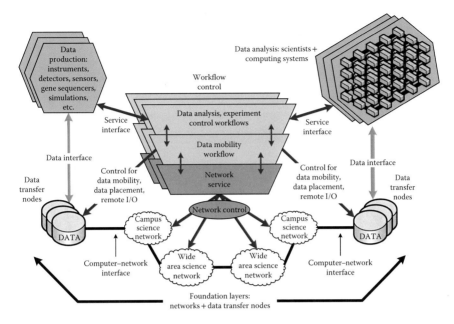

FIGURE 9.3

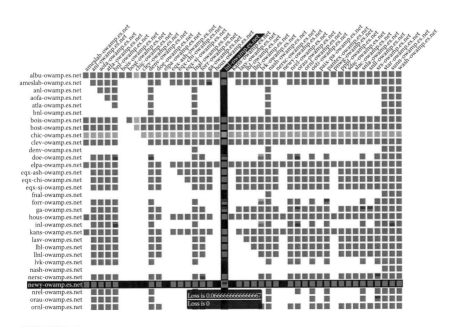

FIGURE 9.4

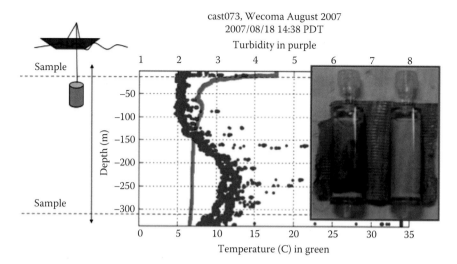

cast073, Wecoma August 2007
2007/08/18 14:38 PDT
Turbidity in purple

Temperature (C) in green

FIGURE 11.2

Data preprocessing steps

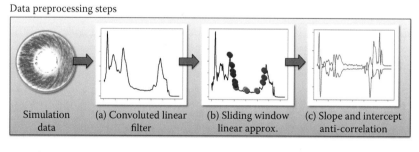

Simulation data | (a) Convoluted linear filter | (b) Sliding window linear approx. | (c) Slope and intercept anti-correlation

Front detection and tracking steps

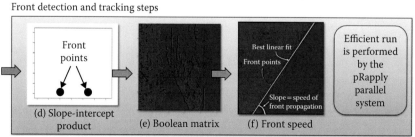

(d) Slope-intercept product | (e) Boolean matrix | (f) Front speed

Front points

Best linear fit
Front points
Slope = speed of front propagation

Efficient run is performed by the pRapply parallel system

FIGURE 13.4

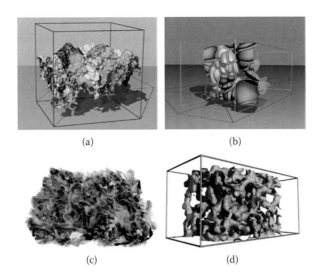

(a)   (b)

(c)   (d)

FIGURE 14.1

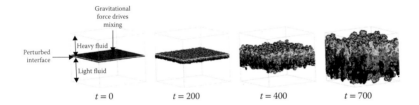

Gravitational force drives mixing

Heavy fluid

Perturbed interface

Light fluid

$t = 0$ $t = 200$ $t = 400$ $t = 700$

FIGURE 14.10

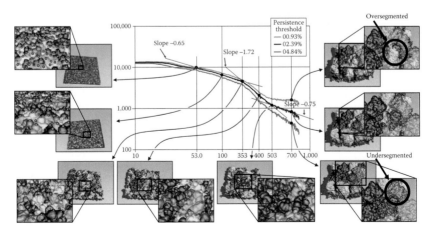

FIGURE 14.12

III

From Challenges to Solutions

CHAPTER 9

Infrastructure for Data-Intensive Science

A Bottom-Up Approach

Eli Dart and William Johnston

CONTENTS

9.1 INTRODUCTION: NETWORKING CENTRAL TO DATA-INTENSIVE SCIENCE

Data-intensive science (DIS) has as one of its core concepts the achievement of scientific discovery through the analysis of large volumes of data. The data sets of DIS are so large that interaction with the data by scientists—collection, transfer, storage, manipulation, analysis, and curation of the data—is a dominant part of the scientific process. The data-intensive paradigm is different from others in that scientific discovery is achieved through data analysis, predominantly using computers. This is not the domain of the lone explorer seeking rare specimens at the ends of the earth but that of teams of scientists pushing the limits of modern computing and networking capabilities to extract discoveries from large, often expensive, and frequently unique scientific instruments that generate data sets on the scale of terabytes to petabytes and beyond. While the majority of productive work is done by scientists using computers to analyze data, a scientist can expend significant effort on just managing data—locating data, caching data, moving data to analysis resources, and so on. There are many reasons to move large data sets or to access them remotely, both of which generate large data flows in the network—these include locations of the analysis resources, instruments, collaborators, and long-term archival storage. In fact, in many cases DIS is not possible without the ability to effectively move large volumes of data. This chapter will examine the infrastructure that is deployed and maintained to support DIS.

The foundation of the infrastructure is a global science network that interconnects the many different institutions where the science is conducted and extends into the site networks of those institutions. Resting on that foundation are the data transfer and analysis resources connected to those site networks, the software tools that control the movement and placement of scientific data by using the capabilities of the underlying infrastructure, and the experimental facilities and computational resources that generate and analyze the data. These interconnected components must work together in a reliable, predictable, and cohesive way for DIS to be successfully conducted. If the infrastructure is configured well and properly maintained, it can be a powerful enabler for DIS.

Here, we take a bottom-up view of the infrastructure needed to efficiently move large data sets. We call it bottom-up because this view is derived from years of operational experience as infrastructure providers engaged in the design, deployment, and operation of networks tailored for science applications. Networks built for science differ in several ways from those built for other applications. However, just as networks are critical for the successful operation of a modern business enterprise, they are also critical for DIS. We will describe the key aspects of network infrastructure that are critical for successful DIS.

9.2 MOTIVATING EXAMPLES

To understand the global scale of the infrastructure necessary to support DIS, it is helpful to consider the scale of some of the experiments being conducted. In order to advance humanity's understanding of the fundamentals of the universe, ever more powerful instruments must be constructed to extract more and more subtle science. As the cost, size, and time to construct the instruments have increased, their number has decreased. An example of such an instrument is the Large Hadron Collider (LHC) at CERN, Geneva, Switzerland, [1], an experimental facility that is unique: a particle accelerator 17 mi. (27 km) in circumference that spans the Franco-Swiss border outside Geneva. With a construction budget of some €7.5 billion (~$9 billion or £6.19 billion as of June 2010) and a construction time of 15 years, LHC is one of the most expensive scientific instruments ever built, and as of now there is only one such experimental facility. Therefore, everyone working in the field must work with the data LHC produces, resulting in collaborations that are truly global in scope. In addition, the detectors at LHC produce a huge volume of data. To extract physics from the data, the data sets must be distributed to analysis teams and computational resources around the globe.

The experiments conducted at LHC provide an example of DIS that cannot be accomplished without the underlying network infrastructure. LHC experiments, particularly the two largest experiments (ATLAS and CMS), require the ability to efficiently process very large volumes of data (petabytes per year) to achieve breakthrough discoveries. Since it is both physically and politically impossible to locate all the computational resources for data analysis at CERN, the data must be distributed throughout the world such that the experiment collaborations can benefit from the resources available to physicists at research institutions regardless of location.

For example, the ATLAS experiment moved almost 7 PB of data between the hundred or so institutions involved in the data storage, curation, and analysis between April and November 2010 [2], the result of sustaining data movement of up to 500TB/day. See Figures 9.1 and 9.2.

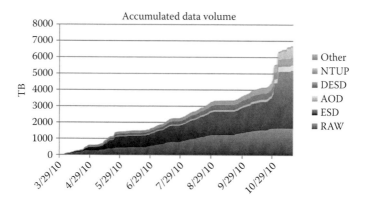

FIGURE 9.1   ATLAS cumulative data volume (in terabytes) by type of data set for a 7-month period. (Courtesy of Michael Ernst, Brookhaven National Laboratory, Upton, New York.)

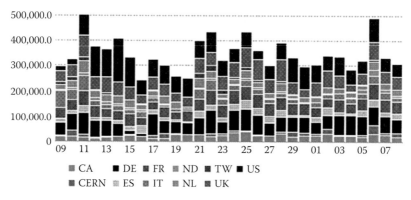

FIGURE 9.2   ATLAS cumulative data volume per day for a typical 1-month period (up to 500 TB/day). (Courtesy of Michael Ernst, Brookhaven National Laboratory, Upton, New York.)

Although previous generations of high-energy physics (HEP) experiments relied, to a lesser or greater extent, on shipping the experiment data on physical media (typically tapes), this is no longer practical due to the volume of data and the human effort required. It would be impossible to conduct the LHC experiments without an international computational and networking infrastructure that distributes the data, performs the analysis, and gathers the results from scientists worldwide.

This is a recurring pattern that applies not only to LHC. In any scientific field where the best instrument in the field is unique for a generation of scientists, all scientists collaborating on the science of the instrument must be able to effectively make use of that instrument to conduct cutting-edge research. Therefore, the network that provides access to the instrument and access to the data from the instrument must reach both the instrument and all scientists in the field. In addition, the scientists must have the means to effectively analyze data and collaborate with each other. This aspect of science does not apply to unique instruments alone. Scientific data sets are often more valuable when their analyses can incorporate other data (e.g., model validation that incorporates experimental or observational data). Large supercomputer centers must exchange data sets as the output from a model run at one center is used as the input to analysis at another center, and data sets of common interest must be distributed to many institutions to facilitate easy access by the scientists in the field. Therefore, in the general case the network must interconnect all aspects of a data-intensive scientific collaboration and be able to effectively meet the needs of all aspects of the collaboration.

As the need for networking has grown, science networks have evolved to meet the needs of science. Since data distribution at the scale required by LHC is only feasible with significant automation, the networks must present consistent services that are of sufficient quality and performance to allow automated processes that use the network services to handle the bulk of the data distribution tasks.

An international collaboration of research and education networks, including the Energy Sciences Network (ESnet), Internet2, and the Regional Optical Networks (RONs) in the United States, CANARIE in Canada, GEANT and the many national research and education networks in Europe, and others from the Americas and Asia, has built such an infrastructure, which provides the following:

1. A consistent set of services providing quality-of-service guarantees to scientific data flows

2. A means by which the experiments can provision end-to-end services from the network edge

3. A test and measurement infrastructure that allows rapid troubleshooting and fault isolation

As we will see, these capabilities are generally applicable to DIS; in fact, the conduct of DIS is very difficult or impossible without them. As noted, the LHC experiments are completely reliant on the network for their success, in the same way that the experiments are reliant on computing for data analysis: without an effective means of distribution and analysis of the data, the science could not be done.

9.2.1 Other Disciplines Following in HEP's Footsteps

Several other fields are following in the footsteps of physics.

The cost of genome sequencing machines is falling dramatically, whereas the volume of data produced is increasing rapidly [3]. The data must be moved from biology labs to data centers for processing and the resultant finished genomes entered into community databases.

Climate scientists must analyze observational data and model data, and these data—comparable in size to the LHC experiments data—are housed at centers around the world. The requirement for productive access to those data sets has resulted in the construction of the Earth System Grid, a global data workflow infrastructure that allows climate scientists to access data sets housed at modeling centers on multiple continents including North America, Europe, Asia, and Australia [4].

New instruments are being deployed at light source facilities that generate data at unprecedented resolutions and rates—the current generation of instruments can produce 300 MB/s or more, and the next generation will be able to produce data volumes many times higher [3].

Future large-scale science projects that are data intensive and that are currently being planned and built include, for example, the international fusion energy prototype ITER and the Square Kilometre Array [5], a massive scale radio telescope that will generate as much data as the LHC or more.

In all these cases, scientific productivity is governed by the ability to analyze large volumes of data with computers, because without the ability to effectively analyze data science is difficult or impossible. Further, since it is often physically, politically, or organizationally impossible to put all the data storage and computational analysis resources necessary for conducting science at the location where the data are generated, plans must be made for the distribution of large volumes of data between instruments, facilities, and analysis resources in the general case.

9.3 CORE SET OF CAPABILITIES THAT ARE NECESSARY FOR DIS

From an infrastructure perspective, DIS relies on a core set of capabilities for success. These capabilities do not stand in isolation; rather, they are complementary and synergistic. They are as follows:

- A federation—because all large-scale science is conducted across many network domains—of high-performance networks that operate in concert to provide high-performance services for science, including end to end service guarantees where required.

- Properly configured end systems that are well matched to the capabilities and characteristics of the network.

- Test and measurement infrastructure that can continuously monitor the network and guard against degeneration in performance due to "soft" failures.

- Software infrastructure, including data transfer tools, network service interfaces, and workflow tools, for automating data-handling and data analysis tasks.

This ensemble of capabilities is typically used as a foundation by a complex data analysis ecosystem that has a workflow of its own.

In Figure 9.3, the general elements of a DIS environment and their relationships are illustrated. A high-performance network provides connectivity between the data transfer nodes (DTNs) at geographically distant institutions. The workflow control software stack in the middle

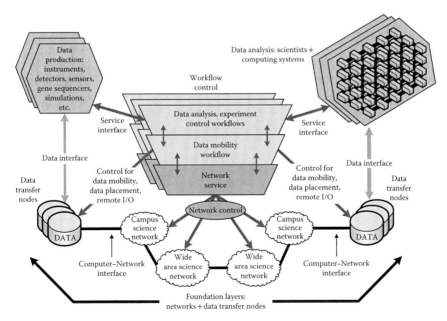

FIGURE 9.3 **(See color insert.)** The "ecosystem" for data-intensive science.

of the diagram (which, as a set of distributed applications, runs on key computers throughout the ecosystem) implements several services for data management and analysis management. The network service layer provides a means by which network services can be abstracted from the configuration details of underlying networks. The data mobility workflow manages the DTNs to exchange data sets or provide streaming access by taking advantage of the network services. The DTNs provide a data interface to the instruments on the left side of the diagram and to the analysis resources on the right side of the diagram. The data analysis workflow requests the data mobility layer to place the data in the proper location for analysis or to stream data from remote data storage nodes, and the analysis infrastructure reads the data through the data interface (e.g., a large file system or a self-describing data stream).

Most of the key elements necessary for supporting DIS can be found in Figure 9.3: the networks at the bottom with the DTNs that provide data mobility service by means of the network, and the layered software infrastructure that provides control, intelligence, and automation. Test and measurement is not present in the diagram, mostly to avoid clutter—the test and measurement infrastructure is part of the network and data mobility layers, and its job is to promote the health of the underlying network and data transfer substrate through active monitoring to detect performance degradation (e.g., due to soft failures that are not visible to conventional monitoring systems). Health and error reports are provided to the system and network engineering groups that manage the infrastructure so that problems can be proactively fixed.

9.4 SOFT FAILURES

The reference to test and measurement in Section 9.3 highlights an important aspect of high-performance networking and system design. As part of their basic functionality, individual elements of a network (e.g., routers, switches, optical systems) have the ability to alert system and network administrators when the element or one of its components fails such that the network element can no longer provide service. Examples of this include raising an alarm when the loss of optical signal is detected due to a cable cut, the failure of an interface in a router, and the loss of connectivity between networks due to a protocol event. We refer to such obvious faults as "hard" failures; they are easy to observe because no network traffic can traverse a hard failure and network devices are good at reporting hard failures when they occur. In contrast, it is not uncommon for a network device or component to perform poorly for high-speed applications while still providing sufficient basic service to avoid triggering an alarm for a hard failure. Examples include packet loss from failing optics (these typically turn into hard failures eventually but can cause significant performance problems for weeks or more before they become hard failures), routers that are misconfigured so that they forward traffic using the low-performance management CPU rather than the high-performance forwarding hardware (this is typically not easily detectable at low data rates), and improperly configured switches or data transfer hosts. These are called soft failures because they cause problems for high-performance applications without causing a hard failure that alerts the network operator or system administrator to their presence. Soft failures can go undetected for years and are a common cause of frustration: the network can appear normal from the perspective of low-bandwidth applications such as web browsing and e-mail and, at the same time, be crippled from the perspective of high-performance data-intensive applications.

By far the most effective way of finding and fixing soft failures is to independently test the infrastructure. If test and measurement hosts are deployed near the data transfer systems, the test hosts can test the entire network path that will be used by production data transfers. If the tests perform well, it is much more likely that the data transfers will perform well; if the tests perform poorly, it is unlikely that the data transfers will do any better than the tests. Obviously, this testing approach only covers the portions of the infrastructure common to the testers and the data transfer hosts, but in practice it is so effective that science networks all over the world (including ESnet) have large test and measurement deployments to periodically test the networks to ensure that they continue to perform well. When the tests begin to exhibit degraded performance, the problem can be investigated proactively and the source of the soft failure located and fixed.

9.4.1 Transmission Control Protocol Considerations

The underlying protocol used for the vast majority of data transfer applications is Transmission Control Protocol (TCP) from the Internet Protocol suite. This is relevant because TCP faces particular challenges when pushed to high performance levels, and support for TCP drives much of the engineering and architecture for networks that support DIS. TCP provides a reliable byte-stream delivery service to networked applications and has been the workhorse transport protocol for client–server interactions on the Internet from the beginning.

TCP is sensitive to packet loss because of changes made to avoid the congestion collapse of the Internet. Although these modifications were necessary at the time (and still are), they make TCP fragile in high-performance environments. The Internet engineering community has been working on improvements to achieve higher performance (either further enhancements to TCP or a high-performance protocol to replace TCP) for many years with limited success. Therefore, in the near to medium term, DIS infrastructure must be built to provide necessary support for TCP such that TCP-based applications can perform well in science environments.

There are two primary causes of poor TCP performance in science networks: (1) the configuration parameters of TCP itself on the data transfer systems, and (2) packet loss in either the local area network (LAN) or the wide area network (WAN) infrastructure. The appropriate configuration of TCP on DTNs can be accomplished by competent system administrators with the help of public knowledge base sites [6]. Packet loss is a much more challenging problem for several reasons, including soft failures as described in Section 9.4 and the number of organizations and devices involved in typical long-distance data transfer. In particular, as the distance between two data transfer hosts as measured by the time it takes for the hosts to exchange packets (termed latency) increases, TCP becomes much more vulnerable to packet loss. Packet loss rates of less than 1% can reduce performance by a factor of 100 or more.

To see why TCP is so sensitive to packet loss, we offer this brief overview: TCP begins a connection in a phase called "slow start," where the data rate is gradually increased. Since TCP interprets packet loss as network congestion, when TCP encounters packet loss it assumes that there is a congested link in the network between the sending and receiving hosts and reduces its sending data rate. This reduction in sending rate varies depending on the algorithm used, but typical reduction values are 20% and 50% (depending on the TCP implementation). After such an event, the sending rate is gradually increased again. If network conditions are such that packet loss is encountered early in slow start or loss events occur frequently, such as with failing optics, misconfigured switches, or other soft failures, TCP can spend all its time reducing its sending rate and slowly ramping the rate back up until it encounters loss again. In these situations, TCP (and by extension the DTNs and the network) typically never comes close to achieving the transfer rate that would be possible in the absence of packet loss. This is a common cause of poor performance.

Packet loss can be difficult to locate without well-deployed test and measurement infrastructure. A particularly challenging aspect of the problem is that because of the way TCP operates, low-latency TCP connections (e.g., for data transfers within a building or between resources in the same laboratory or university) result in different packet behavior compared to long-distance or high-latency TCP connections. The packet behavior of long-distance TCP connections makes them more susceptible to common causes of packet loss. This frequently results in situations where all local tests and measurements indicate that there is no loss and local performance is good while long-distance connections experience poor performance due to packet loss, even if the cause of the loss is local to the site in question. Because of this, test and measurement infrastructure must be deployed at the sites that own the DTNs (and preferably on the same subnet as the transfer nodes) as well as in the networks that connect those sites to collaborating institutions.

Engineering for very low packet loss rates is a key driver for the architectural aspects of science networks that support DIS. High-performance data transfer is one of the critical components of DIS, and a reliable network infrastructure with very low (i.e., as close to zero as possible) packet loss is a necessary part of the foundation for DIS.

9.5 CORE CAPABILITIES IN DETAIL

9.5.1 WANs

WANs interconnect sites, institutions, instruments, facilities, and scientists. Without this interconnectedness, most of the assumptions we make today about the ability to communicate would fall apart. While all WANs use the Internet Protocol suite (IP, TCP, UDP, etc.), the networks built for science are different from the networks of the general Internet in several key ways. The need for these differences largely comes from two drivers, both directly derived from the science that the networks must support. These drivers are the characteristics of network traffic and the structure of scientific collaboration.

Science traffic is different from general Internet traffic in that the data volume carried in a single science connection (e.g., the transfer of a scientific data set) is vastly larger than the data volume carried in a single general Internet connection such as an e-mail, a web page, or a video. Data transfers on the general Internet typically range from a few kilobytes to at most a few gigabytes. In contrast, science data sets typically range from gigabytes to petabytes, with single data transfers commonly moving several terabytes of data. Since the data transfers must occur on timescales that are useful to people, the data transfer rate for science data sets must be much higher than that for general Internet traffic. Today, science data throughput requirements are often well above 1 Gbps, and the throughput requirements increase with the size of the data sets that must be moved. The network traffic profile of a science network is therefore dominated by a relatively small number of very high data rate traffic flows between a small set of end points. Further, a significant fraction of these end points tend to be long distances apart (frequently cross-continental or intercontinental) due to the widely dispersed collaborations of large-scale science.

This traffic profile, which is mostly TCP based, places stringent demands on the networks that must carry the traffic. As noted, since TCP throughput is unstable at high speeds in the presence of even very low packet loss rates, the network must be engineered to make packet loss events extremely rare. This in turn has implications for the capabilities of the equipment used to build the network, including both the LAN and WAN. In particular, routers and switches need deep interface queues,* comprehensive diagnostics, and the ability to accurately count packets for troubleshooting purposes so that the causes of packet loss can be localized and fixed. The sensitivity of high-speed data transfers to packet loss also requires that either science networks be provisioned with sufficient spare capacity to accommodate traffic bursts from large transfers or science networks provide service guarantees to large science flows. In practice, both are necessary.

9.5.2 End-to-End Virtual Circuit Service

The requirement that the reach of a network must extend to the full length and breadth of the scientific collaborations it serves has implications for the services offered by the network. In practical terms, the set of things perceived by its users as the network is the set of things that must all work together to provide a service that is useful to science between sites. Since many independent organizations are involved in the end-to-end network connectivity between scientific resources, these organizations must work together to provide a common set of services. This is a key difference between science networks and the general Internet, because in the former the network

---

* Many of the recommendations given in this chapter only apply to high-performance networks where the end systems are located far apart. These recommendations are not necessarily optimal for general-purpose network traffic.

must be able to provide a consistent service set across many administrative domains without manual intervention. In the general case, science traffic crosses multiple administrative domains. For example, traversing at least five domains is typical: a data transfer may originate from an instrument at a national laboratory; then flow across the network in the laboratory; then traverse ESnet to a peering with Internet2; and then traverse Internet2 to a peering with a regional network that delivers it to the university network, which is responsible for getting the data to the DTN on a scientist's cluster. Note that the data do not stop in the intermediate infrastructures—the data transfer is between the hosts at the ends, and a successful data transfer requires that all the independent infrastructures that support the transfer provide a consistent service for the duration of the transfer.

For science networks to provide services that are more sophisticated than the best-effort delivery of the commodity Internet, they must have a means of signaling to each other so that the attributes of an end-to-end service can be shared between the networks that must cooperate to provide the service. Examples include the quality-of-service parameters necessary to provide bandwidth guarantees, topology information for traffic engineering and interconnection, and virtual LAN (VLAN) ID numbers for service delivery. Each network in the path must implement a portion of the service, and all the pieces must work together so that the service is effectively delivered to the users. This is significantly different from the general Internet, where the service provided is typically limited to basic best-effort IP connectivity.

One platform that provides a sophisticated services suite for interdomain service provisioning in science networks is a network resource reservation system called OSCARS [7–9]. OSCARS provides the infrastructure for a virtual circuit service, including the ability to reserve end-to-end paths through multiple networks that have compatible services providing bandwidth and service guarantees for the virtual circuit. Typical uses for virtual circuits are to provide Ethernet VLANs with guaranteed bandwidths between distant sites for dedicated data transfers, to provide multiple Ethernet VLANs with different network paths so that redundancy can be achieved, and the delivery of traffic to a particular location in a network as needed. The circuits can be scheduled in advance and have a lifetime set at the time of reservation so that the circuit is automatically decommissioned when it is no longer needed. Circuits can also be set up on demand by higher level tools such as data transfer applications or other workflow engines. OSCARS was primarily developed by ESnet and has been deployed in ESnet and many other networks around the world.

OSCARS is the foundation of the ESnet Science Data Network, the Internet2 ION service, the DYNES project [10], and many other U.S. and international networks and services [8]. The wide deployment of OSCARS and compatible systems (such as GÉANT's AutoBAHN) that use the InterDomain Controller Protocol (IDCP) [11] for interdomain signaling, as OSCARS does, allows high-performance science services to be provisioned across multiple networks on multiple continents. LHC experiments make extensive use of this capability to provide primary, secondary, and tertiary paths for data—all with appropriate traffic engineering and bandwidth guarantees—so that science data services are deterministic and are resilient and/or predictable in the face of failures in the network. These sophisticated services allow much of the data distribution for the LHC experiments to be automated.

9.5.3 Monitoring

As noted in Section 9.4, a network test and measurement infrastructure is critical for supporting DIS. The current toolset used by several science collaborations and by the international research and education (R&E) networking community is that of the perfSONAR collaboration [12].

perfSONAR is intended as a significant first step in cross-domain monitoring and testing by both network operators and users. It has been widely deployed* in the international R&E networking community and the networks that support LHC data management and analysis. perfSONAR has succeeded because it is widely deployed in the R&E community and provides an invaluable cross-domain monitoring and testing functionality. It is widely deployed because the design of perfSONAR explicitly allows for federated deployment; that is, domain operators deploy perfSONAR in ways that are consistent with the policies in their domain (e.g., what is tested, what data is distributed and to whom, etc.).

There are three general categories of performance measurement data—active measurements, passive measurements, and network state variables—that can be thought of as data producers. From the network data user's point of view, these data must be available in various ways and must have various services associated with them both to homogenize the information from different networks and to present the data in useful ways. Data should be provided as a data flow or via polling.

---

* In early 2011, there were 353 instance of perSONAR testers measurement archives deployed by 88 institutions in North America and Southeast Asia, based on the list complied at http://ndb1 .internet2.edu/perfAdmin/directory.cgi. Additionally there is a substaintial number in Europe.

Active measurement generators are typically tools like One-Way Active Measurement Protocol (OWAMP) (one-way delay times) and iperf (TCP throughput) running on dedicated hosts. The active measurement generators can be used ad hoc for troubleshooting, and systems also allow them to be configured to run periodic tests to monitor network health and provide a performance baseline against which future measurements can be evaluated.

Periodic throughput tests (e.g., performed several times a day) are typically scheduled for specific paths of interest. Tests are often set up to verify critical backbone links or tests between measurement hosts collocated with DTNs, and the resulting measurements are exported to a measurement archive. The path measurements may be limited to the local domain, or they may probe specific long-distance paths of importance. In any event, all of these measurements can be made available for community access so that a continuous, global picture of network performance can be built up and the overall health of the network monitored continuously. The goal is not to exhaustively test every combination of paths but to establish a meaningful baseline that allows infrastructure providers to measure current performance against past behavior, and to raise alerts when test results deviate from expected behavior.

An example of the power of visualizing network test data is a perfSONAR dashboard application developed by ESnet that is similar to a prototype implementation developed by the ATLAS experiment. Data from active measurements is plotted in a matrix, which uses a simple red/green color scheme to show where potential problems might exist in a network. Since the human eye excels at pattern recognition, a simple display such as this can tell a network engineer a lot about the health of the infrastructure at a glance. In Figure 9.4, a test between two OWAMP hosts is showing packet loss, but only in one direction. Since packet loss causes significant problems for TCP performance, patterns of loss should be investigated if they are persistent. In the picture, there is loss seen by the testers at several locations, all involving the test host at Idaho National Laboratory (INL)—this is a pattern that can be investigated by an engineer (it could be the test host or it could be a network link or device common to some test paths from the INL test host). In addition, the test host in Chicago is having problems (note the rows of orange results) and requires attention by a system administrator. This tool allows a network operator to monitor the infrastructure and fix problems as they occur rather than waiting for users to complain. The end result is a network infrastructure that is perceived as reliable by the applications and collaborations that use it. Reliability and performance consistency are very important for enabling DIS, because

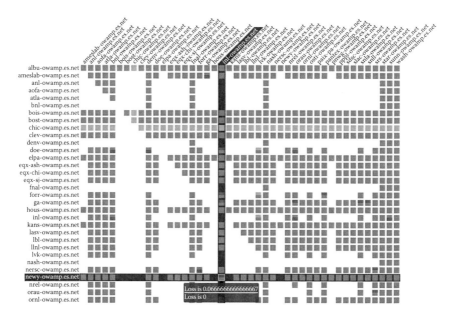

FIGURE 9.4 **(See color insert.)** perfSONAR dashboard display of One-Way Active Measurement Protocol data.

performance problems are significant barriers to productivity: the network must function properly for high-performance applications in the general case. This is an example of the power of a simple tool that is correctly deployed. The tool need not identify the exact nature of the problem to be useful; it just has to tell the owner of the infrastructure where to look. The key is to collect data and present the data in a way that gives network engineering staff actionable information for improving performance. Over time, visualization tools will be enhanced and new tools will be developed. However, it is critical to deploy the infrastructure to collect data; without the data, tools are useless.

The analysis tools, threshold alarms, and visualization tools are data consumers that, in turn, need data that are already transformed in various ways. Therefore, between data producers and data consumers there may be a pipeline of aggregators, correlators, filters, and buffer services that can be regarded as data transformers and data archives.

Further, the services—data producers, consumers, transformers, and archives—are all resources that need to be discovered and almost certainly used within an authentication and authorization framework that maintains the policy prescribed by the network operators that own the measurement data.

9.5.4 Site and Campus Networks

The network at a national laboratory, university, or other research institution provides connectivity to local resources—computers, data, instruments, collaboration systems, and so on. A site network is typically the interface between a WAN and local scientists and resources, and so the site network must deliver the services provided by the WAN to the local resources at the site. In most cases, a site network has multiple missions. These include the mundane business aspects of running a scientific institution, supporting computer security directives for the protection of financial and personnel data (and the avoidance of the embarrassment of news headlines detailing security breaches), and (sometimes as an afterthought) the pursuit of scientific discovery. These missions of a site network are often in direct competition; this is especially the case when the combined missions of business continuity and computer security are considered in contrast to the scientific mission.

In most cases, the needs of the scientific mission (high-speed, low-loss networks that can support large-scale wide area data transfers) are very difficult to satisfy with the devices and configurations typically deployed to build networks for business purposes; these include firewalls, proxy servers, low-cost switches, and so forth. Because of this tension, it is usually desirable to deploy DIS resources in a separate portion of the network that has a different packet-forwarding path and is not subject to the same security policy controls as the institutional business systems. This is feasible from a security point of view because the data transfer systems involved in DIS typically run a well-defined set of special-purpose applications compared with business computers, which run a wide variety of complex user–agent applications like e-mail clients and web browsers.

In traditional network security architectures, there is the notion of a "demilitarized zone" or DMZ network. This portion of a network contains the site resources that are routinely contacted by off-site systems and is normally located at or near the site network perimeter—the connection between the site network and the WAN. Services commonly deployed on the DMZ include authoritative domain name system (DNS) servers, incoming e-mail servers, external web servers, and so forth. Since DMZ resources are assumed to interact with external systems, the security policy for the DMZ is tailored for these functions. For example, the firewall that protects office desktop systems may not permit incoming connections from the addresses assigned to the DMZ (i.e., the DMZ may be considered "external" by the firewall). However, since the DMZ exposes a well-known set of services to the Internet at large, a site host can access those services just like other Internet hosts as long as the site host initiates the connection.

We can use these ideas to design an element of network architecture that explicitly supports DIS services. Much like the traditional DMZ, we can build a portion of the network, at or near the site perimeter, that is specifically configured to support DIS. This "Science DMZ" has several key elements, and the high-performance site networks at many research institutions incorporate these elements in multiple different ways. However, the core set of concepts is the same. These are as follows:

1. A network architecture that supports high-performance science flows

2. Dedicated systems for wide area data transfer

3. Test and measurement systems for performance verification and rapid fault isolation, typically perfSONAR as described previously

4. A security policy tailored for science traffic and implemented using appropriately capable hardware

In some cases, the Science DMZ is a small portion of the site network, consisting of perhaps a single switch, a single DTN, and a perfSONAR host for test and measurement. In other cases, it accounts for the majority of the network infrastructure, with a small portion of the network devoted to office computing behind a small firewall (this model can be a good fit for a mission facility such as a supercomputer center). In either case, science flows are handled by capable hardware that is configured specifically for the support of science. This is illustrated in Figure 9.5. The Science DMZ can also be deployed incrementally. Once a single Science DMZ switch and a perfSONAR host are deployed, additional DTNs for other projects can be added as needed. This model is a good fit for the budgeting realities of a diverse multiprogram institution such as a university or a large laboratory.

The fundamental concept of a Science DMZ is the deployment of DIS resources in a portion of a network that is purpose-built for science, including hardware, configuration, security policy, and security enforcement. The separation of security policy can be very important, since science traffic and resources often have different characteristics than normal enterprise traffic and resources. One example is support for parallel data transfers. The data transfer tools used in science applications typically use parallel data transfer techniques to improve performance, which requires that pools of TCP ports must be open and available between the DTN and the outside world.

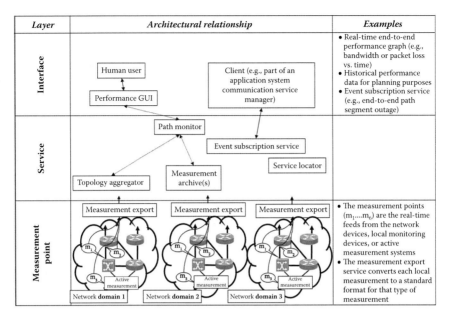

Layer	Architectural relationship	Examples
Interface	Human user ↔ Performance GUI; Client (e.g., part of an application system communication service manager)	• Real-time end-to-end performance graph (e.g., bandwidth or packet loss vs. time) • Historical performance data for planning purposes • Event subscription service (e.g., end-to-end path segment outage)
Service	Path monitor; Event subscription service; Service locator; Topology aggregator; Measurement archive(s)	
Measurement point	Measurement export (Network domain 1); Measurement export (Network domain 2); Measurement export (Network domain 3); m₁...m₆, Active measurement	• The measurement points (m₁....m₆) are the real-time feeds from the network devices, local monitoring devices, or active measurement systems • The measurement export service converts each local measurement to a standard format for that type of measurement

FIGURE 9.5 perfSONAR architecture.

This is often difficult to do because the security policy for a site's business system typically stipulates that all unused ports must be blocked. By placing the DTN on a Science DMZ, the security policy change to enable high-performance data transfers for DIS need not affect the security posture of the site's business systems. Another example is the huge volume of data exchanged between DTNs as science data sets are moved. If there is a requirement that all traffic destined for the site's business systems must be scanned for viruses, the virus-scanning appliance must be able to run at speeds required for science data transfer, which is often many gigabits per second. Since DTNs do not typically run software that is vulnerable to the viruses that affect business systems, the Science DMZ architecture can eliminate the need to scan terabytes of science data for PC viruses. This requires the hosts in a Science DMZ to run a limited set of applications specific to their purpose, but this turns out to be the typical deployment case. Finally, off-the-shelf security appliances typically do not understand the higher level protocols used by science applications (e.g., GridFTP, workflow tools). There is little advantage to pushing terabytes of science data through a security appliance that will not understand the data anyway; all this does is reduce performance while increasing costs. So by deploying applications and security measures where appropriate, the Science DMZ model allows the science infrastructure and the business infrastructure at a site to be optimized separately in terms of performance and security.

Without the separation provided by the Science DMZ model, science traffic and normal business traffic share a common security infrastructure. The common security infrastructure must then be sized to support the high data rates of DIS without causing any packet loss. It must also support science-specific services such as virtual circuits that may carry traffic that off-the-shelf security appliances do not understand. In many cases, the devices that implement sophisticated protocol analytics for the protection of business systems (e.g., virus scanning as described above) do not scale to the data rates required for science traffic, and scientific data flows often do not require sophisticated application-layer analysis by security appliances. By separating the two, the Science DMZ model applies appropriate security policy controls to science traffic and different policies that are appropriate to business traffic, and it avoids the extra costs incurred by conflating the security policies for these different classes of traffic.

9.5.5 DTNs

In addition to the network architecture and the networking hardware that are necessary to support DIS services, the systems used for wide area data transfers typically perform far better if they are purpose-built and dedicated to the function of wide area data transfer. These systems, which we call DTNs, are typically PC-based Linux servers built with high-quality components and configured specifically for wide area data transfer. A DTN also has access to local storage, whether it is a local high-speed disk subsystem, a connection to a local storage infrastructure such as a SAN, or the direct mount of a high-speed parallel file system such as Lustre or GPFS, or a combination of these. The DTN runs the software tools for data transfer to other systems; typical software packages include GridFTP and its service-oriented descendent Globus Online [13], discipline-specific tools such as XRootd [14], and enhanced versions of a default toolset such as SSH/SCP with high-performance patches [15].

DTNs typically have high-speed network interfaces (10 Gbps as of writing this chapter, although experiments with 40 Gbps interfaces are already underway, e.g., at SC11 with 40 Gbps RDMA over a WAN [16]), but the key is to match a DTN to the capabilities of the site and WAN infrastructure. So, for example, if the network connection from the site to the WAN is 1 Gbps Ethernet, a 10 Gbps Ethernet interface on the DTN may be counterproductive.

Finally, the default system configuration options (e.g., network interface tuning options and TCP parameters) are wrong for high-performance data movement. Because of this, there are several knowledge base sites, such as the ESnet "fasterdata" site [6], that maintain system tuning information for science DTNs.

Changes in computing are making the construction of DTNs more challenging. In particular, the shift to multicore processors and the stagnation of the growth rate of per-core clock speed make network-based data transfer more complex. The transfer of a data flow in or out of a process via a network is inherently a serial task, and the multicore world is ill-suited to providing performance increases to single-threaded tasks [17]. The use of parallel transfers, for example, by GridFTP, is one answer to the limits on per-core performance growth. However, this requires that a DTN must be deployed with software that supports parallel transfers, and each connection in the parallel transfer is still limited by the performance of a single CPU core.

In response to this paradigm shift to multicore systems, changes are being made to the ways in which networking is handled at the system level in an attempt to increase performance. In general this is good, but it can have unintended consequences. For example, an advanced network interface technology that steers the processing of a network connection to the same processor core as the process that is analyzing the data from the network connection causes packet reordering as execution of the analysis is moved between processor cores by the operating system scheduler [18,19]. Packet reordering can cause significant degradation in data transfer performance due to the additional processing burden that out-of-order packets place on the receiver. So, the enhancement has the perverse effect of lowering network performance over the wide area.

As data mobility requirements increase and technology continues to change there will be additional challenges to overcome, and these will undoubtedly result in operational issues for production deployments.

9.5.6 Site Technical Support

The combination of requirements and strategic challenges in site/campus networking spaces collectively lead to an important conclusion: the effective support of DIS at the site level combines several areas of expertise, including architectural support for DIS at the network level, the deployment of dedicated network and system resources with specific high-performance configurations, and the navigation of the challenges that stem from fundamental changes in the technology that is available to build the infrastructure. Collectively, these lead to the need for knowledgeable systems and networking personnel at the site. A competent operational team is absolutely critical to the success of DIS at a research institution. Without local human resources who understand the scientific process of the research being conducted and have the ability to effectively map those scientific processes onto the site's local systems and networking infrastructure, DIS will be constantly mired in performance problems with its underlying foundations.

"Streaming data to the cloud" and other similar buzz phrases that imply the outsourcing of complexity have little to offer here. Although cloud services can have significant utility for DIS, "streaming data to the cloud" as opposed to transferring the data to any other place only changes one part of the end-to-end collection of systems and networks that must interoperate effectively. The site infrastructure and the networks in the middle are all still part of the set of things that must perform properly for DIS to be successful, and the local knowledge necessary for success typically cannot be effectively outsourced.

There is simply no substitute for a capable local team that understands the site infrastructure, the needs of the science, and the integration of the two with the services in the wide area.

9.5.7 Data Transfer Tools: Data Mobility Layer

In the data mobility layer, the workflow tools are implemented—ranging from manually invoked commands to sophisticated, automated, and distributed data location management systems.

Data transfer is typically accomplished by software running on computer systems. Ideally, this means proper data transfer tools running on dedicated hardware (a DTN) that is connected to a properly functioning network infrastructure. Software tools are the means by which connections are made between systems to store, move, or stream data. Software builds on the capabilities of the systems on which it is installed, and if the systems (i.e., DTNs) are not configured properly, or are interconnected by networks that have performance problems, even the best toolset will likely have difficulty in achieving consistently adequate results.

However, if installed on properly deployed systems operating in correctly configured networks, the data transfer tools form a data mobility layer upon which higher-level data placement services can be built so that data analysis tasks can have access to the data.

In its simplest form, people must manually issue commands to the software to move individual files between systems. More advanced tools allow a user to tell the data mobility tools what to do (e.g., replicate the specified directory structure and its contents from the DTN at site A to the DTN at site Z) and then go work on other tasks while the data are moved without further manual intervention. This level of automation is provided by toolsets such as Globus Online [13].

In its most sophisticated form, the data mobility layer receives directives from higher level analysis workflows and ensures the data are available to the analysis codes at the appropriate time and location. This is the case, for example, with suites such as the production and distributed analysis (PanDA) data analysis system [20] used by LHC/ATLAS, which has an explicit data mobility system (DDM) (illustrated in Figure 9.6), and with similar systems for the LHC/CMS experiment.

In deploying data mobility tools, one must understand the limitations of the standard toolset. It should be noted that for unauthenticated distribution of smaller data sets, HTTP and anonymous FTP are common; but since they typically offer only read-only access and do not permit on-demand data placement, they are not addressed further. The typical open source data transfer tools that are supported by default by common

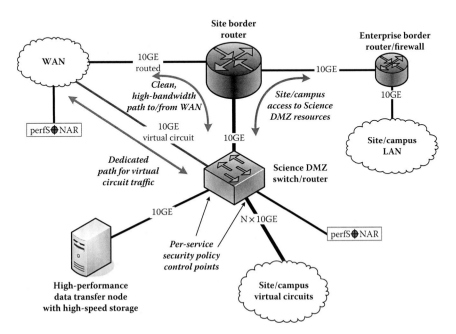

FIGURE 9.6   An example of the Science DMZ model.

open source operating systems are SCP and SFTP from the SSH suite, and rsync [21] (typically using SSH as transport for nonanonymous access). The common thread here is the SSH suite [22], which is no accident—SSH is the standard means by which authenticated system access is granted because its built-in cryptography prevents the transmission of system access credentials in cleartext over the network. Since the SSH suite is already installed on essentially all production systems, it is natural to use the SSH tools (or rsync over SSH) for data transfer—the credentials are the same, the commands are well known, and the software is free.

However, the current SSH toolset has fundamental problems that limit its utility as a high-performance data transfer tool. The SSH toolset has a maximum wide area data rate that is typically less than 50 Mbps, and the maximum data rate decreases as the distance between end points increases. These limitations are outlined and addressed in the work done by the Pittsburgh Supercomputing Center (PSC), Pittsburgh, Pennsylvania, [15]. Even if the high-performance patches from PSC are applied, SSH is still not an effective data transfer tool for throughput requirements beyond 1 Gbps. Other tools, such as GridFTP/Globus Online, are capable of much higher data rates (over 5 Gbps per DTN pair).

The tools used for data transfer are often the primary interface to the network from the perspective of scientists who need to move data between resources to conduct DIS. Whether the tool is SCP from the SSH suite, rsync over SSH transport, BBCP, GridFTP, Globus Online, Xrootd, classic FTP, a higher level tool like PanDA, or any other of a myriad assortment of utilities for transferring data between computers, including remote I/O approaches, the data transfer tool is the typical means by which users are exposed to the network.

It follows that the user interface of a tool or workflow is the primary means by which users are informed of the performance, correctness of operation, and reliability of a network. However, in this circumstance what the user perceives to be the network is really the entire set of software tools, computers, and interconnected networks between the source and destination systems involved in the data transfer. It is critical, therefore, that a substantial data mobility tool set is installed on a well-configured, reliable foundation to manage large science data flows. In current best practice, this means one or more dedicated DTNs that have been properly configured and connected to a well-run site network that is well integrated with the WAN services that provide high-speed connectivity to the DTNs at the far end.

Most data transfer toolsets are unable to effectively provide actionable information about network failures via their user interfaces. This means that unless the underlying foundation for the data transfer tools behaves in a consistent and reliable way, the user of the data transfer tools must have an alternate means of troubleshooting the network. This is an unreasonable expectation to place on most scientific network users, who are sophisticated in the conduct of their science but typically are not skilled network administrators. Therefore, it is critical that the site and wide area infrastructure administrators maintain their networks and systems such that the data mobility toolset performs consistently well from the perspective of the scientists who use it—hence the need for a monitoring infrastructure, such as perfSONAR, for use by the infrastructure administrators. Without a correctly functioning underlying infrastructure, the data mobility layer cannot function reliably and it will therefore be of limited scientific utility.

9.6 KEY ROLE OF SERVICE INTEGRATION

The integration of the technologies and infrastructures described in this chapter can seem a daunting task, and it is typically complex for people who have never done it before. Operational experience indicates that the best way to approach the deployment of a new data mobility infrastructure is to build a prototype and experiment with it in a demanding way.

LHC experiments provide an instructive example. A major contributor to the success of the LHC experiments' data distribution infrastructure was a formal prototyping effort that required the data transfer components (including WANs, site networks, and site data transfer systems) to be tested periodically during their development and deployment. The LHC tests [23], called "data challenges and service challenges," were conducted periodically and at partial scale, but they fully exercised the networks, systems, and software with synthetic data. As the date for production operation approached, the scale of the tests increased until the international data distribution and analysis system was tested at full scale. This approach ensured that the infrastructure was ready when the production operation of detectors began at LHC.

9.7 DATA INTEGRITY

Several mechanisms for ensuring the integrity of data packets are built into the TCP/IP protocol suite. However, these mechanisms were designed decades ago in an environment where data sets were much smaller, CPUs were much less powerful, and network bandwidths were far less than what are routinely available today. As a result, the checksums built into the TCP/IP protocols are reaching the limits of their ability to protect large-scale data sets from corruption. In addition, problems in the hardware of hosts and network devices can result in packet corruption that is not detected by link-layer checksums or the checksums of TCP and UDP [24,25].

As a direct result of this, when GridFTP was designed [26] several strong checksums were made available for integrity checking and these (optional) checks should always be used in high-volume, WAN transfers.

A recent example observed by the authors involved the failure of an interface in a switch at an exchange point. Performance problems were identified, and a troubleshooting effort was underway. While network engineers were localizing the problem, data transfers were continuing to use the link (although with reduced performance). Many of the files that were transferred failed their checksums and had to be transferred again. The offending switch hardware was identified and replaced, and the integrity checks in the data transfer tool did their job. However, this case adds to the evidence documented by Paxson [24] and Stone and Partridge [25], which serves to illustrate that the issue of data integrity is not simply theoretical. As increasingly large data sets are routinely transferred over the network their vulnerability to silent corruption increases, and the data mobility and workflow tools must include mechanisms to safeguard the data in transit.

This is far from being a solved problem. The network is not the only source of undetectable errors when high-data-rate, very-large-volume data flows are involved. For example, high-volume data transfers to and from certain types of disk storage arrays have also exhibited this problem [27].

9.8 SUMMARY

DIS is a critical aspect of modern science, and the network is the foundation on which DIS is built: the very structure of data-intensive scientific collaboration assumes that a high-performance, feature-rich network infrastructure interconnects all the instruments, facilities, analysis centers, institutions, and scientists that collectively make up the scientific enterprise. DIS uses the network very differently from how commodity applications use the Internet.

In the near term, there is work to do, both in core networks and in laboratory and campus networks, in the adoption of performance-based operational models. For research institutions this often means architectural changes along the lines of the Science DMZ, and performance measurement must be integrated into all science networks. In the near to medium term, networking organizations, experiments, computational facilities, and software authors must collaborate to ensure that the data mobility tools, workflow tools, and analysis tools are well matched to the services provided by the network and that the network is providing the appropriate services to scientific applications.

Also, the data scale that comes with the data-intensive paradigm brings the issue of data integrity into sharp focus. Data integrity verification is already used by several scientific communities, and the use of integrity verification will expand as data sets grow.

If the network is treated as an afterthought or is not properly configured, or if the systems that must use the network are themselves not well matched to the network or properly configured, it can be very difficult to reap the benefits of the infrastructure deployed to support DIS. However, the use of appropriate software tools, careful deployment, and proactive maintenance can make the network a catalyst for collaboration and an enabler of scientific discovery when very large data sets are at the heart of science.

ACKNOWLEDGMENTS

This chapter was supported by the director, Office of Science, Office of Advanced Scientific Computing Research, of the U.S. Department of Energy under contract number DE-AC02-05CH11231 with the University of California.

REFERENCES

1. "European Organization for Nuclear Research, The Large Hadron Collider," http://public.web.cern.ch/public/en/LHC/LHC-en.html.
2. Graphs are from the Atlas Dashboard, 2010. Available at http://dashb-atlas -data.cern.ch/dashboard/request.py/site and are courtesy of Michael Ernst, Brookhaven National Lab.
3. "BES (Basic Energy Sciences) Network Requirements Workshop, September 2010." LBNL report number LBNL-4363E, published 2011. Available at http://www.es.net/about/science-requirements/reports/.
4. "About the Earth System Grid," http://www.earthsystemgrid.org/about /overview.html.
5. P. E. Dewdney, P. J. Hall, R. T. Schilizzi, T. Joseph, L. W. Lazio. "The Square Kilometre Array." *Proceedings of the IEEE*, Vol. 97, No. 8, pp. 1482–1496. pages 1482–1496 Available at http://www.skatelescope.org/publications/.
6. "ESnet Fasterdata Knowledge Base," http://fasterdata.es.net/.

7. C. Guok, D. Robertson, M. Thompson, J. Lee, B. Tierney, W. Johnston. "Intra and Interdomain Circuit Provisioning Using the OSCARS Reservation System." In BROADNETS 2006: 3rd International Conference on Broadband Communications, Networks and Systems, 2006 – IEEE. Pages 1–8. 1–5 Oct. 2006, San José, CA, USA. Available at http://es.net/news-and-publications/publications-and-presentations/.

8. N. Charbonneau, V. M. Vokkarane, C. Guok, I. Monga. "Advance Reservation Frameworks in Hybrid IP-WDM Networks." *Communications Magazine, IEEE*, Vol. 49, No. 5 (May 2011). Available at http://es.net/news-and-publications/publications-and-presentations/.

9. W. E. Johnston, C. Guok, E. Chaniotakis. "Motivation, Design, Deployment and Evolution of a Guaranteed Bandwidth Network Service." TERENA Networking Conference 2011, Prague, Czech Republic. Available at https://tnc2011.terena.org/core/presentation/32.

10. "MRI-R2 Consortium: Development of Dynamic Network System (DYNES)" Available at http://www.internet2.edu/ion/dynes.html (2012).

11. InterDomain Controller Protocol (IDCP), http://www.controlplane.net/ (2012).

12. B. Tierney, J. Metzger, J. Boote, A. Brown, M. Zekauskas, J. Zurawski, M. Swany, M. Grigoriev. "perfSONAR: Instantiating a Global Network Measurement Framework." In proceedings of 4th Workshop on Real Overlays and Distributed Systems (ROADS'09) Co-located with the 22nd ACM Symposium on Operating Systems Principles (SOSP), October, 2009, Big Sky, MT, USA. Available at http://es.net/news-and-publications/publications-and-presentations/.

13. "Globus Online – Reliable file transfer – no IT required." https://www.globusonline.org/.

14. "XRootD" http://xrootd.slac.stanford.edu/.

15. C. Rapier, M. Stevens, B. Bennett. "High Performance SSH/SCP -HPN-SSH." November, 2012. Available at http://www.psc.edu/networking/projects/hpn-ssh/.

16. "System Fabric Works: Demonstrating RDMA Protocols Over a 40 Gb/s WAN," https://www.openfabrics.org/resources/document-downloads/presentations/cat_view/57-ofa-documents/23-presentations/39-trade-shows/72-sc11.html.

17. C. Partridge. "The Parallel-to-Serial (and back) problem" in "Forty Data Communications Research Questions." Available at BBN Report-8528. Available at http://www.ir.bbn.com/~craig/BBN-Tech-Report-8528.pdf.

18. W. Wu, P. DeMar, M. Crawford. "Why Does Flow Director Cause Packet Reordering?" CoRR abs/1106.0443: (2011).

19. W. Wu, P. DeMar, M. Crawford. "A Transport-Friendly NIC for Multicore /Multiprocessor Systems." CoRRabs/1106.0445: (2011).

20. W. E. Johnston, C. Guok, J. Metzger, and B. Tierney. "Network Services for High Performance Distributed Computing and Data Management." In The Second International Conference on Parallel, Distributed, Grid and Cloud Computing for Engineering, 12–15 April 2011. Available at http://es.net/news-and-publications/publications-and-presentations/ Ajaccio, Corsica, France.

21. "rsync" http://rsync.samba.org/.

22. "OpenSSH," http://www.openssh.com/.

23. G. Bagliesi, (INFN, Pisa ; CERN); S. Belforte, (INFN, Trieste); K. Bloom, (Nebraska U.); B. Bockelman, (Nebraska U.); D. Bonacorsi, (Bologna U. ; INFN, Bologna); I. Fisk, (Fermilab); J. Flix, (PIC, Bellaterra; Madrid, CIEMAT); J. Hernandez, (Madrid, CIEMAT); J. D'Hondt, (Vrije U., Brussels); M. Kadastik, (NICPB, Tallinn), et al. "Debugging data transfers in CMS." *Journal of Physics: Conference Series*, Vol. 219. (2010) 062055 In: 17th International Conference on Computing in High Energy and Nuclear Physics, Prague, Czech Republic, 21–27 Mar 2009, pp. 062055. Available at http://iopscience.iop.org/1742-6596 /219/6/062055/pdf/1742-6596_219_6_062055.pdf.

24. Paxson, V. "End-to-End Internet Packet Dynamics." *IEEE/ACM Transactions on Networking*, Vol. 7, No. 3, pp. 277–292, (June 1999). Available at ftp://ftp .ee.lbl.gov/papers/vp-pkt-dyn-ton99.ps.gz.

25. J. Stone, C. Partridge. "When the CRC and TCP checksum disagree." *ACM SIGCOMM Computer Communication Review*, Volume 30 Issue 4, October 2000, Pages: 309–319.

26. I. Mandrichenko, FNAL, editor, W. Allcock, ANL, T. Perelmutov, FNAL. "GridFTP v2 Protocol Description." Open Grid Forum draft standard. May 4, 2005. Available at www.ogf.org/documents/GFD.47.pdf.

27. D. Foster. Personal communication, CERN to WEJ. October, 2010.

CHAPTER 10

A Posteriori Ontology Engineering for Data-Driven Science

Damian D. G. Gessler, Cliff Joslyn, and
Karin Verspoor

CONTENTS

10.1 PROBLEM OVERVIEW

Science—and biology in particular—has a rich tradition in categorical knowledge management. This continues today in the generation and use of formal ontologies. Unfortunately, the link between hard data and ontological content is predominately qualitative, not quantitative. The usual approach is to construct ontologies of qualitative concepts and then annotate the data to the ontologies. This process has seen great value, yet it is laborious and the success of ontologies in managing and organizing the full information content of the data is uncertain. An alternative approach is the converse: use the data itself to quantitatively drive ontology creation. Under this model, one generates ontologies at the time they are needed, allowing them to change as more data influences both their topology and their concept space. We outline a combined approach to achieve this, taking advantage of two technologies, the mathematical approach of formal concept analysis (FCA) and the semantic web technologies of the web ontology language (OWL).

Biology has a rich tradition in classifying knowledge that extends back to Aristotle (384–322 BCE) (Aristotle 350 BCE) and more recently to Linnaeus (Linné 1735), Whittaker (1969), and Woese (Woese et al. 1977; Woese et al. 1990). Although these latter examples are primarily taxonomic, knowledge organization today witnesses efforts across the breadth of biology in the creation of controlled vocabularies, ontologies, and knowledge bases (e.g., see online resources at BioPortal,* Open Biological and Biomedical Ontologies [OBO†], and the National Library of Medicine‡).

Biological ontologies, such as those hosted at BioPortal, tend to be predominantly axiomatically weak, deep subsumption hierarchies. "Axiomatically weak" means a preponderance of axiomatic subsumption statements (i.e., subclass of "is a"), and "deep" means successive static relations creating long, transitive chains of class subsumption. These ontologies are noted

---

* http://bioportal.bioontology.org.
† http://obo.sourceforge.net.
‡ http://www.nlm.nih.gov.

for their relatively few semantic relationships other than subsumptive relations between concepts, relative to the number of classes (del Vescovo et al. 2011).

For perspective, note that the classification system of the Library of Congress (LC) of 21 top-level classes and a few hundred subclasses pales in comparison with the 33,000+ entries in the three major ontologies of Gene Ontology (GO),* not to mention the over 26,000 descriptors and 177,000 supplemental headings in Medical Subject Headings (MeSH†) and the 1 million biological concepts in the Unified Medical Language System's Metathesaurus (UMLS‡). While biological classes tend to be axiomatic and static (i.e., humans conceived the class, named it, and placed it in the ontology in some relation to other classes), LC's classification system classifies any publication in the realm of human knowledge by dynamically generating a part of every item's identifier (the so-called Cutter number). This is equivalent to dynamically generating a class (the set of all publications—e.g., physical copies of a book—that share the same Cutter number) by applying an algorithm to a representative publication's properties to generate the class name.

There is, though, a similarity between the LC system and modern biological ontology creation: the static component of the LC classification—the 21 classes and their subclasses—was built in the late nineteenth and early twentieth centuries using the same methodology that predominates in biology today, a basic methodology of an iterative loop of observation → induction → model creation → deduction. In this methodology, classifiers (1) observe individuals, conceptualized as instances of more general classes; (2) propose the existence of classes, often striving for a certain notion of orthogonality; (3) propose a model (a metaset of classes and relationships between classes) for the assignment of any assemblage of individuals into classes; and (4) implement the model on new individuals, thereby classifying those individuals according to the rules of class assignment. The process is then repeated in refining the classification. The methodology has been productive in biology, but the process is laborious and prone to artifacts, as we discuss in Section 10.2.1. In this chapter, we discuss two approaches that yield a different, and data driven, approach to ontology creation.

---

* http://www.geneontology.org
† http://www.nlm.nih.gov/pubs/factsheets/mesh.html
‡ http://www.nlm.nih.gov/research/umls/about_umls.html#Metathesaurus

10.2 VISION FOR THE FUTURE

Our vision for the future is a more integrative, scientific information environment. To achieve this, our research and technological approach has been to enable computers to contribute a greater role upstream in data description, discovery, invocation, and response phases. This helps to better position scientists for downstream tasks, such as the interpretation and analysis of data integration products, an area where the human scientific mind is vastly superior to its machine analogs. But deployment of computers upstream means that computers need to be able to access and *assess* data and services with greater degrees of automation. Achieving this requires a computational semantics (from the Greek *semaino*, which means "to mean," and *semantikos*, which means "significant"). A computable semantics is achieved in two tiers: the first uses a nonvariant structural framework with a rigid formal semantics. We use the technologies of Resource Description Framework (RDF) + Web Ontology Language (OWL) + Simple Semantic Web Architecture and Protocol (SSWAP) (Gessler et al. 2009) to produce a semantic web services framework. The second tier uses domain-specific ontologies. We can think of RDF, OWL, and SSWAP as the semantic framework for a web-based computable language; the content—the concepts that actually capture domain-specific statements and their derived inferences—are captured by the use of domain-specific ontologies.

Current-day implementation of domain-specific ontologies—for example, biological ontologies such as those at BioPortal—presents a unique set of challenges for achieving this vision. They can contribute more to biological integration if we can address the following needs: (1) in ontology creation, to ensure that an ontology truly does cover the concept space, minimally and efficiently, and is less susceptible to the conceptual biases of the creators (even if such biases are unintentional and accepted by the scientific paradigm of the day); (2) in ontology maintenance, to have formal support for collaborative editing and quality control, whereby the repercussions of changes to an ontology could be systematized and tested; (3) in ontology annotation, to allow the definition of ontological classes in a logical formalism to inform and drive the process of annotation, that is, assigning individuals to classes; and (4) in ontology deployment, to better enable ontologies to inform the discovery and engagement of semantic web services, thereby contributing to dynamic, on-demand data discovery and integration.

10.2.1 Ontology Engineering: Creation and Maintenance

Ontology creation and maintenance are activities of ontology engineering (Gómez-Pérez, Fernández-López, and Corcho 2003). Although there is no single formula for ontology engineering (Fernández-López and Gómez-Pérez 2002), thoughtful practices have delineated a general process of an environment study, a feasibility study, specification, conceptualization, formalization, implementation, maintenance, and ontology support activities (formula and terminology from Corcho, Fernández-López, and Gómez-Pérez 2007; Fernández-López et al. 1997; see also Sure, Tempich, and Vrandecic 2006). The literature on ontology engineering is well developed. Yet, while surveys of practitioners in the field show general support for formalized processes, 80% of ontology engineering projects reported by Simperl and Tempich (2006) did not follow a specific methodology. Despite the lack of widely adopted methodologies, this general process of human inspection driving ontology creation is widely practiced. We call this general process, as outlined earlier, *a priori* ontology engineering, in reference to the fact that practices across methodologies share the fundamental assumption that ontology engineering should proceed from the informed, human analysis of a problem and that a goal is a formalization of this human conceptualization.

Arguably the most prominent biological ontology is GO (Ashburner et al. 2000). GO is manually constructed and maintained and appears to follow the pattern uncovered by Simperl and Tempich (2006): a formal process is followed, but it is not a scientific methodology as demonstrated by Alterovitz et al. (2010). This work used information theoretic analyses to report topological and structural inefficiencies in GO. Based on this, Alterovitz et al. (2010) proposed a small, restricted set of new classes and relationships. Although their approach was deemed successful in achieving both an improvement in information theoretic measures and operational changes (they proposed 14 recommendations to the Gene Ontology Consortium, many of which have been adopted), the observed effect on a test case of gene expression results is disturbing: "… these changes significantly affected the functional interpretations of 97.5% ($P < .001$) of the experimental gene signatures and altered the resulting set of GO categories by 14.6% on average (p. 130)." In other words, after they improved the information theoretic properties of the ontology by repositioning classes in the subsumption hierarchy, scientific interpretations on data annotated to the ontology changed in 97.5% of the cases. This is disturbing because if such a heavily used ontology—one subject to extensive human inspection and professional use over many years—could be subject to improved topological changes resulting in sweeping reclassifications of sample data, then we are left to wonder to what degree the ontology is organizing knowledge and to what degree it is simply organizing.

The truth is perhaps at neither extreme, but the results suggest that a more empirical process can yield a more robust and actionable knowledge representation. We also note that although there is a compelling basis for information theoretic approaches in ontology engineering, we must still ask how we are to validate their reclassifications. Are there noninformational theoretic criteria that should also be applied? The answer is yes, since we expect exceptions to information-maximizing and entropy-minimizing routines, for example, because of low sample sizes or systematic ignorance of the true, a priori, information model. It appears that the former was explicitly recognized by the Gene Ontology Consortium, as one of the terms recommended to be repositioned in the ontology was not moved due to the low number of extant gene annotations (Alterovitz et al. 2010). The latter—systematic ignorance of the true, a priori, information model—is always implicitly assumed whenever we apply theoretical models to real-world data.

We can conclude that GO is susceptible to alternative topologies, yielding new scientifically relevant classifications, but we cannot conclude yet that these new classifications are optimized over a universe of constraints relevant to knowledge representation. In a way, then, we have the worst of both worlds: our annotation of data to concepts is vulnerable to the topology of the concepts, but we have no guarantee on the "correctness" of the concept topology or concept universe itself.

10.2.2 Ontology Annotation

In biology, gene annotation is the process of assigning a gene—a delineated segment of DNA for which there is evidence that it is transcribed—to a function. Gene annotation is one of the first steps in understanding what a gene does and what specific genes are associated with various biological processes. Genome annotation meetings are often called *jamborees*. The name reflects the activity: knowledgeable, domain experts collaboratively and manually assign genes to functional categories, weighing a variety of qualitative and quantitative evidence to make assignments. Computational assignment is also well practiced, often based on transfer annotation, but the signal-to-noise ratio is low and computational assignments are accepted as putative until they are supported by experimental evidence.

Currently, for the model plant *Arabidopsis thaliana* (perhaps the best studied model plant) less than 20% of predicted genes have laboratory-based experimental evidence (Goff, pers. comm.). Yet experimental evidence is a weak bar, because experimental evidence is open world: just because a gene is implicated in process *X*, rarely does that mean it is not possible to be implicated in process *Y*. So in gene annotation, negation-as-failure does not apply. Ontology *realization*—the process of assigning individuals to ontological classes—is an activity of gene annotation. However, axiomatically weak subsumption hierarchies are poorly suited to represent negative assertions. For example, many biological ontologies have an expressivity akin to RDF schema (RDFS) semantics for which there exists no notion of negation, inconsistency, or nontrivial satisfiability (all classes are trivially satisfiable in RDFS because there are no negation semantics [Cuenca Grau 2007]). Thus when there is experimental evidence that a gene does not belong to a class, it is difficult or impossible to represent that information in the ontology and the information is lost. Biological ontology annotation is a labor-intensive, human-dominated activity that tends to capture positive class extension (the set of individuals belonging to a class) based on a human interpretation of the data, but it misses extensions that could be derived by negative assertions or inference-derived inconsistencies.

10.2.3 Ontology Deployment

Biological ontologies are recalcitrant to the standards and best practices of the World Wide Web. This limits their contribution to reuse, as well as to emerging technologies such as semantic web services. The problem is threefold.

First, biological ontologies exist primarily as creations of domain-specific human knowledge. As such, they adhere to no web standards per se—they are technology independent; so, for example, there is nothing inherent about the terms, topology, or naming conventions that makes the ontologies web aware. Compare this, for example, with OWL DL (McGuinness and van Harmelen 2004): although the abstract syntax and semantics of OWL DL is also technology independent—it is driven by the axioms and theorems of first-order description logic—implementation standards are tightly linked to web technologies such as international resource identifiers (IRIs) and RDF.

Second, biological ontology engineering is monolithic and opaque to high-throughput, cross effort integration. In general, terms are not formally defined in relation to other terms in other ontologies: terms are de novo to separate ontologies, even if they are derivatives or informally dependent on preexisting concepts. For example, the GO term "Biological Process" (GO:0008150) is a term in GO and the Cell Cycle Ontology (CCO) term Biological Process (CCO:U0000002) is a distinct term in CCO, despite the fact that CCO is by design developed and built from other ontologies, including GO (Antezana et al. 2009). Regardless of how similar or dissimilar the terms' semantics are, they reside in different ontologies as separate, distinct concepts. This creates broad challenges to ontology alignment, because instead of the ontology authors themselves directly using terms from other ontologies, or introducing terms via formal semantic relations to terms in other ontologies, terms are tied tightly to ontology ownership where alignment then proceeds ex post facto. This is evident at repositories of ontology alignments* where secondary information resources such as these mappings are used to describe formal (and informal) alignments across ontologies. Compare this to the IRI and RDF basis of OWL, where it is natural to leverage across domains at the point of term declaration under a formal semantic (e.g., http://thisWebSite.org/thisOntology/thisTerm rdfs:subClassOf, http://thatWebSite.org/thatOntology/thatTerm).

Third, standard ontology packaging practices are not web savvy. Standard practice is to bundle all terms of an ontology into a single file. OWL version of CCO is over 300 MB of uncompressed RDF/XML. CCO terms cannot be independently dereferenced on the web with persistent uniform resource locators (URLs). Access to ontologies is often via non-RESTful (REST stands for Representational State Transfer) user interfaces (Fielding 2000). Even when access is programmatic and RESTful (e.g., BioPortal REST services† or EBI services‡), the returned content may be idiosyncratic (e.g., a nonformal semantics, such as arbitrary XML or HTML). Terms do not use uniform resource identifiers (URIs) for universal addressing; for example, the GO term "Biological Process" exists as a GO concept independent of any web address (GO:0008150); as an OBO term in its 19.4-MB OBO file§ in the 20.6-MB file on the GO website¶ (and

---

* e.g., http://obofoundry.org/index.cgi?show=mappings.

† http://rest.bioontology.org.

‡ http://www.ebi.ac.uk/QuickGO.

§ http://obo.cvs.sourceforge.net/viewvc/obo/obo/ontology/genomic-proteomic/gene_ontology_edit.obo.

¶ http://www.geneontology.org/ontology/obo_format_1_2/gene_ontology_ext.obo.

other places); as a RESTful, HTML page;* and so forth. When ontologies do use URIs, they may fail to leverage the underlying web capability of dereferencing, for example, the adult mouse brain (ABA) term (#ENT1)† does not resolve to a representation of the term's semantics but to the human readable home page of the *Allen Reference Atlas*. The term's semantics are defined at a separate place, in the ontology itself.‡ The use of the delimiting hash (#) fragment identifier (e.g., #ENT1), instead of a slash (e.g., /ENT1), guarantees that no server can satisfy per-term requests, because the hash is strictly a client-side secondary resource reference (RFC 3986, Section 3.5§). Thus, any server must return the content associated with the reference prior to the hash (e.g., the entire ontology) even if the user agent requests just a single term.¶ These and other factors conspire to render ontologies as artifacts of human knowledge organization, rather than web technologies.

10.2.4 Vision for the Future: Dynamic, Restful, Data-Driven Ontologies

How can we address the limitations mentioned in Section 10.2.3? They are limitations in both ontology engineering and web deployment. We want ontologies to be more quantitative and "objective," that is, subject to algorithmic quantification on the utility of their conceptualization. And we want them to be more informatically aware, such that they leverage the world's dominant distributed informatic infrastructure, that is, the web, to the benefit of data integration. There are ongoing efforts to make biological data interoperable in the context of the semantic web, for example, Bio2RDF (Belleau et al. 2008) and KaBOB (Bada, Livingston, and Hunter 2011); this is in part to address the limitations of current approaches to ontology representation.

Our approach is to invert traditional ontology engineering into what we call "a posteriori" ontology engineering for data-driven science. In a posteriori ontology engineering, one starts with no preconceived concepts. One canvasses the *measurement* technologies of the science to delineate a list of measureable *properties*. Empirical measurement commences, and properties of individuals are assigned observed *values*. Informally, a *class* is defined as "the set of all individuals which share the same properties and

---

* http://www.ebi.ac.uk/QuickGO/GTerm?id=GO:0008150.
† http://mouse.brain-map.org/atlas/index.html#ENT1.
‡ http://rest.bioontology.org/bioportal/ontologies/download/40133 (API key needed).
§ http://www.apps.ietf.org/rfc/rfc3986.html#sec-3.5.
¶ http://mouse.brain-map.org/atlas/index.html#ENT1.

property values." For continuous values, shared property values encompass binning at an arbitrary resolution. We call this an *informal class*. The network of individuals (datum items) clustered according to shared property values creates a subsumption network, or more formally a lattice: a subclass is a class of individuals that have all the same shared properties and values of its super class, and may be more. This corresponds with an intuitive notion of a class as a set of individuals sharing common properties and values. As a second step (see Section 10.4.3), the definition of a class is formalized to be "the set of all individuals which share necessary and sufficient conditions." We call this a *formal class*. For example, naively, the class of eukaryotes is the set of all individuals such that their cells have the property *hasNucleus* with the value *true*, or the cellular nuclear envelope has *Number of cellular membranes* with the value 2, and so forth. Animals have the properties of eukaryotes and may be more (e.g., are heterotrophic); thus animals are a subclass of eukaryotes in a subsumption lattice built from observed properties and their values.

It may first appear that creating the list of properties has simply shifted the axiomatic class creation of a priori ontology engineering from classes to properties, but this is not the case. The property selection is driven by the measurement methodology—a reflection of the practice and technology of the day. This decouples preconceived notions of what concepts should exist to an empirical canvassing of what properties are being measured. Existential arguments are precluded because the act of measuring itself instantiates the reality of the measurement property and value. Properties may be grouped into a priori subsumption chains, but this is not necessary. We will see that subsumption relations on the data itself are the *result* of data analysis, rather than the a priori scaffold of knowledge organization. Annotation (ontology realization) is the algorithmic process that creates a subsumption lattice, rather than the product of human assignment. A posteriori ontology engineering uses the data itself to create the ontology; thus the ontology is the product, not the assumption. Such an ontology is both temporal and marginal: it is based on the data used to construct it. New data may change both the topology and the realization of the ontology. But the change occurs under the influence of the new data and thus should be a refinement toward a more encompassing model of nature, rather than an undirected change in organization.

We examine a process to achieve a posteriori ontology engineering. The first step uses FCA to create the subsumption lattice from a collection of data with arbitrary measured properties. This creates unnamed, informal classes, that is, groupings of individuals according to shared properties and values. The second step is to analyze the resultant informal classes for groupings that are deemed scientifically relevant. This is a manual step reliant on human assessment. From this, property restrictions are formalized in OWL, `owl:Restrictions`, which are used to create necessary and sufficient conditions for formal, named classes. The use of necessary and sufficient conditions yields class definitions, and thus concepts are created only to the degree that data support their existence. The existence of formal classes allows new, unclassified data to be annotated to a formal knowledge representation framework based on their observed property values. Lastly, the ontology is deployed web aware, using separately dereferenceable URIs and OWL + SSWAP.

10.3 RELATIONSHIP TO OTHER TECHNOLOGIES

10.3.1 Formal Concept Analysis

A key technology in the order theoretical approach is FCA (Ganter and Wille 1999). FCA produces mathematical methods to represent the hierarchical relationships and implications present among relational data, which are represented as sets of objects and their properties. Within mathematics, FCA derives from the algebraic theory of binary relations and complete lattices. Within computer science, FCA is increasingly applied in conceptual clustering, data analysis, information retrieval, knowledge discovery, and ontology engineering (Ganter, Stumme, and Wille 2005).

FCA defines a formal *context* as a mathematical structure, $K = (G,M,I)$, where G is a set of *objects* (individuals), M is the set of *attributes* (properties) of those objects, and $I \subseteq G \times M$ is an *incidence relation*. The expression $(g,m) \in I$ means that the object g has the attribute m. A formal context can be visualized by a two-dimensional table called a *cross table*, where the presence of a cross in a cell indicates that the object on that row has the attribute on that column. A formal *concept* is a pair of sets (combination of objects and attributes) such that every object has every attribute and every attribute is present on every object. There are often multiple concepts for a given context. We will provide a detailed example in Section 10.4.1.

FCA as a fundamental technique for both construction and integration of ontologies has been known for some years. Bain (2002, 2003) investigated using FCA and inductive logic programming as a means to both identify and create concepts from a formal context. Later, Akand, Bain, and Temple (2007, 2010) extended the approach to GO such that genes could be annotated to derived classes composed of combinations of classes. Cimiano, Hotho, and Staab (2005) provided an example of how formal contexts can be constructed from the output of text parsers. The resulting concept lattices compared favorably to the reference ontologies constructed manually on the same domains. Joslyn, Paulson, and Verspoor (2008) used a similar approach, analyzing a larger general language corpus and focusing specifically on taking advantage of linguistic relations among nouns and verbs to investigate the hypothesis that semantic generality of terms, as represented by hierarchical relations among them, can be determined from the analysis of their shared linguistic contexts. Formica (2006, 2008) first used human-curated ontologies to influence FCA concept similarity measures and then extended the method to substitute an information content metric on the concepts, removing the need for human-curated expertise entirely.

10.3.2 Web Ontology Language

A second technology important for data-driven, a posteriori ontology engineering is OWL (McGuinness and van Harmelen 2004). OWL is the World Wide Web Consortium (W3C)-recommended technology for distributed computational logics on the web. W3C is the voluntary sanctioning body of the World Wide Web. OWL is built on URIs, the more expressive RDF, and the helper technologies of RDFS and XML schema definition (XSD). With OWL, we address both the construction of necessary and sufficient named classes and the deployment of the ontology onto the web (via OWL's tight linkage to the web technologies of URIs, RDF, RDFS, and XSD).

Unrestricted OWL Full is a higher order description logic. As a description logic, its entities are "things" (individuals), relations between things (called properties or predicates), and sets of things (called classes). Individuals are akin to FCA objects as are properties to FCA attributes. OWL, like RDF, allows the expression of properties of classes, classes of properties, and so forth. OWL posses the following important properties: (1) completeness (all truths can be derived), (2) validity and soundness (no falsehoods are derived), (3) monotonicity (inferred truths cannot be later proved not true), (4) nontrivial consistency (no contradictions), and (5) nontrivial satisfiability (logical possibility of a class containing at least one individual). But as a higher order logic it is known to be undecidable, and thus computational inference algorithms are not guaranteed to finish in finite time with finite resources. Yet a few key restrictions on OWL's use and expressivity yield it first order (called OWL DL), and as such OWL DL gains decidability. Complexity remains high (finite time could still be a long time in worse-case scenarios), but in practice many if not the majority of biological ontologies can be reasoned over in a few seconds (del Vescovo et al. 2011).

OWL's semantics allows one to construct classes via numerous means: axiomatically (simply assert the class); by identity (equivalence and difference); by set operations (intersection, union, complement, enumeration); by specific value, or universal or existential restrictions on properties (*hasValue*, *for all* ∀, and *there exists* ∃); by cardinality constraints on properties (minimum, maximum, and exact); by property characteristics (reflexivity, symmetry, transitivity, and chaining); and so on. This expressivity allows one to state the necessary and sufficient conditions of classes (ontological concepts) based on individuals' observed properties and yields the resultant ontology amenable to computational analysis. An analysis using OWL DL in biology is overall highly favorable, with noted exceptions where DL's fragment of first-order logic (FOL) cannot capture the full breadth of biological expressivity (Stevens et al. 2007).

10.4 CURRENT STATE OF THE ART

FCA creates a concept lattice based solely on objects and their attributes. Objects—data—are grouped by virtue of their shared attributes (properties), and thus these collections of shared properties are essentially unnamed classes, which grow more general (encompassing more objects) as one ascends the hierarchies. Thus, a concept lattice maps to a subsumption lattice. Similar to annotations in manually constructed ontologies such as GO, the process yields data assigned to, or annotated to, classes. But unlike GO, where a class may exist independent of the scientific data (e.g., posited by humans as a placeholder), in FCA the classes are derived automatically and solely from experimental properties of the data. Furthermore, all classes can be proven to be logically consistent, complete, and "minimal," in the sense that if there is no data driving a formal concept's creation there is no consequent class. Most significantly, while there is not necessarily any explicit hierarchical structure in the formal context K, concept lattices derive hierarchical relations among the attributes implicit in the structure of their objects. The resulting concept lattices represent ontological subsumption hierarchies derived from the data. The conspicuous missing element in FCA is that classes remain unnamed (Bain 2002, 2003).

10.4.1 Formal Concept Analysis Illustration

Consider Table 10.1, a cross table showing a brief excerpt of data given by Sjoblom et al. (2006). In this table, the objects, G, on the rows are genes or open reading frames (ORFs)—sequences of DNA that are putative, yet not confirmed, genes. (For ease of presentation, we use the word gene to refer to both genes and ORFs). An 'X' appears in the column under the attribute "Breast" if that gene is associated with breast tumors according to the criteria of Sjoblom et al. (2006; espec. Supplementary Material) and the attribute "Colorectal" if it is associated with colorectal tumors. The remaining attributes relate to the cancer mutation prevalence (CaMP) score. The CaMP score is proportional to "... the probability that the number of mutations actually observed in a gene is higher than that expected to be observed by chance given the background mutation rate" (Sjoblom et al. 2006, p. 270). A score of CaMP < 1 is interpreted as evidence that the gene is not implicated in cancer; CaMP ≥ 1 is a decision point implicating a gene in cancer. Out of the total 13,023 genes examined by Sjoblom et al. (2006), 189 have CaMP > 1 and these comprise a class called "CAN-genes" (candidate cancer genes). The use of CaMP score in Table 10.1 illustrates both how nontrivial scientific measurements (complex summaries) may be used as FCA attributes and how continuous variables may be binned on scientifically relevant thresholds into a discrete cross table.

Identifying relevant patterns such as which collections of genes are highly, somewhat, or not implicated in breast, colorectal, or both forms of cancer involves sorting the table by rows and columns multiple times to move checkboxes together. Some patterns are obvious: all genes are associated with colorectal cancer. Others are more subtle: every gene associated with breast tumors is also associated with colorectal tumors, yet it does not induce disease. This simple example is illustrative, but as the number of objects and attributes increases (the real data set involves

TABLE 10.1   Cross Table Summarizing Five Attributes of Two Genes and an ORF

	Tumor		CaMP		
	Breast	**Colorectal**	**<1**	**≥1**	**≥1.5**
SKIV2L	X	X	X		
C6orf29		X		X	
SLC29A1		X		X	X

Source: Sjoblom T et al., *Science,* 314(5797):268–274, 2006. With permission.

thousands of genes) the complexity of permuting and sorting such tables to find all patterns is combinatorially overwhelming for manual inspection.

The formal context K on objects (rows: genes) and their attributes (columns: tumor types and CaMP levels) summarizes all possible permutations of rows and columns to identify *maximal rectangles* of checkboxes in its creation of a concept lattice, a semantic hierarchy that dually catalogs both the collections of objects that have certain attributes and the collections of attributes that hold for certain objects. An example concept lattice is shown in Figure 10.1. In the lattice, nodes contain information about precise facts in Table 10.1, such as the fact that SKIV2L is the only gene that is both implicated in breast tumors and not disease related. Similarly, one can examine attributes such as CaMP $\geq$ 1 to show that C6orf29 is disease implicated, but traversing downward we see that so is SLC29A1 (so that C6orf29 has only CaMP $\geq$ 1, and not CaMP $\geq$ 1.5). Going back the other way, both genes are associated with colorectal tumors and only colorectal tumors. Finally, information is available about what pairs or groups of objects and attributes are in common. For example, to see what SKIV2L and SLC29A1 have in common we traverse upward from both until we arrive at their *join* at "Colorectal" tumor: not only are both genes implicated in colorectal tumors (though only SLC29A1 is additionally

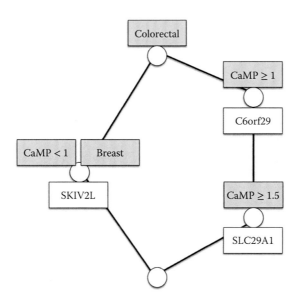

FIGURE 10.1   Concept lattice for Table 10.1.

implicated as disease associated) but also that is *the only thing* they have in common.

The example also shows the marginal effect of limited data sets. Among the genes present in the formal context of Table 10.1 and the concept lattice of Figure 10.1, genes involved in breast tumors "imply" genes involved in colorectal tumors, that is, breast tumor is below (is subsumed by) colorectal tumor. Yet this is not true for the full data set (data not shown). Thus as more data is added, scientific conclusions may also alter, reflecting the new data's contribution. Yet in marked distinction with the situation described by Alterovitz et al. (2010) and GO, subsequent changes to a resultant ontology built from the formal context necessarily reflect the *data's impact* on new knowledge implications, rather than causing a reorganization of a priori concepts.

10.4.2 Hybrid Example Combining Data and an A Priori Ontology

A priori domain ontologies operate by coding the fundamental concepts and semantic relations of a domain and, as such, they may still contribute a scaffold for knowledge representation. FCA can leverage data against this scaffold to further refine the ontologies. Figure 10.2 shows an example of a small portion of GO, which currently holds over 36,000 such categories. Each functional category is adorned with some of the gene products that perform those functions. For example, the gene *Mcmd4* in mice performs (nonexclusively) the function "DNA ligation," and thence by subsumption the functions "DNA-dependent DNA replication," and so forth. Such structures need to be maintained manually, which is both difficult and error prone.

In addition to pure induction of taxonomic structures from underlying relational data, FCA can be used to combine hand-crafted ontological structures with automatically constructed hierarchical information (see the studies of Kaiser, Schmidt, and Joslyn [2008a, b] and Guo et al. [2011] for a more complete formal mathematical consideration). For example, extant GO annotations can be used as attributes on the data. Consider again the GO fragment from Figure 10.2, and now use it to construct a formal context by making the objects (rows) the gene products (e.g., *Mcmd4*) and the columns GO functional categories (e.g., "DNA ligation"). While these relationships alone will determine a hierarchical structure using FCA, the subsumption implications in GO that incorporate its manually constructed taxonomic structure are represented in the columns.

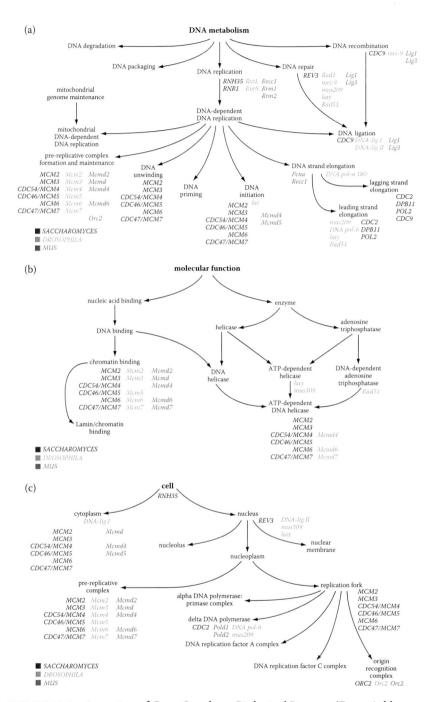

FIGURE 10.2 A portion of Gene Ontology, Biological Process. (From Ashburner, M., et al., *Nature Genet.*, 25(1), 25–29, 2000.).

The result is an adjustment of the original hierarchy, as shown in Figure 10.3. Note, for example, that "DNA initiation" is now both a child of "DNA dep. DNA replication" and a parent of "DNA unwinding," whereas before it was only a child of "DNA dep. DNA replication." This is because of the great amount of annotation overlap between these categories. Similarly, "DNA repair" is now a parent of "DNA recombination," because of the common annotation with the Lig 1 and Lig 3 genes for mouse.

FCA has limitations. Subsumption only occurs to the extent that objects share attributes; thus, it is trivially uninformative if objects use different attribute tags for equivalent attribute concepts. If objects represent ambiguous concepts, they will be associated with sets of attributes that conflate multiple meanings. Thus, FCA benefits from the consistent use of a controlled vocabulary. Most FCA algorithms are optimized for discrete

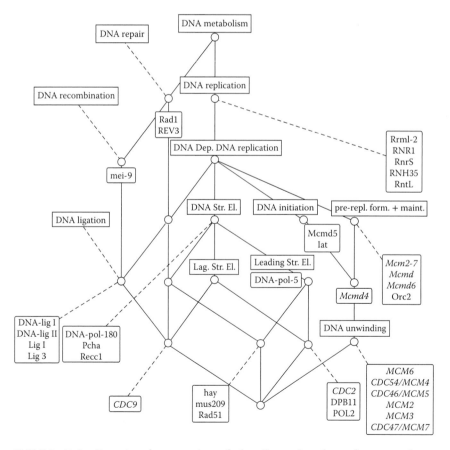

FIGURE 10.3 Functional categories of the Gene Ontology fragment from Figure 10.1 adjusted by their genomic annotations.

characters. Continuous values can be addressed by first binning values into a finite set of value ranges or by ranking. Either transformation loses information. Additionally, as the number of individuals and attributes grows it is computationally expensive to compute the subsumption lattice. Finally, subsumption implies necessary but not sufficient conditions, so FCA alone is not sufficient to generate formal class definitions. Although none of these limitations appear to be fatal to the use of FCA in ontology engineering, they do hint that more research is warranted before FCA can be deployed in a production environment.

10.4.3 Mapping to Classes

From a data-driven subsumption lattice, we now seek to map necessary relations to formal necessary and sufficient class definitions. We proceed by examining each class from the FCA subsumption lattice and its shared properties (attributes). Our goal is to identify those classes that capture concepts that we value sufficiently so as to name and construct necessary and sufficient conditions. This may be done manually, or by identifying topological or informationally rich parts of the concept lattice (Bain 2003; Formica 2008). The process is nondestructive (we do not dismantle the source subsumption lattice); it may be quantitative (e.g., we may use threshold criteria such as the presence/absence of properties, the size of class extension, or the position of the class in the lattice), and it may be qualitative (studying a specific disease drives our relative valuation on which classes to name). The process may be done statically (once, to produce an authoritative model of the data) or dynamically (as affected by specific questions on the data).

10.4.4 Using OWL

Once key concepts are identified, class definition constructs of OWL can be used to create class restrictions. To create necessary but not sufficient conditions, we use subsumption, that is,

```
<owl:Class rdf:about = "&dataOntology;Eukaryota">
    <rdfs:subClassOf>
        <owl:Restriction>
            <owl:onProperty rdf:resource =
              "&propertyOntology;hasNucleus"/>
            <owl:hasValue rdf:datatype =
              "&xsd;boolean">true</owl:hasValue>
```

```
          </owl:Restriction>
        </rdfs:subClassOf>
</owl:Class>
```

The above W3C-recommended RDF/XML snippet serialization declares the class `Eukaryota` to be a subclass of the anonymous class of all individuals that have the `hasNucleus` property with a value of `true` (see the work of Stevens et al. [2007] for exceptions). Informally, this can be serialized in Notation 3 (N3) without loss of information:

```
dataOntology:Eukaryota
    a owl:Class ;
    rdfs:subClassOf [
        a owl:Restriction ;
        owl:hasValue "true"^^xsd:boolean ;
        owl:onProperty propertyOntology:hasNucleus
    ] .
```

To create necessary and sufficient conditions (a formal class definition), we use equivalence:

```
dataOntology:Eukaryota
    a owl:Class ;
    owl:equivalentClass [
        a owl:Restriction ;
        owl:hasValue "true"^^xsd:boolean ;
        owl:onProperty propertyOntology:hasNucleus
    ] .
```

Some care needs to be exercised when using equivalence because it is possible that the restrictions will be uncovered by a DL reasoner to be equivalent to `owl:Thing`—the top OWL concept to which all concepts, properties, and individuals belong. Informationally this is redundant, whereas computationally it can be expensive: equivalency is transitive, and thus equivalence to `owl:Thing` can generate excessive statements that unnecessarily burden computational analysis. Equivalence to `owl:Thing` is informationally vacuous and should be removed.

Yet the key motivation for definitions (necessary and sufficient conditions), versus just subsumption (necessary conditions), is that new, unannotated data can be assigned to such classes. Thus, one uses an initial set of experimental data to construct the FCA concept lattice and the resultant ontology to annotate new data under the logical requirements for class extension. At some point, this new data can also be iterated back to the FCA step to revise the underlying concept lattice.

From just a few named classes, one can define new classes based on logical relations. For example, consider the property `hasNucleatedCell`, which is an object property that connects a subject individual to an object datum—in this case, the representation of a `NucleatedCell`, that is, a cell with a nucleus. Thus, we may define

```
dataOntology:Eukaryota
    a owl:Class ;
    owl:equivalentClass [
        a owl:Restriction ;
        owl:onProperty propertyOntology:has
          NucleatedCell ;
        owl:someValuesFrom dataOntology:NucleatedCell
    ].
```

The existential quantifier *there exists* ∃ is modeled with OWL construct `owl:someValuesFrom`. Thus, any individual with the observed property `hasNucleatedCell` with a value belonging to the class `NucleatedCell` will be assigned to the class `Eukaryota`; that is, it is a eukaryote. We can force the range of `hasNucleatedCell` to always be `NucleatedCell` with the following property definition:

```
dataOntology:NucleatedCell
    a owl:Class.

propertyOntology:hasNucleatedCell
    a owl:ObjectProperty ;
    rdfs:range dataOntology:NucleatedCell.
```

The objects of `hasNucleatedCell` will be inferred to belong to the class `NucleatedCell` if they are not already so declared. Axiomatic assignment to classes can occur on both the range and the domain. Depending on our data collection quality control, if we want to state a priori that anything with the `hasNucleatedCell` property is axiomatically a eukaryote then we can enforce this by adding a global domain:

```
dataOntology:NucleatedCell
    a owl:Class.
```

```
dataOntology:Eukaryota
    a owl:Class.

propertyOntology:hasNucleatedCell
    a owl:ObjectProperty ;
    rdfs:domain dataOntology:Eukaryota ;
    rdfs:range dataOntology:NucleatedCell.
```

Axiomatic domains and ranges should be used judiciously because they force a reasoner to deduce that a datum belongs to the respective class regardless of other statements. Alternatively, we may choose to define eukaryote (and thus its class extension) solely a posteriori, in which case we would not put a domain on `hasNucleatedCell` but would add a universal quantifier *for all* ∀ to `Eukaryota`, for example,

```
dataOntology:Eukaryota
    a owl:Class ;
    owl:equivalentClass [
        a owl:Class ;
        owl:intersectionOf ([
            a owl:Restriction ;
            owl:onProperty propertyOntology:
              has NucleatedCell ;
            owl:someValuesFrom dataOntology:
              NucleatedCell
        ] [
            a owl:Restriction ;
            owl:allValuesFrom dataOntology:
              Nucleated Cell > ;
            owl:onProperty propertyOntology:
              has NucleatedCell
        ])
    ].
```

In this case, only individuals that have one or more `hasNucleatedCell` property instances of which all values belong to the class `NucleatedCell` would be assigned to the class `Eukaryota`. Conversely, any individual assigned to the class `Eukaryota` is inferred to have at least one instance of the property `hasNucleatedCell`, and its value must be a datum belonging to the class `NucleatedCell` (even if that datum is not identified in the knowledge base).

OWL DL reasoning is designed under the open world assumption (OWA). As such, we cannot syntactically restrict the number of properties or the number of instances of a property. We can, however, achieve its logical equivalency with cardinality semantics. For example,

```
dataOntology:Nucleus
    a owl:Class ;
    rdfs:subClassOf [
        a owl:Restriction ;
        owl:cardinality "1"^^xsd:nonNegativeInteger ;
        owl:onProperty propertyOntology:hasNucleolus
    ] .
```

The earlier discussion states that individuals that belong to the class Nucleus also belong to the class of all individuals that have at least one instance of the property hasNucleolus. The cardinality is 1, and thus if an individual has more than one instance of the property the reasoner will infer that the various objects that are the values of the property instances are semantically equivalent, that is, subject to an owl:sameAs relation connecting them. This will have cascading inference effects in the knowledge base. If other data contradict this, then the entire ontological model (knowledge base) is rendered inconsistent. Even if other data does not contradict this, inferring that two objects (the values of multiple hasNucleolus instances) are the same may yield surprising results. OWL DL's guarantees of completeness, validity, and monotonicity mean that factual errors cannot exist in a consistent knowledge base subject solely to first-order reasoning. But they may be uncovered when new data is added, a consequence that would lead to ontological inconsistency. This disruption is considered a good thing, because as more data is added and subject to a new round of ontology engineering, the resulting consistent ontology is a growing systematic endorsement of a true and consistent model of nature, whereas an inconsistent ontology forces resolution of the conflicting statements. Thus, there is a characteristic of convergence toward truth—a characteristic that is shared by the scientific method, even if it is only imperfectly realized.

Once the formal OWL ontology with named classes is constructed, we can subject it to tests for redundancy (two or more named classes inferred to be related by owl:equivalentClass), satisfiability (no class is equivalent to owl:Nothing), and closure (all data is realized to at least one class). We can use the formal ontology to the exclusion of the subsumption lattice, or we can use it as a marginal conceptualization on a subset of data. If the ontology includes necessary and sufficient conditions, we can use it to annotate new data or use new data to drive a new subsumption lattice to drive a new ontology.

Procedurally, the subsumption lattice drives the initial candidate classes to name and define; subsequent modeling allows us to derive formal named classes based on observed property conditions. If we do not axiomatically create classes de novo and assign individuals ex situ of the data, then we practice pure a posteriori, data-driven engineering. There are cases where we may value a priori knowledge, especially in cases of a priori *synthetic* statements in the Kantian sense (Kant 1787). For example, metadata and higher order logics may lead us to propose axiomatic classes that exceed the expressivity of OWL DL. In these cases, we pursue a *hybrid* of a priori and a posteriori engineering.

10.4.5 World Is Not First Order

An important caveat is that the success of this approach relies on the power and applicability of first-order description logics. First-order description logics are more expressive than the deep, static subsumption hierarchies so common in biology today. But our broader conceptualization of the world is not first order: *understanding* is reliant on higher order relations where the world cannot be neatly demarcated into mutually exclusive and exhaustive entities of individuals, properties, and classes. Even when given first-order constraints, OWL 1.1 DL has a complexity of NEXPTIME complete [NEXPTIME stands for nondeterministic exponential time: $O(2^{p(n)})$] (Cuenca Grau 2007), which is "harder" than NP complete (NP stands for nondeterministic polynomial time). OWL 2 DL is exponentially harder still (N2EXPTIME) (Kazakov 2008). Language-weakening constructs such as OWL 2 EL and OWL 2 QL can guarantee PTIME (polynomial time) complexity, but this comes at a cost (e.g., removing expressivity for the inverse relation or the universal quantifier *for all* ∀). Thus, we do not naively claim that FCA + OWL DL a posteriori ontology engineering is sufficient for all knowledge representation; rather, we offer it as an approach with quantifiable tractability for data-driven science.

10.4.6 Making Ontologies Web Savvy

SSWAP (Gessler et al. 2009) uses OWL ontologies as the foundation of a semantic web services platform. SSWAP supports the refactoring of ontologies into terms with separately dereferenceable URIs, such that dereferencing terms returns OWL DL statements about each term.[*] Current research is moving in two directions: (1) delivering on guarantees on completeness (del Vescovo et al. 2011), and (2) establishing a resolver service so that the original ontologies will not need to be separately refactored.

SSWAP is the underlying technology for The iPlant Collaborative's semantic web platform.[†,‡] It addresses limitations discussed in Section 10.2.3 by using ontology terms as resolvable URIs under a semantic web services protocol. Thus, ontologies—the conceptual constructs that describe data—are made instrumental in service description, discovery, invocation, and response. Semantic web services become the engagement layer for the underlying data and transformations upon them.

iPlant's semantic web platform uses transaction-time reasoning to match data types with service requirements, thereby allowing the construction of semantic pipelines of one service to the next. This matching is achieved with OWL reasoning. The entities of reasoning are service descriptions and their data, which are described with biological (or any third-party) ontologies. Thus, biological ontologies are being brought to bear on semantic web services a very real production environment. The missing link is that these biological ontologies are not themselves a posteriori reflections on data but exist as a priori knowledge management constructs. This makes data annotation burdensome and error prone, thus hindering the recruitment of data into the larger, semantic web service framework.

10.5 CONCLUSIONS AND THE PATH FORWARD

"Prediction is very difficult, especially about the future" (attributed to Niels Bohr [1885–1962]). We do not know how knowledge representation will evolve. We do know that current practices have limitations, and we discuss an approach to better formalize and quantify scientifically driven knowledge representation. Data-driven, a posteriori ontology engineering does not replace the a priori ontological construction of concepts from the empirical realization of individuals to classes (annotation). Rather, it advocates using measureable properties on individuals to simultaneously drive

---

[*] http://sswapmeet.sswap.info.

[†] http://www.iplantcollaborative.org/discover/semantic-web.

[‡] http://sswap.info.

the concept space, its topology, and its realization. Challenges remain, in the implementation of FCA, in its mapping to FOL, and in the ability of FOL DL to suitably capture the relevant facts and relations. Yet the basic position of letting the data dictate the knowledge representation has promising characteristics. The most fundamental is a computationally tractable monotonicity, such that by adding more data we become increasingly confident that the resultant model gains comprehensive breath. It is a characteristic of the scientific endeavor—the pursuit, explicit or otherwise—of unifying theories, of a model or set of models that are internally consistent and that together cover the data-driven concept space with an explanatory and predictive power.

A litmus test for any knowledge representation is the degree to which it enables actionable, evidence-based decision making. Data-driven, a posteriori ontology engineering attempts to enable this by allowing the data to define the conceptualization, rather than vice versa. One may posit that when data are rare, deduction is underpowered and induction is a necessary risk. Today, especially in biology, we are entering the age of commoditization of data generation: data are not rare. This is especially evident in areas such as high-throughput DNA sequencing (e.g., see the study by Nowrousian [2010]). With data, deduction gains power, whereas induction and abduction (positing explanatory hypotheses based on informed analysis of the data) gain focus. Thus, data-driven, a posteriori ontology engineering is aimed at extracting the signal from the data, utilizing scientific data and measurements on them to drive a new understanding of the data's interconnectedness and ultimately their integration.

ACKNOWLEDGMENTS

This chapter is based on the work supported by the National Science Foundation (NSF) under grant #0943879 and the NSF Plant Cyberinfrastructure Program (#EF-0735191).

ABBREVIATIONS

DL: description logic

FCA: formal concept analysis

FOL: first-order logic

GO: Gene Ontology

IRI: international resource identifier

LC: Library of Congress

MeSH: Medical Subject Headings

ORF: open reading frame

OWL: web ontology language

REST: representational state transfer

RDF: resource description framework

RDFS: RDF schema

SSWAP: simple semantic web architecture and protocol

UMLS: Unified Medical Language System

URI: uniform resource identifier

URL: uniform resource locator

W3C: World Wide Web Consortium

XML: extensible markup language

XSD: XML schema definition

REFERENCES

Akand E, Bain M, Temple M. 2007. Learning from ontological annotation: An application of formal concept analysis to feature construction in the gene ontology. In: *Proceedings of the Third Australasian Ontology Workshop (AOW-2007)*, Gold Coast, Australia. CRPIT eds. T. Meyer and A. Nayak, Vol. 85, 15–23.

Akand E, Bain M, Temple M. 2010. Learning with gene ontology annotation using feature selection and construction. *Applied Artificial Intelligence*, 24:5–38.

Alterovitz G, Xiang M, Hill DP, Lomax J, Liu J, Cherkassky M, Dreyfuss J, Mungall C, Harris MA, Dolan ME, Blake JA, Ramoni MF. February 2010. Ontology engineering. *Nature Biotechnology*, 28(2):128–130. PMID: 20139945.

Antezana E, Egaña M, Blondé W, Illarramendi A, Bilbao I, De Baets B, Stevens R, Mironov V, Kuiper M. 2009. The cell cycle ontology: An application ontology for the representation and integrated analysis of the cell cycle process. *Genome Biology*, 10:R58. doi:10.1186/gb-2009-10-5-r58.

Aristole. 350 BCE. On the Parts of Animals. Trans. by William Ogle. Kegan Paul, Trench & Co., London, 1882.

Ashburner M, Ball CA, Blake JA, Botstein D, Butler H, Cherry JM, et al. 2000. Gene ontology: Tool for the unification biology. Gene Ontology Consortium. *Nature Genetics*, 25(1):25–29.

Bada M, Livingston K, Hunter L. 2011. An ontological representation of biomedical data sources and records. In: *Proceedings Bio-Ontologies SIG at ISMB 2011*, Vienna, Austria, 75–78.

Bain M. 2002. Structured features from concept lattices for unsupervised learning and classification. In: *AI 2002: Proceedings of the 15th Australian Joint Conference on Artificial Intelligence*, eds. B. McKay and J. Slaney, LNAI 2557, 557–568, Berlin: Springer.

Bain M. 2003. Inductive construction of ontologies from Formal Concept Analysis. In: *AI 2003: Advances in Artificial Intelligence*. 16th Australian Conference on AI, Perth, Australia, December 3–5. Lecture Notes in Computer Science Volume 2903, http://link.springer.com/chapter/10.1007%2F978-3-540-24581-0_8, 88–99.

Belleau F, Nolin MA, Tourigny N, Rigault P, Morissette J. 2008. Bio2RDF: Towards a mashup to build bioinformatics knowledge systems. *Journal of Biomedical Informatics*, 41:706–716.

Cimiano P, Hotho A, Staab, S. 2005. Learning Concept Hierarchies from Text Corpora using Formal Concept Analysis. *Journal of Artificial Intelligence Research*, 24:305–339.

Corcho O, Fernández-López M, Gómez-Pérez A. 2007. Ontological Engineering: What are Ontologies and How Can We Build Them? In: *Semantic Web Services: Theory, Tools and Applications* IGI Global, ed. J Cardoso, 44–70.

Cuenca Grau, B 2007. OWL 1.1 Web Ontology Language. Tractable Fragments. http://www.webont.org/owl/1.1/tractable.html.

del Vescovo C, Gessler DDG, Klinov P, Parsia B, Sattler U, Schneider T, Winget A. 2011. Decomposition and modular structure of BioPortal ontologies. In: *Proceeding of International Semantic Web Conference ISWC 2011*, Bonn, Germany, LNCS 7031, 130–145.

Fernández-López M, Gómez-Pérez A, Juristo N. 1997. METHONTOLOGY: From Ontological Art Towards Ontological Engineering. Spring Symposium on Ontological Engineering of AAAI, 33–40, Stanford, CA: Stanford University.

Fernández-López M, Gómez-Pérez A. 2002. Overview and analysis of methodologies for building ontologies. *The Knowledge Engineering Review*, 17:129–156. doi:10.1017/S0269888902000462.

Fielding RT. 2000. Architectural styles and the design of network-based software architectures. Doctoral dissertation, University of California.

Formica A. 2006. Ontology-based concept similarity in Formal Concept Analysis. *Information Sciences*, 176:2624–2641.

Formica A. 2008. Concept similarity in Formal Concept Analysis: An information content approach. *Knowledge-Based Systems*, 21:80–87.

Ganter B, Stumme G, Wille R. eds. 2005. *Formal Concept Analysis: Foundations and Applications*, Springer-Verlag, Berlin Heidelberg.

Ganter B, Wille R. 1999. *Formal Concept Analysis*, Springer-Verlag, Berlin Heidelberg.

Gessler DDG, Schiltz GS, May GD, Avraham S, Town CD, Grant D, Nelson RT. 2009. SSWAP: A Simple Semantic Web Architecture and Protocol for semantic web services. *BMC Bioinformatics*, 10:309. doi:10.1186/1471-2105-10-309.

Gómez-Pérez A, Fernández-López M, Corcho, O. 2003. *Ontological Engineering*. Springer Verlag, Berlin Heidelberg.

Guo L, Huang F, Li Q, Zhang GQ. 2011. Power contexts and their concept lattices. *Discrete Mathematics*, 311(18–19):2049–2063.

Joslyn C, Paulson P, Verspoor KM. 2008. Exploiting term relations for semantic hierarchy construction. In: *Proceedings of the International Conference on Semantic Computing (ICSC 08)*, 42–49, IEEE Computer Society, Los Alamitos CA.

Kaiser T, Schmidt S, Joslyn C. 2008a. Concept lattice representations of annotated taxonomies. *Concept Lattices and their Applications, Lecture Notes in AI*, eds. S. B. Yahia, E. M. Nguifo, R. Belohlavek, Vol. 4923, 214–225, Berlin: Springer-Verlag.

Kaiser T, Schmidt S, Joslyn C. 2008b. Adjusting Annotated Taxonomies. *International Journal of Foundations of Computer Science*, 19(2):345–358.

Kant I. 1787. *Critique of Pure Reason*. Second edition. Translated by Norman Kemp Smith. England: Palgrave Macmillan.

Kazakov Y. 2008. RIQ and SROIQ are Harder than SHOIQ. In: *Proceedings of the 21st International Workshop on Description Logics* (DL2008), Dresden, Germany, May 13–16, 2008.

Linné, C. von. 1735. *Systema Naturæ per Regna Tria Naturæ, Secundum Classes, Ordines, Genera, Species, cum Characteribus Differentiis, Synonymis, Locis* [System of Nature, in Three Kingdoms of Nature, with Classes, Orders, Types and Species, with Differences of Character, Synonyms, Places] Joannes Wilhelm de Groot for Theodor Haak, Leiden.

McGuinness DL, van Harmelen F. 2004. OWL Web Ontology Language. Overview. http://www.w3.org/TR/owl-features.

Nowrousian M. September 2010. Next-generation sequencing techniques for eukaryotic microorganisms: Sequencing-based solutions to biological problems. *Eukaryotic Cell*, 9(9):1300–1310.

Simperl EPB, Tempich C. 2006. Ontology engineering: A reality check. R. Meersman, Z. Tari et al. (Eds.): OTM 2006, LNCS 4275, Springer-Verlag Berlin Heidelberg, 836–854.

Sjoblom T, Jones S, Wood LD, Parsons DW, Lin J, Barber TD, Mandelker D, Leary RJ, Ptak J, Silliman N, Szabo S, Buckhaults P, Farrell C, Meeh P, Markowitz SD, Willis J, Dawson D, Willson JK, Gazdar AF, Hartigan J, Wu L, Liu C, Parmigiani G, Park BH, Bachman KE, Papadopoulos N, Vogelstein B, Kinzler KW, Velculescu VE. 2006. The consensus coding sequences of human breast and colorectal cancers. *Science*, 314(5797):268–274. Epub 2006 Sep 7. PubMed ID: 16959974.

Stevens R, Aranguren ME, Wolstencroft K, Sattler U, Drummond N, Horridge M, Rector A. 2007. Using OWL to model biological knowledge. *International Journal of Human-Computer Studies*, 65:583–594.

Sure Y, Tempich C, Vrandecic D. 2006. Ontology engineering methodologies. In: *Semantic Web Technologies: Trends and Research in Ontology-based Systems*, 171–190. eds. J. Davies, R. Studer and P. Warren, Chichester, UK: John Wiley & Sons, Ltd.

Whittaker RH. 1969. New concepts of kingdoms of organisms. *Science*, 163: 150–161.

Woese CR, Balch WE, Magrum LJ, Fox GE, and Wolfe RS. 1977. An ancient divergence among the bacteria. *Journal of Molecular Evolution*, 9:305–311.

Woese CR, Kandler O, and Wheelis ML. 1990. Towards a natural system of organisms: Proposal for the domains Archaea, Bacteria, and Eucarya. *Proceedings of the National Academy of Sciences*, 87:4576–4579.

CHAPTER 11

Transforming Data into the Appropriate Context

Bill Howe

CONTENTS

Data acquired from sensors, experiments, or simulations or identified in external databases must be transformed into the right context to make discoveries. These transformations rely on establishing a common structure and semantics tailored for the given task. Top-down approaches to data integration that rely on establishing universal consensus about the meaning and interpretation of data—global ontologies, schemas, and metadata standards—are incomplete. At the frontier of research, such universal consensus is elusive, by definition. If the data are understood well enough to construct a permanent global schema, then it would not be the subject of research. We argue that there will always exist data that are not "born into compliance" with any such schema or standard. Heterogeneity, ambiguity, and quality issues cannot be "designed away" and must be tolerated at runtime by any practical and comprehensive scientific data management system. Consider search engines: Google and Bing do not refuse to index sources that are not compliant with W3C standards, and neither can, say, a geo-catalog service afford to refuse to import data that are not compliant with OGC standards [1].

In this chapter, we explore these issues, emphasizing bottom-up, "pay-as-you-go" approaches [2] that can tolerate heterogeneous, nonstandards-compliant data sources while still exploiting structure, patterns, and quality metadata when they exist. For example, a dataset may initially be represented as an opaque file admitting only simple operations: lookup by name, owner, date, or operations based on file extension. Later, text may be extracted, affording keyword search as a new operation. Similarly, a dataset from which rows and columns can be extracted admits structured query in a suitable language (e.g., SQL). A dataset from which coordinate pairs can be extracted admits basic two-dimensional visualization. A dataset from which locations on the Earth's surface can be extracted (e.g., [latitude, longitude] pairs) admits a mapping service. We advocate systems that can span this spectrum of capabilities rather than imposing strict format and metadata conditions on all data before admission into the system. Moreover, such a system should facilitate "upward mobility" of data, incrementally adding structure and semantics as they become apparent to admit higher-level services.

The goal of such systems should be to support *transformation* whenever possible rather than just search and retrieval. By "transformation," we intend tools that permit manipulation, restructuring, and semantic enrichment on-the-fly. For example, computationally expressive query languages such as SQL, XQuery, or SPARQL can be used to reorganize tabular, tree, and graph data, respectively, and the OPeNDAP protocol used in the Earth Sciences [3] can be used to restructure NetCDF files. In contrast, many scientific data systems settle for a "fetch and compute" model, requiring users to download data to their local environment before performing any nontrivial transformation or analysis tasks. As data sizes grow, this approach cannot scale. Instead, we advocate "pushing the computation to the data" rather than pulling the data to the computation.

This approach requires universal access to significant computing resources: clusters of computers and appropriate software to apply them to scientific data transformation and query problems. The advent and rapid growth of cloud computing plays a critical role in this vision; we will discuss its role.

Beyond (1) designing services to make fewer assumptions about structure and semantics and (2) pushing the data to the computation, a critical bottleneck is to reach a new class of users and allow them to be productive in this environment. Tools designed for IT professionals need to give way to tools designed for data analysts and domain experts. Moreover, the rate at which data is increasing is outpacing "reactive" software that can only respond to explicit requests by users. Instead, "proactive" software that can generate candidate results for review by scientists must become the norm.

This chapter is organized as follows. We will give an overview of the problem through a series of example application scenarios. We then describe a vision for the future, where data at any scale are universally accessible and manipulable in a common environment, and the software assists in focusing user attention. Next we describe relevant techniques and technologies that serve as "building blocks" in realizing this vision, followed by a discussion of our recent research on a database-as-a-service for science that attempts to integrate some of these building blocks. Finally, we will offer some conclusions and next steps.

11.1 PROBLEM OVERVIEW

Data volumes are doubling every year. While Moore's law has allowed capacity and some measures of performance to keep pace with this growth, the imbalances described by Amdahl's laws have become serious

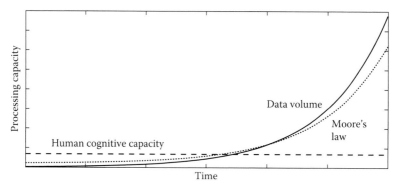

FIGURE 11.1 An illustration of a critical challenge of data-intensive science developed by Cecilia Aragon. Data volumes grow exponentially, and computing resources have arguably kept pace. However, human cognition has remained essentially flat. This gap represents the dominant challenge for data-intensive science in the long tail.

bottlenecks [4]. But the most significant bottleneck to data-intensive science is not between two levels of the memory hierarchy, but between the computer and the user. As data and computing grow exponentially, human cognition has remained essentially flat. Cecilia Aragon illustrates this idea as a plot similar to Figure 11.1 [5].

The effect of this gap on science is difficult to underestimate. At the UW eScience Institute, we routinely engage with researchers who tell a story of the form "In my dissertation, I analyzed N data points over five years. Last week, I collected $100N$ data points." At this scale, it is not just the algorithms that need to scale. The entire process of science needs to evolve. Consider the following examples, drawn only from oceanography and microbial biology:

Example: Environmental metagenomics: One researcher in environmental microbiology collects water samples by passing ocean water through a 0.2-μm filter, trapping particulate matter (Figure 11.2). These filters are then treated chemically and frozen at sea. Back at the lab, the samples are broken open, heated, and processed using polymerase chain reaction (PCR), which amplifies the nucleic acid present in the sample. The amplified samples are then sequenced, and the resulting reads are analyzed to determine a population profile. This profile helps answer two questions: "who is there?" and "what are they doing?"—that is, what organisms are present in the sample and what genes are they expressing?

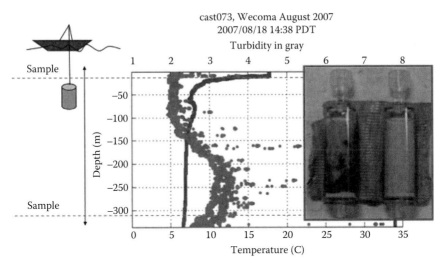

FIGURE 11.2 **(See color insert.)** Oceanographic metagenomics involves sequencing entire microbial populations under different environmental conditions. Here, two samples are collected at different depths in the water column from a single *cast* of a conductivity temperature depth (CTD) sensor package. The sample near the surface has significantly more particulate matter, as is visible once the water is passed through a 2-µm filter (at right). These samples are frozen at sea, sequenced on shore, and computationally compared to correlate environmental conditions with population characteristics. This area of research may involve significantly more data and more samples than genomics techniques involving a single organism.

Her dissertation involved the analysis of five such samples, characterizing a particular organism and its calcium "plates." The bottleneck to this work was the cost and time required to sequence the samples, and the process led to a new understanding of the formation of these plates. In her work as a postdoc, the cost and time of sequencing had dropped by orders of magnitude. As a result, there was significant interest in characterizing *entire microbial populations*, as opposed to an individual organism, across a variety of gradients: salinity, temperature, depth, diurnal cycles, distance from shore, nutrients, oxygen, and more. This kind of science required an enormous number of samples. Students were recruited to accompany the scientists on research cruises to help process samples at sea. Very quickly, and for the first time, the bottleneck became the *analysis* of data rather than the *acquisition* of data. The problem was not (only) one of computing performance, but one of new "data-intensive" interfaces and applications. In particular, at the Center for Coastal Margin Observation and Prediction [6], the new approach was to build a database and associated web application to enable students, researchers, and collaborators to perform ad hoc query and analysis over the data as opposed to relying on a single researcher to personally (and manually) inspect the sequence data, search public databases for genes and other features, and prepare visualizations. Without allowing multiple "sets of eyes" access to the data and providing access to scalable algorithms for processing it (through a relational database in this case), the value of the data, per byte collected, would have approached zero.

Example: In situ microscopy: We encounter similar stories from other researchers, even in other subfields of oceanography. The images in Figure 11.3 are acquired by a device called a flowcam [7]. This device is deployed in situ in the ocean and it automatically acquires images of (large) microscopic organisms and applies some image analysis techniques to recognize the organism based on its morphology. So the result of deploying this device is, potentially, a continuous stream of tagged images of organisms. Previously, acquisition of such images required a vessel expedition (at a cost of $20,000 per day or more), manual sampling at sea, and significant lab-based microscopy work after the cruise, all to collect a handful of images. With the level of automation offered by the flowcam device, the way in which the data are managed and analyzed is fundamentally transformed. Researchers draw inferences algorithmically from huge sets of related images, build three-dimensional reconstructions of organisms, and search for rare events.

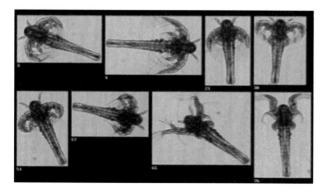

FIGURE 11.3 Images from a flowcam device, which can automatically acquire and tag images of organisms in situ, replacing a labor-intensive, manual process. The increase in data acquisition motivates new science as well as new tools.

Example: Observational oceanography: Multiple groups at the University of Washington are involved with environmental observation and modeling [8]. Datasets in this domain are fields defined over four-dimensional space-time of the form $(x, y, z, t) \rightarrow (v_0, v_1, ..., v_n)$. These datasets vary in which quantities are being measured or simulated, but also in which dimensions, if any, are fixed and implicit. For example, an immobile buoy equipped with a conductivity temperature depth (CTD) sensor at 1 m below the surface has the form $(t) \rightarrow$ (*salinity, temperature, elevation*), since (x, y, z) need not be recorded. A profiling station can move vertically in the water column, but not horizontally, producing datasets of the form $(z, t) \rightarrow$ (*salinity, temperature, elevation*). Autonomous underwater vehicles (AUVs) are free to swim in any direction, so all attributes must be recorded. Simulations may involve one-dimensional, two-dimensional, or three-dimensional spatial domains. The dependent variables $(v_0, v_1, ..., v_n)$ are also a source of heterogeneity. In the past, each project might focus on the deployment, maintenance, and data analysis associated with a single sensor type, often deployed in only a single area. Increasingly, projects are funded to synthesize measurements across multiple sensor types, across scales ("river to ocean"), and in multiple regions [6,9]. Individual researchers on such projects work with hundreds of data sources when they may have previously been accustomed to working with one or two.

Having encountered this situation in multiple contexts, we began to ask researchers how much time they spend "handling data" as opposed to "doing science." The mode answer among graduate students and postdocs has been 9 to 1. This number should give pause: taxpayer money flows through federal funding agencies to pay researchers to spend 90% of their time doing something that they do not consider to be science!

11.1.1 Long Tail Science

The problem we have described is especially acute in the *long tail* of science: the small labs and individual researchers who collectively produce the majority of scientific output. The term was used by Jim Downing in 2008 [10] to differentiate it from "big science," which is frequently used to describe the trend toward larger projects involving hundreds of principal investigators (PIs) and thousands of staff and students (e.g., the Large Hadron Collider [11]). Typically, big science projects have big IT budgets. Such projects are staffed with proficient programmers with access to significant computing resources. But in the long tail, researchers may have only recently begun using computation as an integral part of their science and have little support from programmers or system administrators.

The cost to computer scientists in engaging with long tail scientists on a lab-by-lab basis is prohibitive. Instead, there is a need to understand and attack only the cross-cutting requirements, while also considering deployment models that can reach those with a limited technical background. For example, a database product optimized for an array data model might be useful for long tail science [12], but the installation, configuration, schema design, and application development are at least as difficult as they are with a conventional database, and would therefore have little impact in the long tail. Instead, a web-based application with the database product deployed in the back end is more useful in the long tail. In general, there is a need to build centralized, multi-tenant tools and services that can serve the needs of hundreds or thousands of labs simultaneously. For example, myGrid [13], CrowdLabs [14], and the SQLShare system described in this chapter exhibit these properties.

11.1.2 How Big Is Big?

There is no absolute scale at which the challenges associated with data-intensive science are triggered. We find that our collaborators may find it as difficult to move from datasets that fit in memory to datasets that do not (e.g., 20 MB to 20 GB) as it is to move from a single server to a cluster of servers (e.g., 20 GB to 20 TB). In the long tail, researchers who need to "graduate" from files and spreadsheets to scalable databases or rewrite a script to exploit multicore parallelism are more common than those who need to manage petabytes. The potential impact of a solution that serves hundreds of labs is likely greater than the impact of a solution that serves a single lab with hundreds of times the data.

Michael Franklin, a professor at Berkeley and Director of the AMP Lab [15], defines "big data" as any size large enough to to make it difficult for analysts to complete their tasks [16]. For many researchers, this condition may be encountered at 100s of MB—some statistical techniques are fundamentally exponential, and even the approximations are high-degree polynomial (e.g., Markov chain Monte Carlo techniques for molecular sequence analysis, or de novo assembly of short-read sequences).

11.1.3 Problem Summary

We find that the ratio of time spent on science to the time spent performing such manipulation is approaching zero. We observe an increasing need for individual researchers to manipulate a large number of heterogeneous datasets to make discoveries. No assumptions can be made about the source, format, or semantics of the datasets; these issues must be addressed as part of the researcher's task. Any of these datasets may be too large for processing in main memory, making it difficult to rely on hand-coded scripts. Furthermore, the researchers we focus on are in the *long tail*, where they do not necessarily have access to programming support or significant computational resources. In this scenario, it is not enough to assume that data will be discovered, downloaded, and processed locally. The system must provide server-side, scalable transformation capabilities. In Section 11.2, we will articulate requirements for a system that addresses this problem.

11.2 VISION FOR THE FUTURE: COMPUTING WITH BAGS

Having found that our collaborators report spending up to 90% of their time "handling data" as opposed to "doing science," our vision for the future is to reduce this "data overhead" to zero—computing should never be the bottleneck to discovery. We distill four characteristics of an effective, comprehensive scientific data management system; the BAGS requirements:*

1. *Bottom-up*: There are no preconditions on the format, structure, or semantics of the data. Any data source may participate in the system, but those with richer and more reliable structure and metadata afford additional services. As additional structure is inferred and additional metadata attached over time, additional service becomes available—the system supports "incremental semantic enrichment." Datasets can be considered at multiple levels of abstraction simultaneously—as a file, as a table, as a time series, and as a trajectory. This approach is in contrast to top-down approaches that define a "master" standard or schema to which all data must conform.

2. *Algebraic*: Whenever possible, operations are composable and expressions are manipulable by the system. The system is free to optimize and evaluate the expression in any manner it sees fit. User-defined functions, when they are necessary, can be annotated with algebraic properties to afford optimization. Operations are algebraically

---

* The acronym connotes a loose and permissive structure, which is a hallmark of our approach.

closed whenever possible—the results of a transformation can be further transformed by the same operations.

3. *Global*: All data are addressable, at any granularity. There is no (logical) distinction between local data and remote data, "inside" and "outside" the system, or client and cloud. The internal components of a dataset (i.e., the header, a column, a row, a pixel) are similarly addressable—no dataset is an opaque stream of bytes.

4. *Speculative*: With nonzero probability, results are ready before anyone requests them. With access to all data, and some level of interoperability achieved, the system is equipped to *proactively and automatically derive new data products* for review by researchers. The choice of which operation f applies to which dataset X is informed by user preferences, heuristics, and statistical learning. For example, the system may learn that many users have applied the operation `trim _ reads` (a common transformation of *omics sequences) to datasets of type `fasta` (a common file format for sequence data) and decide to proactively apply the operation to a new `fasta` dataset upon upload. Similarly, entire workflows that perform specific tasks, predefined by analysts for use in one context, can be automatically applied to new datasets in other contexts. We refer to this capability as *semantic speculative execution*.

We consider these desiderata in more detail in the following subsections.

11.2.1 Bottom-Up: Inclusive Rather Than Exclusive

An intuitive approach to reducing data heterogeneity is to design and impose standards on how data can be represented, shared, described, and manipulated. In practice, such standards are at best incomplete—there will always be science data "born in the wild" that does not comply with any given standard. There are several reasons why this should be the case:

- A standard represents a shared consensus on the structure of semantics of data. But at the frontier of research, such a shared consensus is elusive by definition: if everyone agreed on a common interpretation of the data, then it would not be the subject of research.

- Given a single dataset, no one level of abstraction, and therefore no one standard, is suitable for all tasks. For example, measurements from a CTD sensor may simultaneously be viewed as a file, a table, a sequence of generic measurements, a sequence of salinity and temperature measurements, a time series, and many more. In high-performance simulation, a common problem is the design of an on-disk representation for snapshot and output files. The format that is efficient and convenient to write (time-ordered, column-oriented, and space-partitioned) is not always convenient nor efficient to read during analysis (e.g., large-scale multivariate visualization). Mass conversion of files post hoc is not often feasible, so typically software is written that can directly manipulate the "inconvenient" formats [17].

- The boundary between prototype and production is thin in scientific data analysis. Researchers use custom scripts coupled to third-party tools to construct new pipelines. Once these pipelines produce the desired result, there is little incentive to rewrite the application to produce standards-compliant XML instead of a convenient text format. It may be tempting to argue that scientists simply need to be more diligent, but since compliance will never be 100%, there will always be a need to tolerate this kind of heterogeneity. Also, the "freedom" to work with data in arbitrary ways using arbitrary tools is a unique and powerful characteristic of computational science—chemists cannot typically reengineer a centrifuge to meet new requirements in the way that a computational chemist can reengineer an algorithm. Policies that constrain this freedom should be applied conservatively.

- Flexible tools that use "lowest common denominator" data models, and tolerate or even embrace heterogeneity, are pervasive; it is difficult to ask scientists to give them up. Consider search engines: they do not refuse to index web content that does not conform to HTML 5. They attempt to extract whatever text they can to afford keyword search or attempt to extract geo-coordinates to position the resource on a map. Filesystems manage data as streams of bytes and support only read and write operations. Spreadsheets remain extremely popular in science because of their flexibility—anything that can be cast as rows and columns is potentially suitable. In contrast, relational databases impose significant structural constraints on their input and, unsurprisingly, remain underused in science [4,18,19]. The typical value proposition of a relational database imposes a significant barrier to entry: Once you clean, organize, normalize, and reformat your data, the database management system (DBMS) is an appropriate tool to manage it. This precondition is wholly prohibitive in science applications. In Section 11.4, we consider a different way of using relational databases that relaxes these constraints.

This argument suggests that the opportunity cost of designing standards to eliminate heterogeneous data is high, as such standards can never offer a complete solution. We argue that much of this effort is better spent on developing new tools (and democratizing existing tools) that can tolerate heterogeneity.

11.2.2 Algebraic: Reasoning about Computation

Codd observed in 1970 that the operations offered by all database products at the time could be reduced to compositions of about six different operations. If these operations could be formalized and made explicit, he reasoned, most of the work in accessing and manipulating data could be off-loaded to the database system instead of the programmer. Thirty years later, relational databases are an $18 billion dollar industry that powers most of the web.

The key insights of this model are *physical and logical data independence* and *algebraic cost-based optimization*. Physical data independence refers to the level of indirection between the physical representation on disk and the representation presented to the user. Logical data independence refers to the ability to present the underlying data through multiple logical interfaces (i.e., *views*). As Codd argued [20], "Activities of users at terminals and most application programs should remain unaffected when the internal representation of data is changed and even when some aspects of the external representation are changed." Algebraic cost-based optimization refers to the ability of the system to transform the user's query into an equivalent expression that is easier to evaluate. The intuition is that the database, and not the user, is best equipped to evaluate the query.

These techniques have been applied in other contexts besides relational databases, including stream processing [21–24], trees [25,26], graphs [27,28], and meshes [29]. More recently, large-scale data-intensive computing systems designed for sheer scalability and fault-tolerance such as MapReduce/Hadoop [30,31] have been retrofitted with algebraic reasoning capabilities [32–34] and data independence features such as views and incremental type inference [35]. This trend suggests that database features are being incrementally reincorporated into so-called NoSQL systems.

Although scientific computations cannot always be distilled into a composition of simpler operators [36], we find that the "irreducible" component of most data processing pipelines is only a small part of the overall computation. Filtering, joining, grouping, and sorting, operations at which this algebraic approach excels, tend to dominate the overall computation.

Relational databases, perhaps the most compelling example of these techniques, still remain remarkably underused in science. In Section 11.4, we will consider how to deliver this functionality in more effective ways.

11.2.3 Global: Universal Addressability

Artificial technical barriers complicate data sharing. External collaborators do not have access to your laboratory server. Datasets exist only on your former postdoc's laptop, which has not been turned on in 6 months. The data you need exists somewhere on the shared filesystem, but the organization of the directory structure housing it has been lost. Your inbox is the safest place to store data, thanks to routine backups and basic search capabilities.

Consider the requirements for a system that eliminated these barriers. All data would exist in a common logical namespace that transcended physical location. New data could be registered with the system easily, perhaps implicitly as a side effect of handling. Multiple access methods would be supported: keyword search, lookup by name, browsing by social network, usage history, popularity, similarity ("more like this"), and more. Explicit data movement would never be required. Implicit data movement, initiated by the system automatically in response to user tasks, would be optimized appropriately; data transfer would be minimized by moving computation to the data rather than moving data to the computation.

The idea of such a global address space was a primary motivation for the work that led to the Internet [37], but the fluid movement of both code and data between any two nodes—automatically in response to user requests— is not yet realized. Perhaps, the most common modality for data analysis on the Internet might be described as a "fetch-and-compute" model, where the analyst finds and downloads all relevant data (perhaps through remote interfaces) and performs the computation locally. Research on web service composition, orchestration, and choreography [38–40] provides somewhat more freedom, where data (i.e., the result of a service call) can be transferred directly between two remote sites. This capability opens a variety of optimization opportunities that have to be explored in the literature [38]. But web services are not typically portable; they can only be executed on the host that provides them. A third capability, which we advocate in this chapter, is to move the computation to the data rather than move the data to the computation.

These three models are illustrated in Figure 11.4. Three datasets, A, B, and C are located in three different hosts, `local`, `site1`, and `site2`, respectively. Each site is also equipped with a set of operations that it can perform. The host `site1` can apply two functions $g(*)$ and $h(*)$ to data, while both `site2` and `local` can only apply $f(*,*)$. The goal is to compute $f(g(B),C)$. We are told that the size of $g(B)$ is much larger than the size of either C or B. Three possible methods of computing the desired result are labeled (1) *Fetch*, (2) *Web Services*, and (3) *Cloud*. The *Fetch* method requires copying C and $g(B)$ to one's local environment. The *Web Services* method affords an optimization: $g(B)$ can be sent directly to `site2` through web service orchestration techniques, avoiding the cost of sending C to the local environment. Further, `site2` is perhaps more likely than `local` to have the computing capacity needed to evaluate $f(g(B),C)$ efficiently. The *Cloud* method affords another optimization: the function $f(*,*)$ can be sent to `site1`, along with C, and evaluated there. This mobility of both data and operations is the critical feature we see for a platform for large-scale data-intensive science.

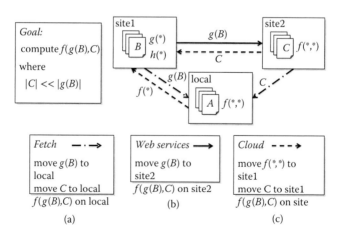

FIGURE 11.4   (a) An illustration of three methods of evaluating a distributed computation. The *Fetch* method requires all data to be downloaded locally before performing the computation. (b) The *Web Services* method affords an optimization, moving $g(B)$ to `site2`. (c) The *Cloud* method moves the computation $f(*)$ to `site1`, avoiding any expensive data transfer.

We labeled the third method to indicate a possible implementation that relies less on distributed, autonomously administered web services and more on centralized, scalable, elastic, and generic computing resources. Under this model, `site1` and `site2`, and possibly even `local`, can be considered separate logical environments within a shared physical environment such as Amazon Web Services [41]. In this context, movement of data can be handled automatically by the cloud manager, and it is feasible to move the computation (embodied as a virtual machine) nearer to the data. Private, local data that has not been uploaded to the cloud must also be addressable in a common namespace, but work remains to be done in the area of "client + cloud" architectures to span the gap between local and shared contexts [42].

An important source of difficulty in the example in Figure 11.4 is the binary operation $f(\star,\star)$. When a binary operation must be applied to arguments in different locations, there is no general way to avoid copying one of the arguments. Such operations are frequently overlooked in many service-oriented architectures [43] and often among users of NoSQL systems [44–46]. But as data-intensive science becomes increasingly integrative and comparative, these operations will be the norm rather than the exception. For example, the inability to conveniently and accurately compare two ocean models (through *regridding* [29]) is frequently considered to be a critical roadblock to progress [47]. Approaches to this problem have typically focused on developing standards for representing the data on disk along with application programming interfaces (APIs) for extracting subsets. Although such approaches help make data addressable, comparative operations—a form of spatial join, typically—can only be performed locally after downloading both datasets. The programming effort involved in implementing this join locally on a case-by-case is considerable; as a result, direct comparisons of ocean models are rare.

11.2.4 Speculative: Proactive Derivation of Data Products

As we have argued, the bottleneck to discovery in data-intensive science is rarely the availability of computational resources, but instead the relatively flat increase of human cognition (Figure 11.1). As data volumes and processing capacity grow exponentially, the opportunity cost of sitting idle waiting for the human operator to express the next query becomes increasingly wasteful. A system that can proactively mine the data, recommending interesting results for further analysis and review by the scientists, is becoming necessary. This capability is needed to focus the scientist's attention on the most promising results and thereby optimize for the limited resource. There simply is not enough time for an individual to personally scrutinize all the data they collect today (see the examples in Section 11.1). We refer to this capability as *semantic prefetching*.

We envision the following four levels of semantic prefetching:

1. Prefetch data that already exist based on a model of the user's task. This capability includes ranking results by similarity based on the user's recent browsing history ("more like this"), by results of interest to one's social network, by global popularity, and by compatibility with explicit user preferences.

2. Prefetch data that do not yet exist, but could be derived by applying an operation f to a dataset D. This capability involves ranking (operation, dataset) pairs using similar techniques as level 1.

3. Prefetch data that could be derived by applying a tree of operations, where each leaf is a single dataset. This capability is simply the transitive closure of level 2. We differentiate it because the potential search space is exponentially larger.

4. Prefetch data that cannot be derived from the current holdings, but could be derived by collecting more data D^* and applying some f. This level requires that data acquisition itself be at least partially automated.

This last level warrants the following examples.

- Observational oceanographers sometimes make use of the concept of *vessels of opportunity*—vessels that are not under direction of the chief scientist, but that are in the right place at the right time to take an important measurement or sample. Empowering the system to automatically identify these opportunities and issue the request could significantly increase the value of data collected. Currently, scientists may not identify these situations until it is too late to exploit them.

- Citizen science projects provide volunteers with enough training and equipment to collect data on behalf of a research project. For example, volunteers for the NatureMapping project [48] record observations of wildlife species in populated areas, dramatically improving the quality of the range maps used to inform public policy. A system that could proactively identify regions and species for which little data exists and issue standing requests for additional observations could amplify the effectiveness of these projects and make the experience more rewarding for the volunteers.

- The term *adaptive sampling* refers to the capability of some sensors to receive commands while operating autonomously in the field. For example, an AUV may adapt its trajectory based on commands issued from shore, or an atmospheric radar may rotate its antenna to face an incoming storm [49]. A system that can direct these resources automatically based not only on current observations but also on the value of the potential derived products could increase the return on investment of deployment.

For all four levels, the search space is enormous and cannot be searched directly. Instead, we need to identify promising results by modeling importance to the user. While quantifying importance is difficult, a simplifying factor is that it is not necessarily important that the model is accurate when first deployed, as long as it is equipped to incorporate human feedback and learn statistically what is important and what is not. We envision an interface where scientists can browse the data products derived the previous night over their morning coffee, selecting a few for further review while rejecting the majority. This interaction provides a strong signal on which to base a ranking algorithm.

11.3 RELATIONSHIP TO OTHER TECHNOLOGIES

The vision we have articulated is not out of reach. The foundational components are already emerging. Cloud computing provides a platform for scalable, ubiquitous, democratized access to arbitrary storage and computational resources. Scientific workflow systems are raising the level of abstraction for expressing computational pipelines, which affords the kind of reasoning required to automate their use. Scalable databases and related data-intensive computing systems are experiencing a renaissance.

11.3.1 Cloud Computing

The cloud offers a compelling alternative to deploying data management services and applications for use by the long tail. Besides a basic economic argument stemming from the economy of scale achieved by commercial cloud vendors, the cloud offers novel usage modalities for collaboration and sharing:

- Providing a collaborator with access to your data requires simply changing the permissions—no data need be moved, no software exchanged, and no firewalls punched.

- A virtualized shared development environment, outside the security jurisdiction of any one collaborator's institution, can be provisioned in a few minutes. This kind of shared environment cannot be set up without reciprocal exchange of accounts—something system administrators are typically resistant to.

- Virtualization supplies the code mobility motivated by the example in Figure 11.4.

- The cloud helps distribute the cost of data hosting among those who consume the data. Amazon will host a public dataset at no cost if it generates sufficient traffic. In this way, the cost of hosting an important dataset can therefore be shifted to the consumer rather than the provider. Even ignoring Amazon's free hosting program, the data provider need not provision a parallel computational environment required to operate on large datasets. This cost is borne by the consumers.

- There is a trend toward developing *data markets* [50], enabled by cloud computing. Such a facility may allow scientists to charge directly for access to their data.

11.3.2 Service-Oriented Architectures

Web services have had a significant impact on the way distributed applications are architected. The benefits in this context include the potential for automatic composition, orchestration, and choreography of distributed computations [38,43,40]. However, current efforts relying on service oriented architectures (SOAs) for science (c.f., IOOS [43]) over-rely on top-down standards that are designed primarily for professional software engineers rather than working scientists. Further, SOAs do not exploit a critical advantage of scientific data management: it is rarely updated destructively. This fact allows centralized architectures that copy the data to a single (large-scale) platform such as the cloud to become feasible.

11.3.3 Scientific Workflow

Workflow systems propose to raise the level of abstraction for computational tasks by providing a library of typed, reusable components, some visual programming interfaces, and a strong emphasis on provenance, sharing, and reproducibility. Computations are typically modeled as directed graphs capturing data flow.

The Kepler project [51,52] has provided a foundational platform with which to study the research and practice of scientific workflow systems and has enjoyed many users of success.

The myGrid project [13] emphasizes sharing of workflows, but offers no support for data-intensive (i.e., parallel) computation and does allow seamless sharing of data. HubZero [53] is a platform for collaboration and sharing tools, especially simulations. The "autocompletion" techniques for mashups [54] and workflows [55] can be used to derive a chain of steps from a corpus, but assume that precise, accurate metadata exists. HubZero provides community features such as user ratings and comments, but its primary research contribution is a mechanism for wrapping a script with a web-based graphical user interface (GUI). Workflow provenance systems track executions of tasks for forensic analysis, documentation, and reproducibility [56–60].

We assume that provenance will be captured and stored using these techniques, but we use the provenance in a novel way—to inform semantic prefetching (see Section 11.2).

The weakness of current workflow systems is that the data is a second class citizen; the systems are designed around the *Fetch* model we described in Section 11.2.

11.3.4 Databases

Relational databases offer the algebraic languages and optimization capabilities we seek. However, scalability has historically been limited to a few nodes (although this is changing due to pressure from Hadoop-based systems). Large-scale database deployments typically required a full-time database administrator for reliable operation. Such a resource is prohibitively expensive in the long tail of science. Moreover, relational databases overemphasize the importance of a fixed, carefully engineered schema for performance and to ensure data quality. In Section 11.4, we describe a system that relaxes this constraint while preserving the use of the relational model, algebra, and query languages.

Database-as-a-service systems [61–63] are gaining popularity due to their ability to lower the cost of deploying and managing a database.

Maintenance tasks and basic server administration are handled by the cloud provider.

11.3.5 Dataspaces

The concept of dataspaces [64,2,65,28] provided a model for our interest in querying ad hoc data without relying on a predefined schema [28]. However, the incremental, "pay-as-you-go" process for data curation and knowledge engineering activities emphasized in the dataspace does not go far enough. In many applications, we find an urgency in extracting results from ad hoc datasets *right now*. We therefore emphasize speculative, proactive processing of data to help focus users' attention and reduce the end-to-end time to insight.

11.3.6 Data Integration

Data integration techniques (c.f., [66,67]) will play a significant role in the proposed vision, but the problem as it is typically formulated overemphasizes the existence of well-defined schemas.

11.3.7 Visualization

The VizDeck system (see Section 11.4) builds on seminal work on automatic visualization of relational data using heuristics related to visual perception and presentation conventions [68,69]. More recent work on intelligent user interfaces attempts to infer the user's task from behavior and use the information to recommend visualizations [70]. Dörk et al. derive coordinated visualizations from web-based data sources [71]. Mashup models have been studied in the database community [54,72,73], but do not consider visualization ensembles and assume a pre-existing repository of mashup components.

11.4 STATE OF THE ART

In this section, we describe current work on components of the vision we have described, focused on leveraging cloud computing to deliver easy and powerful transformation and visualization capabilities to researchers in the long tail of science.

11.4.1 SQLShare: Smart Services for Ad Hoc Databases

In the SQLShare project, we consider the management of *ad hoc databases*. An ad hoc database is a collection of tables with possibly unknown relationships gathered to serve a specific, often transient, often urgent, purpose. Consider the following examples:

- A scientist assembles an ad hoc database of recent experimental results to prepare a paper or proposal.

- Emergency workers responding to a natural disaster assemble an ad hoc database from lists of addresses of nearby schools, locations of resources (e.g., ambulances), and contact information for emergency workers.

- A consulting business analyst assembles an ad hoc database from a set of financial spreadsheets provided by management for a short-term engagement.

- A paralegal assembles an ad hoc database of call logs, transaction records, and other potential evidence to prepare for litigation.

Spreadsheets remain more popular than relational databases in such scenarios, despite advantages of set-based manipulation, declarative query, and scalability.* But we find that the size and number of datasets involved in a typical ad hoc database is outgrowing the "manual" interaction modalities offered by spreadsheets. An analyst can comfortably use copy-and-paste, sorting, and explicit chart creation to manipulate one or two spreadsheets with O(100) rows each, but these techniques are useless when working with hundreds of spreadsheets with O(10k) rows each.

We find that the key barriers to relational database management systems (RDBMS) adoption in this context include the following:

- The costs associated with installation, configuration, and tuning of commercial database management systems are prohibitive, especially without access to dedicated IT staff or resources.

- The initial investment required to design and populate a database is also costly. Developing a definitive database schema for an urgent project, where knowledge is undergoing daily or hourly revision, is difficult.

---

* Enrique Godreau of Voyager Capital estimated that 90% of all business data is maintained in spreadsheets [74].

- The corpus of data for a given project accretes over time, with many versions and variants of the same information and little explicit documentation about connections between datasets and sensible ways to query them.

- We corroborate findings that novices can become productive with SQL without formal training [48,60,75], but we find that these users need a set of relevant, high-quality example queries to get started.

- Interactive query must be augmented with decision-making tools such as visualization and statistical tests. We use the term *data product* to refer to any digital artifact derived from data in an ad hoc database—query recommendations, visualizations, mashups, and statistical models.

In each of these cases, the scarce resource is user attention. While data sources have exploded in size and complexity, human cognitive capacity remains essentially constant. The problem is exacerbated in the context of an ad hoc database, where time and resources are constrained. As a result, an ad hoc database management system must seek to eliminate or automate as many of these tasks as possible—setup, ingest, query synthesis, and data product creation. For those tasks that cannot be eliminated, the system must optimize for user attention by *proactively* creating and recommending candidate data products for review by the user. We consider two such services in this section.

Are relational models and languages fundamentally unsuitable for the requirements of ad hoc databases? We performed a preliminary experiment to answer this question. In the context of a collaboration with microbiologists accustomed to working with spreadsheets, we provided a simple cloud-based service for working with ad hoc databases called SQLShare* [18] and found it to have a transformative effect.

SQLShare allows users to upload their data and immediately query it using SQL—no schema design, no reformatting, and no DBAs. Queries can be named, saved, shared, and commented on—anything you can do with a YouTube video, you can do with a saved query. Each saved query is also registered as a view, allowing composition and reuse.

At an early demonstration of our platform, the results of a simple SQL query written "live" in less than a minute caused a postdoc to exclaim

---

* http://escience.washington.edu/sqlshare.

"That took me a week!"—meaning that she had literally spent a week manually cleaning and pre-filtering a handful of spreadsheets, then using copy-and-paste to compute what was essentially a join. Within a day, the same postdoc had derived and saved several new queries. The experience was not isolated: the director of her lab has contributed several of her own SQL queries. She has commented that the tool "allows me to do science again," explaining that she felt "locked out" from personal interaction with her data due to technology barriers, relying instead on indirect requests to students and IT staff. She is not alone—several researchers we have surveyed informally have reported that the ratio of time they spend "manipulating data" as opposed to "doing science" is up to 9 to 1.

In the remainder of this section, we sketch our model for ad hoc databases and two "smart" services aimed at minimizing user attention through automation and speculative data product generation. We conclude with a discussion of the relevant related work that can be applied to this topic as well as some next steps for our own research.

11.4.1.1 Data Model

We model an ad hoc database as a graph $(D \cup P, \Sigma)$, where D is a collection of *datasets*, P is a collection of *data products*, and edges in Σ represent a dependency relationship used to reason about workflow provenance [76]. That is, if a data product p is derived from a dataset d, then $(p, d) \in \Sigma$. A dataset is a triple (R, n, M) where R is a relation, n is the name of the relation, and M is a set of metadata represented as key-value pairs. We explicitly include metadata as a part of the model to support services that intentionally refer to sets of relations—search, permissions, ownership, download, and annotation. For example, one can download all relations where source = myworkflow. Similarly, a data product is a triple (X, n, M), where X is an uninterpreted binary object.

The relations can be queried as usual using SQL. A view $Q(D)$ introduces a new relation into the system as well as a dependency edge for every relation mentioned in Q. A data product represents (typically) a nonrelational digital artifact such as a visualization, the result of a statistical test, or a trained model.

No schema: We do not allow *CREATE TABLE* statements; tables are created directly from the columns and types inferred in (or extracted from) uploaded files. Just as a user may place any file on a filesystem, we intend for users to put any table into the SQLShare "tablesystem," not just those that comply with a predefined schema.

Tolerance for structural inconsistency: Files with missing column headers, columns with nonhomogeneous types, and rows with irregular numbers of columns are all tolerated. We find that data need not be pre-cleaned for some tasks (e.g., counting records) and that SQL is an adequate language for many data cleaning tasks.

Append-only, copy-on-write: We do not allow destructive updates. Users insert new information by uploading new datasets. These datasets can be appended to existing datasets if the schemas match. Name conflicts are handled by versioning—the conflicting dataset is renamed, and the dependency graph is updated appropriately.

Simplified views: We find views to be underused in practice. We hypothesize that the solution may be as simple as avoiding the awkward CREATE VIEW syntax. In SQLShare, view creation is a side effect of querying—the current results can be saved by simply typing a name. This simple user interface (UI) adjustment appears to be effective—267 views have been registered in the system from just a handful of users.

Unifying views and tables: Our data model consists of a single entity: the *dataset*. Both logical views and physical tables are presented to the user as datasets. By erasing the distinction, we reserve the ability to choose when views should be materialized for performance reasons. Since there are no destructive updates, we can cache view results as aggressively as space will allow. However, since datasets can be versioned, the semantics of views must be well defined and presented to the user carefully. We find both snapshot semantics and refresh semantics to be relevant, depending on the use case. Currently, we support only refresh semantics.

11.4.1.2 Automatic Starter Queries

Analysts who assemble ad hoc databases frequently do not have significant SQL expertise, but we find that providing a rich set of examples is sufficient to empower nonexperts to use SQL for data analysis [18]. This finding should not be surprising: Many public databases include a set of example queries as part of their documentation [77,78], suggesting that the strategy is effective. We adopt the term *starter query* to refer to a database-specific example query, to distinguish it from an example that merely illustrates SQL syntax abstractly.

In this context, starter queries must be derived from the data alone—ad hoc databases have no explicit foreign keys (initially) from which to derive join relationships [79], and no query logs from which to mine frequent query patterns [80].

Analysts use starter queries in several ways. They *browse* them to learn basic idioms of SQL (joins, in particular, are often frequently unfamiliar). They *execute* them to explore the data itself. They *modify* them by adding or removing *snippets* [81]: predicates in the WHERE-clause, tables in the FROM-clause, and columns in the SELECT-clause. They *compose* them to derive new queries—each saved query is automatically registered as a view and is referenceable as a table. They *share* them during collaboration—each query (and its result) has a unique URL that can be bookmarked or e-mailed to colleagues, who can then add comments, derive their own queries, and so on. Our preliminary results suggest that this query-oriented collaborative analysis is an effective model for working with ad hoc databases. To bootstrap this model, we need only "seed" the collaboration with a set of starter queries. In our existing system, these example queries are provided by database experts.

When we first engage a new potential user with our current SQLShare prototype, we ask them to provide us with (1) their data and (2) a set of questions, in English, for which they need answers. This approach, informed by Jim Gray's "20 questions" requirements-gathering methodology for working with scientists [4], has been remarkably successful. Once the system was seeded with these examples, our users were able to use them to derive their own queries and become productive with SQL.

Our approach to the problem of automatically deriving starter queries for an ad hoc database is to (1) define a set of heuristics that characterize "good" example queries, (2) formalize these heuristics into quantities we can calculate from the data, (3) develop algorithms to compute or approximate these features from the data efficiently, (4) use examples of "starter queries" from existing databases to train a model on the relative weights of these features, (5) evaluate the model on a holdout test set, and (6) deploy the model in the production SQLShare application. In this context, we are given just the data itself: In contrast to existing query recommendation approaches, we cannot assume access to a query log [81], a schema [7], or user preferences [80]. We explore heuristics for identifying four idioms: union, join, select, and group by. To illustrate the approach, we describe our method for join.

Detecting joins: To detect join candidates, we derive a scoring function by combining the following set of heuristics:

1. A foreign key between two columns suggests a join.

2. Two columns that have the same active domain (i.e., the same set of unique values) but different sizes suggest a 1:N foreign key and a

good join candidate. For example, a fact table has a large cardinality and a dimension table has a low cardinality, but the join attribute in each table will have a similar active domain.

3. More generally, two columns with a high similarity offer evidence in favor of a join.*

4. If two columns have the same active domain and that active domain has high entropy (large numbers of distinct values), then this is evidence in favor of a join. Conversely, if both attributes have small entropy, then this is evidence against a join.

The preceding join heuristics 1 to 4 all involve reasoning about the union and intersection of the column values and their active domains. Heuristic 1 identifies foreign key relationships. Given two columns x and y (modeled as bags), a foreign key relationship relationship exists if $x \subset y$. Heuristic 2 adds the condition that $|x| \ll |y|$. Heuristic 3 relaxes the strict subset condition and invokes Jaccard similarity with bag semantics: $\frac{|x \cap y|}{|x \cup y|}$. Heuristic 4 sets conditions on the relative sizes of the active domains: $|\pi(x)| \ll |x|$, where $\pi(x)$ indicates the set derived by removing duplicates from x. For example, a fact table has a large cardinality and a dimension table has a low cardinality, but the join key will typically have a highly similar active domain. These heuristics can therefore be distilled to a set of features that can be used to train a model. For each pair of columns x, y in the ad hoc database, we extract each feature in Table 11.1 for both set and bag semantics. We extract these features for a training set derived from examples of starter queries available on the web and then test the model on a different set.

TABLE 11.1   Features Extracted to Estimate the Joinability

Feature	Expression
Max/min cardinality	$\max/\min(|x|,|y|)$
Cardinality difference	$\mathrm{abs}(|x| - |y|)$
Intersection cardinality	$|x \cap y|$
Union cardinality	$|x \cup y|$
Jaccard similarity	$\dfrac{|x \cap y|}{|x \cup y|}$

---

* An important exception is the case of an "autoincrement" column that is sometimes used as a key and may have no relationship to another autoincrement column.

In preliminary experiments, we use an alternating decision tree (ADTree) [82] on the example queries from the Sloan Digital Sky Survey [78] and then test the model on the example queries from a completely different database, the Gene Ontology Database [77]. Our results are promising: We achieved a recall and precision of over 90% [18].

11.4.2 Self-Organizing Dashboards

To support visual data analytics over ad hoc databases, we are developing VizDeck, a web-based visualization client that uses a card game metaphor to assist users in creating interactive visual dashboard applications in seconds with zero programming. VizDeck generates a "hand" of ranked visualizations and UI widgets, and the user plays these "cards" into a dashboard template, where they are automatically synchronized into a coherent web application that can be saved and shared with other users (Figure 11.5).

FIGURE 11.5 After selecting the desired thumbnails, the user can interact with the dashboard they have created. In this dashboard, the user can select the type of microorganism at the upper left and review the effect. In particular, the clusters in the scatter plots at upper right and lower correspond to microorganism populations.

By manipulating the hand dealt—playing one's "good" cards and discarding unwanted cards—the system learns statistically which visualizations are appropriate for a given dataset, improving the quality of the hand dealt for future users.

Users are increasingly relying on exploratory visual analytics to quickly identify patterns and generate hypotheses for deeper investigation. With our collaborators, we originally proposed to use sophisticated tools such as Tableau [83], but found that scientists were not able to self-train quickly and reverted back to simpler tools such as Excel. The reason seems to be a "blank canvas" effect—even though Tableau and similarly powerful tools make it easy to craft a complex visualization, the user still has to have a preformed idea of what they want to see before they begin (which attributes, which chart type, etc.).

We consider a different approach. Informed by the statistical properties of the underlying dataset, we proactively generate and rank candidate visualizations (and UI widgets) called *vizlets* and display them to the user in a grid. These vizlets can be browsed, discarded, or *promoted* as part of a coordinated, interactive, shareable dashboard designed to tell a particular scientific story. To construct a coordinated dashboard, the user need not perform any explicit configuration; the constituent vizlets automatically respond to events triggered by other vizlets.

Dashboards can be saved and shared among collaborators simply by exchanging URLs. Server-side, a saved dashboard, is represented as a *replay log* of the actions a user took to construct it. Besides affording a useful undo/redo feature, this technique allows a collaborator to review the steps taken by the original author to create the dashboard, which we anticipate will improve cross-training and communication between users.

11.5 CONCLUSIONS AND PATH FORWARD

We have discussed problems and future solutions associated with data transformation in the sciences. Observing that the overhead of working with data dominates the science being done with it, we analyzed the problem in terms of several examples in oceanography and microbiology. We rejected solutions that rely on top-down global consensus on the structure or semantics of data, finding that such constructs are too rigid in practice and ignore the "forward edge" of science data—the data that do not conform to any accepted model, by definition. However, when such structure and semantics exist, they should be exploited. We cannot settle for opaque, uninterpreted files, but nor can we do nothing until comprehensive schemas are defined and universally applied. Instead, we need to take the "middle path" that offers appropriate services across the spectrum of "data maturity." Luckily, many of these services already exist. We have powerful tools for working with documents, tables, trees/XML, graphs, images, and meshes. These tools need to be offered in a centralized platform and linked together.

We argued that the fetch-based model, where researchers are assumed to gather all the data they need, download it to their local computing environments, and process it there, is infeasible. Data sizes are becoming too large, and researchers do not always possess the programming skill required to answer their questions. Instead, we must push the computation closer to the data, implying that both data and computation must be mobile (Figure 11.4)

We distilled the problem into four desiderata for a comprehensive data transformation system, represented by the acronym BAGS: Bottom-up, Algebraic, Global, and Speculative. The data models must be designed *bottom-up* so as not to exclude the exceptional datasets that are not well behaved with respect to any given standard. The interfaces needed to express these transformations must be *algebraic* so that they can be composed to express complete pipelines and so the system can evaluate them in multiple ways, tailoring the execution to the properties of the data and the current load of the system. The system must support *global* addressing—physical locations must be transparent to users, and importing or registering data within the system must be automated whenever feasible. We advocate "client + cloud" approaches that span the gap between the two platforms [42]. Finally, we argued that scientists' attention is the limiting resource, rather than computational power. This "attention gap" (Figure 11.1) is a prominent concern—there is an acute need to offload more work to the computer to make up the gap. The vision is one of scientists reviewing hypotheses and candidate results found by the computer, selecting promising leads for further analysis and eventual publication.

11.5.1 Path Forward

The components of the proposed vision are already emerging. Cloud computing has been proven in the marketplace and is here to stay. In 10 years, it will seem like a quaint idea to purchase one's own hardware. This trend will enable new usage modalities and drive costs down. More importantly, this trend will increasingly lead to data being "born" in the cloud. There will no longer be a discussion of how, when, and whether to migrate one's data to the cloud; it will already be there. As a result, the data will never move; computations, represented as virtual machines, will be launched near to the data, operate on it to derive new results, and then shut down. This trend is already under way: In China, an enormous sequencing center is simultaneously building a state-of-the-art data center, recognizing that the data acquisition and data management can no longer afford to be separate activities [84].

REFERENCES

1. Open geospatial consortium. http://www.opengeospatial.org/.
2. M. J. Franklin, A. Y. Halevy, and D. Maier. From databases to dataspaces: A new abstraction for information management. *SIGMOD Record*, 34(4): 27–33, December 2005.
3. OPeNDAP. Open-source project for a network data access protocol. http://opendap.org/, 2005. Viewed on August 2006.
4. J. Gray, D. T. Liu, M. A. Nieto-Santisteban, A. S. Szalay, D. J. DeWitt, and G. Heber. Scientific data management in the coming decade. *CoRR*, abs /cs/0502008, 2005.
5. C. Aragon. Scientist-computer interfaces for data-intensive science. Microsoft eScience Workshop, 2010.
6. Center for Coastal Margin Observation and Prediction. http://www .stccmop.org.
7. F. I. Technologies. Submersible flowcam—In-situ monitoring. http://www .fluidimaging.com/products-submersible.htm.
8. Pacific northwest environmental forecasts and observations. http://www .atmos.washington.edu/mm5rt/.
9. B. Gegosian. Ocean observing initiative, 2005. http://www.oceanleadership .org/ocean_observing.
10. P. Murray-Rust and J. Downing. Big science and long-tail science. http:// blogs.ch.cam.ac.uk/pmr/2008/01/29/big-science-and-long-tail-science/, term attributed to Jim Downing.
11. Large Hadron Collider (LHC). http://lhc.web.cern.ch.
12. P. G. Brown. Overview of SciDB: Large scale array storage, processing and analysis. In *Proceedings of the 2010 International Conference on Management of Data*, SIGMOD '10, pages 963–968, New York, NY, USA, 2010. ACM.
13. myGrid. http://www.mygrid.org.uk.
14. P. Mates, E. Santos, J. Freire, and C. T. Silva. CrowdLabs: Social analysis and visualization for the sciences. In J. B. Cushing, J. C. French, and S. Bowers, editors, *SSDBM*, volume 6809 of *Lecture Notes in Computer Science*, pages 555–564, 2011. Springer.
15. M. Franklin. Algorithms, machines, and people. http://amplab.cs.berkeley.edu/.
16. M. Franklin. Making sense at scale with algorithms, machines, and people. Keynote, Microsoft Cloud Futures Workshop, http://research.microsoft .com/apps/video/?id = 150301.

17. B. Howe and D. Maier. Logical and physical data independence for file-based scientific applications. *IEEE Data Engineering Bulletin*, 27(4):30–37, 2004.

18. B. Howe, G. Cole, E. Souroush, P. Koutris, A. Key, N. Khoussainova, and L. Battle. Database-as-a-service for long tail science. In *SSDBM '11: Proceedings of the 23rd Scientific and Statistical Database Management Conference*, 2011.

19. M. Stonebraker, J. Becla, D. DeWitt, K.-T. Lim, D. Maier, O. Ratzesberger, and S. Zdonik. Requirements for science data bases and SciDB. In *Fourth Biennial Conference on Innovative Data Systems Research (CIDR)—Perspectives*, 2009.

20. E. F. Codd. A relational model of data for large shared data banks. *Communications of the ACM*, 13(6):377–387, 1970.

21. D. J. Abadi, D. Carney, U. Çetintemel, M. Cherniack, C. Convey, S. Lee, M. Stonebraker, N. Tatbul, and S. Zdonik. Aurora: A new model and architecture for data stream management. *VLDB Journal: The International Journal on Very Large Data Bases*, 12(2):120–139, September 2003.

22. Aleri. http://www.aleri.com/index.html.

23. StreamBase. http://www.streambase.com/.

24. S. Zdonik, N. Jain, S. Mishra, A. Srinivasan, J. Gehrke, J. Widom, H. Balakrishnan, M. Cherniack, U. Cetintemel, and R. Tibbetts. Towards a streaming SQL standard. In *Proceedings of the 34th International Conference on Very Large DataBases (VLDB)*, August 2008.

25. S. Boag, D. Chamberlin, M. Fernandez, D. Florescu, J. Robie, J. Siméon, and M. Stefanescu. XQuery 1.0: An XML query language. W3C Working Draft, June 2001.

26. P. Fankhauser, M. Fernandez, A. Malhotra, M. Rys, J. Siméon, and P. Wadler. The XML query algebra, February 2001. http://www.w3.org/TR/2001/WD-query-algebra-20010215.

27. A. Harth and S. Decker. Optimized index structures for querying RDF from the web. In *LA-WEB*, page 71, Washington, DC, USA, 2005. IEEE Computer Society.

28. B. Howe, D. Maier, N. Rayner, and J. Rucker. Quarrying dataspaces: Schemaless profiling of unfamiliar information sources. In *Data Engineering Workshop, 2008. IEEE 24th International Conference on ICDEW 2008*, pages 270–277, 2008.

29. B. Howe and D. Maier. Algebraic manipulation of scientific datasets. In *VLDB '04: Proceedings of the 30th International Conference on Very Large Data Bases*, Toronto, Ontario, CA, 2004.

30. J. Dean and S. Ghemawat. MapReduce: Simplified data processing on large clusters. In *Proceedings of the 6th USENIX Symposium on Operating Systems Design & Implementation (OSDI)*, 2004.

31. Hadoop. http://hadoop.apache.org/.

32. R. Chaiken, B. Jenkins, P.-A. Larson, B. Ramsey, D. Shakib, S. Weaver, and J. Zhou. Scope: Easy and efficient parallel processing of massive data sets. In *Proceedings of the 34th International Conference on Very Large DataBases (VLDB)*, pages 1265–1276, 2008.

33. Hive. http://hadoop.apache.org/hive/.

34. C. Olston, B. Reed, U. Srivastava, R. Kumar, and A. Tomkins. Pig latin: A not-so-foreign language for data processing. In *SIGMOD'08: Proceedings of the ACM SIGMOD International Conference on Management of Data*, pages 1099–1110, 2008.

35. A. Abouzeid, K. Bajda-Pawlikowski, D. J. Abadi, A. Silberschatz, and A. Rasin. HadoopDB: An architectural hybrid of MapReduce and DBMS technologies for analytical workloads. In *Proceedings of International Conference on Very Large Databases*, Lyon, France, 2009.

36. Y. Kwon, D. Nunley, J. P. Gardner, M. Balazinska, B. Howe, and S. Loebman. Scalable clustering algorithm for N-body simulations in a shared-nothing cluster. Technical Report UW-CSE-09-06-01, Department of Computer Science and Engineering, University of Washington, June 2009.

37. B. M. Leiner, V. G. Cerf, D. D. Clark, R. E. Kahn, L. Kleinrock, D. C. Lynch, J. B. Postel, L. G. Roberts, and S. S. Wolff. A brief history of the internet. *Computer Communication Review*, 39(5):22–31, 2009.

38. D. Berardi, D. Calvanese, G. D. Giacomo, R. Hull, and M. Mecella. Automatic composition of transition-based semantic web services with messaging. In *Proceedings of the International Conference on Very Large Databases*, pages 613–624, 2005.

39. R. Hull. Web services composition: A story of models, automata, and logics. In *Proceedings of International Conference on Web Services*, 2005.

40. C. Peltz. Web services orchestration and choreography. *Computer*, 36(10): 46–52, 2003.

41. Amazon Web Services (AWS). http://aws.amazon.com.

42. K. Grochow, B. Howe, M. Stoermer, and E. Lazowska. Client + cloud: Seamless architectures for visual data analytics in the ocean sciences. In *Proceedings of the 22nd International Conference on Scientific and Statistical Database Management (SSDBM '10)*. IEEE Computer Society, 2010.

43. The integrated ocean observing system. http://ioos.gov/.

44. Apache Software Foundation. Apache Cassandra. http://cassandra.apache.org/.

45. A. S. Foundation. CouchDB. http://couchdb.apache.org/.

46. Yahoo! Research. PNUTS—Platform for Nimble Universal Table Storage. http://research.yahoo.com/node/212.

47. Contributors. Ugrid interoperability group. http://groups.google.com/group/ugrid-interoperability.

48. K. Dvornich. Query NatureMapping data using SQLShare. http://naturemappingfoundation.org/natmap/sqlshare/NM_sqlshare_1.html.

49. B. Plale, D. Gannon, J. Brotzge, K. Droegemeier, J. Kurose, D. Mclaughlin, R. Wilhelmson, S. Graves, M. Ramamurthy, R. D. Clark, S. Yalda, D. A. Reed, E. Joseph, and V. Chandrasekar. CASA and LEAD: Adaptive cyberinfrastructure for real-time multiscale weather forecasting. *Computer*, 39(11):56–64, 2006.

50. M. Balazinska, B. Howe, and D. Suciu. Relational data markets in the cloud. In *Proceedings of the 37th VLDB Conference*, 2011.

51. The Kepler Project. http://kepler-project.org.

52. B. Ludäscher, I. Altintas, C. Berkley, D. Higgins, E. Jaeger-Frank, M. Jones, E. Lee, J. Tao, and Y. Zhao. Scientific workflow management and the Kepler system. *Concurrency and Computation: Practice & Experience*, 18(10): 1039–1065, 2006.

53. M. McLennan. The hub concept for scientific collaboration, November 2008.

54. S. Abiteboul, O. Greenshpan, T. Milo, and N. Polyzotis. MatchUp: Autocompletion for mashups. In *Proceedings of International Conference on Data Engineering*, pages 1479–1482, 2009.

55. D. Koop, C. E. Scheidegger, S. P. Callahan, J. Freire, and C. T. Silva. Viscomplete: Automating suggestions for visualization pipelines. *IEEE Transactions on Visualization and Computer Graphics*, 14(6):1691–1698, 2008.

56. O. Biton, S. Cohen-Boulakia, S. Davidson, and C. Hara. Querying and managing provenance through user views in scientific workflows. In *Proceedings of International Conference on Data Engineering*, 2008.

57. S. Miles and L. Moreau (organizers). First provenance challenge. http://twiki .ipaw.info/bin/view/Challenge/FirstProvenanceChallenge, 2006.

58. J. Freire, S. Miles, and L. Moreau (organizers). Second provenance challenge. http://twiki.ipaw.info/bin/view/Challenge/SecondProvenanceChallenge, 2007.

59. C. Silva, J. Freire, and S. P. Callahan. Provenance for visualizations: Reproducibility and beyond. *IEEE Computing in Science & Engineering*, 9(5):82–89, 2007.

60. Y. L. Simmhan, B. Plale, and D. Gannon. Query capabilities of the karma provenance framework. *Concurrency and Computation: Practice and Experience, Wiley InterScience*, 20(5):441–451, 2008.

61. Amazon Relational Database Service (RDS). http://aws.amazon.com/rds/.

62. Database.com. http://www.database.com.

63. Microsoft SQL Azure. http://www.windowsazure.com/en-us/home/features /data-management/.

64. X. Dong and A. Halevy. Indexing dataspaces. In *SIGMOD*, pages 43–54, New York, NY, USA, 2007. ACM Press.

65. A. Halevy, M. Franklin, and D. Maier. Principles of dataspace systems. In *PODS*, pages 1–9, New York, NY, USA, 2006. ACM Press.

66. P. A. Bernstein and S. Melnik. Model management 2.0: Manipulating richer mappings. In *SIGMOD Conference*, pages 1–12, 2007.

67. J. Madhavan, P. A. Bernstein, and E. Rahm. Generic schema matching with cupid. In *VLDB '01: Proceedings of the 27th International Conference on Very Large Data Bases*, 2001.

68. J. Mackinlay. Automating the design of graphical presentations of relational information. *ACM Transactions on Graphics*, 5:110–141, 1986.

69. S. F. Roth and J. Mattis. Data characterization for intelligent graphics presentation. In *CHI '90: Proceedings of the SIGCHI Conference on Human Factors in Computing Systems*, pages 193–200, New York, NY, USA, 1990. ACM Press.

70. D. Gotz and Z. Wen. Behavior-driven visualization recommendation. In *Proceedings of the 14th International Conference on Intelligent User Interfaces*, IUI '09, pages 315–324, New York, NY, USA, 2009. ACM.

71. M. Dörk, S. Carpendale, C. Collins, and C. Williamson. Visgets: Coordinated visualizations for web-based information exploration and discovery. *IEEE Transactions on Visualization and Computer Graphics*, 14:1205–1212, November 2008.

72. H. Elmeleegy, A. Ivan, R. Akkiraju, and R. Goodwin. Mashup advisor: A recommendation tool for mashup development. In *ICWS '08: Proceedings of the 2008 IEEE International Conference on Web Services*, pages 337–344, Washington, DC, USA, 2008. IEEE Computer Society.

73. R. J. Ennals and M. N. Garofalakis. Mashmaker: Mashups for the masses. In *SIGMOD '07: Proceedings of the 2007 ACM SIGMOD International Conference on Management of Data*, pages 1116–1118, New York, NY, USA, 2007. ACM.

74. E. Godreau. quoted in a talk by Ed Lazowska, 2009.

75. B. Howe, N. Khoussainova, G. Cole, and L. Battle. Automatic starter queries for ad hoc databases. In *Proceedings of the SIGMOD Conference*, 2011.

76. S. B. Davidson and J. Freire. Provenance and scientific workflows: challenges and opportunities (tutorial). In *SIGMOD'08: Proceedings of the ACM SIGMOD International Conference on Management of Data*, 2008.

77. Gene ontology. http://www.geneontology.org/.

78. Sloan Digital Sky Survey. http://cas.sdss.org.

79. D. X. Yang, C. M. Procopiuc, and D. Srivastava. Summarizing relational databases. *Proceedings of the VLDB Endowment*, 2(1):634–645, 2009.

80. J. Akbarnejad, G. Chatzopoulou, M. Eirinaki, S. Koshy, S. Mittal, D. On, N. Polyzotis, J. S. Vindhiya Varman. SQL QueRIE Recommendations. In *Proceedings of the 2009 Edition of the 36th International Conference on Very Large Databases (VLDB 2010)*, Demo paper. Vol. PVLDB 3, Issue 2, pages 1597–1600, September 2010.

81. N. Khoussainova, Y. Kwon, M. Balazinska, and D. Suciu. Snip Suggest: A context-aware AQL autocomplete system. In *Proceedings of the 37th VLDB Conference*, 2011.

82. Y. Freund and L. Mason. The alternating decision tree learning algorithm. In *International Conference on Machine Learning*, 1999.

83. Tableau. http://www.tableausoftware.com/.

84. E. Callaway. Genome giant offers data service. *Nature*, 475(7357):435–437, 2011.

Bridging the Gap between Scientific Data Producers and Consumers

A Provenance Approach

Eric G. Stephan, Paulo Pinheiro,
and Kerstin Kleese van Dam

CONTENTS

12.1 INTRODUCTION

Technological advancements such as the World Wide Web brought together business, engineering, and scientific communities, creating collaborative opportunities that were previously unimaginable. This was due to the novel ways technology enabled users to share information that would otherwise not be available. For science in particular, this means that data and software that previously could not be discovered without direct contact with data or software creators can now be downloaded with the click of a mouse button, and the same products can now outlive the life span of their research projects. While in many ways these technological advancements provide benefit to collaborating scientists, a critical producer–consumer knowledge gap is created when collaborating scientists rely solely on web sites, web browsers, or similar technology to exchange services, software, and data. Without some provided context from producers, collaborating scientific consumers have no inherent way to trust the results or other products being shared, and producers have no way to convey their scientific credibility.

Provenance originates from the French term provenir (to come from), which is used to describe the curation process of artwork as art is passed from owner to owner. Data provenance, in a similar way, allows producers to pass data context to consumers by providing vocabularies and methodologies for collaborators to share the origin of anything, both digital and nondigital. Peter Buneman [BUN2001] was one of the first in the database community to coin the term data provenance as it pertained to describing the origin of microbiology data being shared on the web. Yet in some ways, many years later through all the advances of provenance, evidence of this knowledge gap continues to persist creating many exciting provenance contribution opportunities for both the researcher and the domain practitioner. For science, provenance can fill the knowledge gap by

- Answering questions about the methodologies used to arrive at a result along with the expected level of imprecision

- Establishing credibility by citing the credentials of who published the results

- Documenting the lineage of sources used to generate a particular product, or providing a timeline of what took place.

For datasets, the scientific producers may be anyone who created them or modified their contents. It is assumed that the producers fully understand what they produced because they chose the methods and created the context that was used to derive the data and software. Using data as an example, the producer created the process for data creation, controlled the precision used to collect the results, and knew of any uncertainty that are naturally a part of any scientific result. During dataset creation/modification, the producers hopefully took careful ledger notes, performed instrument/sensor calibration, and recorded any observations about inaccuracies introduced during the experimental or computational process. The producers need to encode this knowledge as provenance so that consumer communities understand and thus accept the scientific results.

12.2 CLIMATE SCIENCE PROVENANCE EXAMPLE

The following example shows how provenance is used to bridge the knowledge gap between climate scientists performing atmospheric simulation studies to identify levels of uncertainty based on parameter settings and collaborating researchers who use these published datasets to rank levels of uncertainty against their own simulation results.

12.2.1 Background

Atmospheric climate researchers rely on models such as the Climate Atmosphere Model (CAM) [CAM] to perform research such as air temperature and precipitation variability studies (Figure 12.1).

Scientists found that changes of one physical input parameter can potentially have significant impact on the overall results. Identifying, understanding, and ranking sources of uncertainty is a vital part the knowledge scientists need to convey when sharing their results with the research community. To perform uncertainty quantification (UQ) studies on CAM, a researcher prepares hundreds or thousands of CAM simulation runs for their particular study. The UQ study uses a range of values to alter

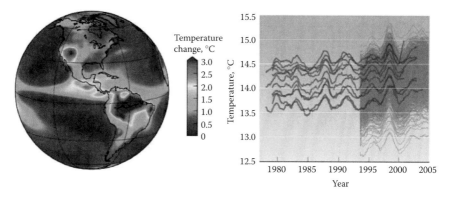

FIGURE 12.1   Visualization and analysis of temperature variability in CAM UQ study. https://str.llnl.gov/JulAug10/klein.html.

a parameter value for each simulation run. After the runs are complete, scientists use visualization and analytical techniques to comparatively assess the different levels of uncertainty experienced across all the simulation runs. The earth sciences communities have developed grids where scientists can upload their results and share them with researchers worldwide. Without provenance, a knowledge gap exists between the UQ study producer and anyone wanting to access those same results.

To convey knowledge about the UQ study, the scientist producing it is faced with some decisions about what to convey to the prospective collaborators. The following scenario summarizes "Bob" the scientist's knowledge about the study.

Bob is an atmospheric scientist at the Pacific Northwest National Laboratory (PNNL) working with a team of National Center for Atmospheric Research (NCAR) climate researchers performing an UQ study on the CAM version 3.0. His fellow climate researchers request Bob to benchmark changes in the results due to setting parameter A in the CAM input deck to minimum, average, and maximum values. Bob runs each CAM job. Each CAM job produces a dataset in the form of netCDF files along with a CAM historical log recording native provenance. Bob merges the results of all three CAM simulations and publishes them as a new dataset to an earth system grid data node. Bob writes an article based on his UQ results in a climate journal.

12.2.2 Scoping Provenance by Consumer Provenance Questions

To convey knowledge, Bob needs to have some understanding of communities accessing his data and what types of provenance-related questions they will have. Fortunately, Bob has an active collaboration with climate scientists Jun and Candis and, drawing on past interactions, knows the following provenance-related questions are typical inquiries:

- Jun wants additional background material on the UQ study. His question is: "Who provided the UQ study criteria to Bob, and what were the criteria?"

- Candis wants to know: "What parameter did Bob alter in his UQ study to produce his published dataset?"

For provenance to bridge knowledge gaps between collaborators, it needs to answer who-, what-, when-, where-, which-, why-, and how-related questions. Jun poses two who- and why-related questions driving the UQ study. Candis poses a what-related question.

12.2.3 Building Consumer-Oriented Provenance

The producer, Bob, must construct answers to what-, why-, and who-related questions to support his user community—Jun and Candis. Provenance vocabularies, such as W3C PROV [PROV2012], provide many capabilities that might seem overwhelming at first. To overcome this concern, W3C PROV provides a core set of terms that do not require a steep learning curve providing users a means to start modeling quickly. Once the user become confident with the core vocabulary terms, it is easy to incorporate more sophisticated parts of the language. Using PROV, the provenance model can be built using only the vocabulary terms: Entities, Activities, Plans, and Agents (Figure 12.2). Entities are any physical, digital, or conceptual thing and can answer what, where, and/or which questions. Activities create entities and modify the attributes of existing entities to create new ones.

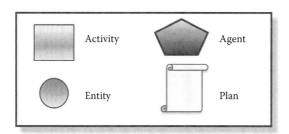

FIGURE 12.2   Legend used to graphically model provenance. Directional arrows that convey relationships between model terms are not shown: *wasGeneratedBy* describes something produced by an Activity, *used* describes input to an Activity or Agent, *wasAssociatedWith* and *actedOnBehalfOf* correspond to Agents.

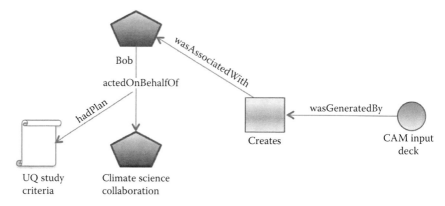

FIGURE 12.3   Provenance model depicting with whom Bob was collaborating to determine the UQ Study Criteria.

They can answer how and/or when questions. Agent is a person, instrument, organization, or piece of software and can answer who and/or what questions. Finally, Plan is a specialized entity that is needed to answer why.

Jun's question "Who provided the UQ study criteria to Bob and what were the criteria?" requires knowledge related to Figure 12.3 (which depicts the relationships between the UQ study, the collaborators, and Bob conducting the study) based on these two paraphrased knowledge statements:

1. Bob is a PNNL atmospheric scientist working with a team of NCAR climate researchers.

2. The joint PNNL-NCAR team is performing a UQ study.

Candis' question "What parameter did Bob alter in his UQ study to produce his published dataset?" requires that Bob convey knowledge related to these three paraphrased knowledge statements:

1. Bob merges the results of all three CAM simulations and publishes them as a new dataset (Figure 12.4a), each CAM job produces a dataset in the form of netCDF files along with a CAM historical log recording native provenance.

2. To perform the UQ study, Bob must prepare and schedule each CAM job (Figure 12.4b).

3. Bob sets parameter A in each CAM input deck to minimum, mean, and maximum values (Figure 12.4c).

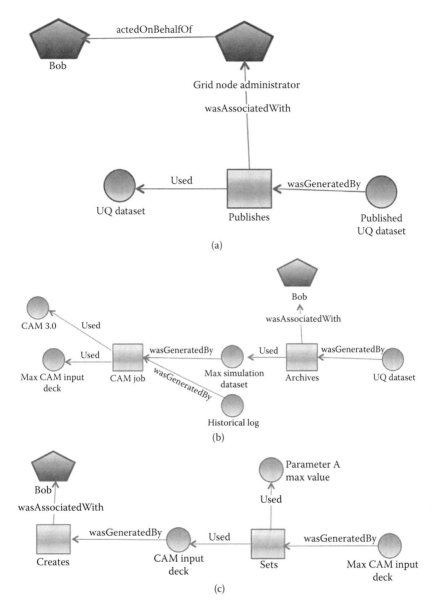

FIGURE 12.4 (a) Provenance model depicting a grid node administrator acting-OnBehalfof Bob to publish his UQ dataset. (b) Provenance model depicting part of the pipeline producing the UQ dataset that is generated from CAM results. The CAM jobs, activities, and simulation dataset entities for the minimum and mean values Bob also used in the study are not depicted. (c) Provenance model depicting part of the pipeline showing parameter A being modified by the set activity for the max CAM input entity. Similar activities and input decks generated for the minimum and mean values Bob also needed to use in the study are not depicted.

The provenance model developed so far covers the basics of the knowledge statement Bob provided about his UQ study. It is consumer-oriented in part because it only conveys the knowledge Bob's collaborators require. To make it more scientifically oriented, Bob will need to extend the provenance model with additional details to the climate community. PROV provides support by allowing users to add customized knowledge to the provenance model. This knowledge can be in the form of simple key value attributes to provide names and directory locations of files, start and end date/time stamps for activities, or more elaborate structures such as specializations of entities, or attributes providing additional details such as a journal article citation using the Dublin Core element set, or details of the UQ study plan.

Once Bob defines his provenance model, the model needs to be saved in a persistent form either in an XML or in an OWL document, or within a database so he can make the knowledge available for Jun and Candis to browse, search, or use derivation to find answers to their questions.

12.2.4 Answering Provenance Questions

Because Bob relied on a publically available formal provenance vocabulary, he provides Jun and Candis a standard representation that describes the origins of his dataset. For Jun to get answers to his questions, he realizes that his questions are regarding who, inferring use of Agents. One obvious Agent is Bob, of whom Jun is already aware, but Jun needs to know who collaborated with Bob. Searching the PROV relation actedOnBehalfOf between Bob and climate science collaboration provides Jun with this answer. Performing a second search on the Agent climate science collaboration, we find this Agent has a PROV relation hadPlan that points to a UQ study criteria plan.

Because the input deck is not published with the UQ study, Candis cannot determine which parameter was being varied and studied. Candis asks a more complex question: "What parameter did Bob alter in his UQ study to produce his published dataset?" With a provenance model, her question relies first on performing a search for the entity UQ dataset and then on walking the relations wasGeneratedBy/used back through the provenance until she finds entity max CAM input deck and examines the activity associated with the wasGeneratedBy relation activity set queries that traverse provenance of cited UQ studies by traversing wasGeneratedBy/used edges back to the input deck of each CAM run.

Relying on a formal provenance vocabulary allows users who store their provenance in a database to search across provenance collected from multiple studies. This is useful if Candis, after examining the entity input deck, finds an error in the file invalidating the study findings. She could then search across her provenance store for any other entity CAM simulation run or plan UQ study that relied on the same input deck to alert the producers of this error.

12.3 BRIEF HISTORY OF RECORDING SCIENTIFIC HISTORY

Although the style of recording provenance might be new to a scientist working through the previous provenance example, from a scientific perspective the discipline of recording knowledge on the origin or history of scientific results has been a practice put into place for millennia. Findings were recorded and procedures described in scrolls and logbooks, and calibration records were kept on analog instruments. Before the existence of a provenance research community, there have been many different efforts to capture provenance in different forms. Some were community based, but most were native forms of provenance using native vocabulary representations.

Evidence of different efforts to make provenance available has been found as early as the late 1980s [GRA1988] when historical information was needed to describe geospatial dataset origin and history. In the 1990s, data provenance was first explored using XML documents [BUN2001] that could trace and convey the processing history of the origin of molecular biology data. The Dublin Core Metadata Initiative (DCMI) [DCMI] emerged as a widely adopted standardized metadata vocabulary. The DCMI had wide appeal in the scientific community to cite data sources. The Collaboratory for Multiscale Chemical Science (CMCS) [MYE2005] was an early adopter of the DCMI and used it to describe shared scientific resources. However with the exception of DCMI, most attempts by scientific projects to record knowledge in databases and on the World Wide Web relied on local terms and structures that supported the data producers, leaving data consumers with the task of interpreting its meaning.

The era of scientific workflows ushered in new interest by scientists seeking to tie together many disparate applications and data sources and have a record of what transpired during the workflow execution. Today many different workflow products exist; these include Kepler [BOW2006], MeDICi [GOR2008], and VisTrails [CAL2006]. While workflows were originally conceived for managing business processes, workflows were increasingly valued to run complex computational and experimental pipelines. Because workflows were adaptable and reconfigurable, they were used to perform complex analyses by coupling together geographically distributed disparate data sources and applications. Listeners were added to collect provenance based on logged workflow events, and the provenance was represented in a form native to the workflow engine.

While native forms of provenance provided certain benefit, consumers often found themselves in a bit of a quandary when attempting to integrate different native forms of provenance.

Without standard provenance vocabulary terms, sharing results and databases became difficult because it was cumbersome to distinguish between provenance-related knowledge and any other metadata or referential information. Even when provenance was distinguishable, each community had their own formats, used their own local structures and conventions, and needed to be expressive based on their particular needs making any hope for settling on one standard impractical. The lack of formalization also led to another issue of trust. Without a formalized approach, it was difficult for scientists to claim they had a complete understanding of what transpired or what scientific protocols were followed.

12.4 PROVENANCE COMMUNITY CONTRIBUTIONS

Universities made many initial contributions to the theory and application of provenance [SIM2005]. Inference Web [INF] provided a semantic provenance language, Application Programming Interface (API) and tools supporting interoperable explanation of sources, assumptions, and learned information and answers as an enabler for trust between collaborators. The International Provenance and Annotation Workshop (IPAW) [MOR2008] began a series of challenges to explore how different native forms of provenance could be exchanged or integrated. Many of the initial workshop participants were scientists and computer scientists wanting to better support E-Science. By 2008, the Open Provenance Model (OPM) interchange language [MOR2011] was developed by IPAW participants so users with their own native provenance could have a common format to compare and exchange provenance with one another.

In 2009, the W3C launched an incubator group composed of universities, industry partners, and research institutions along with members of IPAW and Inference Web. They documented many of the noteworthy provenance efforts [INC2010]. Based on the findings in 2011, the W3C established a Provenance Working Group [PROV2012] to develop a new W3C provenance interchange language.

12.4.1 Provenance Representation

There are many different ongoing provenance efforts; a comprehensive list of known provenance vocabularies is identified [INC2010] by the W3C Incubator Group. This section shares three provenance representations: Proof Markup Language (PML), OPM, and PROV.

Inference Web is a consortium of researchers who developed PML [PIN2006]. It is a semantic web-based representation for exchanging explanations, including provenance information— annotating the sources of knowledge, justification information—annotating the steps for deriving the conclusions or executing workflows, and trust information—and annotating trustworthiness assertions about knowledge and sources. PML was initially developed in 2002 in response to multiple government projects including DARPA Agent Markup Language (DAML), DARPA Rapid Knowledge Formation, and ARDA Aquaint (http://kdi.pnnl.gov/projects/kani.stm). PML's initial purpose was to explain results from hybrid theorem provers, for example, Stanford's JTP (Java Theorem Prover) and SRI's SNARK (SRI's New Automated Reasoning Kit), in support of the Inference Web, a web-based question-answering architecture [MCG2004]. PML supports arbitrary logical data and inference steps including extraction of data from nonlogical sources, conversion to logical forms, classification, and first-order inferences. The work described in the DARPA proposal used PML because of its ability to explicitly represent alternative justifications for a conclusion. In 2004, PML classes were fully implemented as part of the World Wide Web Consortium's standard Ontology Web Language [SMI2004], and PML data are therefore expressible in the RDF/XML syntax. PML was split into three modules, which are referred to as PML2 [MCG2007] to reduce maintenance and reuse cost. PML is used to build OWL documents representing both proofs and proof provenance information.

OPM [MOR2011] is a product developed by IPAW. It is a stable and viable solution expressed both as an XML schema and in OWL. The original intent of OPM was to be an interchange language between two different provenance representations. OPM is also an open model providing extensibility so users can add their own customized attributes and specialization classes to further describe provenance being exchanged. The OPM community also provides java APIs supporting XML and RDF representations.

The W3C Provenance Working Group [PROV2012] is developing a language "that defines a language for exchanging provenance information" about web documents, data, and resources between applications. It currently is working on five specifications including: PROV Primer [PRI2012], PROV Ontology (PROV-O), PROV Data Model (PROV-DM), PROV Notation (PROV-N), PROV Constraints, PROV Access, and Query. The working group draws on many standard bodies of work [CTF2011] and existing provenance vocabularies. These vocabularies include Dublin Core, PML, and OPM.

12.4.2 E-Science Provenance Software and Architectures

To make provenance a more viable commodity to users and software developers, the provenance community performed extensive work writing provenance vocabulary APIs; defined approaches to storing provenance; developed methodologies for browsing, querying, and using derivation; and created new approaches to using provenance for visualization and analysis.

Provenance implementations will vary greatly based on the projects' needs. For any provenance architecture, the basic elements needed to support the four phases in the provenance lifecycle are creation, recording, querying, and managing. The requirements matrices listed in Table 12.1 can be used to scope infrastructure needs. For some users, adopting an existing provenance vocabulary for their standard representation, incorporating a provenance vocabulary API, and leveraging a database (e.g., RDF triple store) for provenance storage may suffice. For team or enterprise solutions, a more substantial provenance architecture is needed (Figure 12.5) and described in Miles et al. [MIL2007]. A technical report on RDF-based Provenance stores can be found in Paulson et al. [PAU2008].

TABLE 12.1 Example Provenance Architecture Requirements Matrix

Key	E-Science Provenance Requirements	Project Provenance Scope?
R1	Record (and query) arbitrary information about individual processes, data that moves between processes, and the relationships between them	
R2	Record enough information to enable references of data regardless of size or location	
R3	Extract and record customized file metadata for context searching	
R4	Record only the provenance from significant events and the processes and data that led to the identification of the event	
R5	Identify processes, experiments, or data as a collection of related work and allow users to record arbitrary annotations and define new relationships.	
R6	Record provenance of high-throughput pipelines with minimal impact on performance	
R7	Determine who ran a particular process, under what conditions, and which settings were used	
R8	Determine if an analysis or experiment has previously been run	
R9	Identify data generated from a particular process	
R10	Retrieve information to be presented for application specific views	
R11	Identify contextual information and results from access to dynamically changing data sources and versions used in an analysis	
R12	Examine full derivation of the result or significant event	
R13	Determine where a process/data was used for data that should be regenerated because of an algorithm or data source change	
R14	Query for derivation graph, filtering on level of detail	
R15	Compare multiple runs of the same workflow execution (differentiated by data source or software module versions) to analyze the effects of the changes	
R16	Retrieve process documentation to reenact an experiment or workflow using new inputs or parameters	

Source: Miles S, P Groth, M Branco, L Moreau, *Journal of Grid Computing,* 5(1):1–25, 2007. With permission.

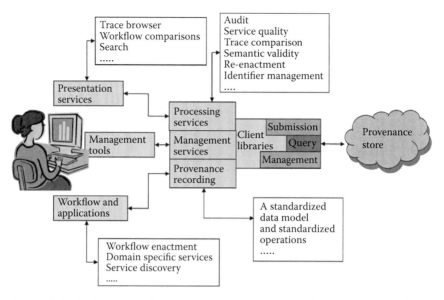

FIGURE 12.5 Conceptual provenance architecture of creation, recording, management, and access.

12.5 SCIENCE PROVENANCE CHALLENGES AND FUTURE PROVENANCE SCIENCE SOLUTIONS

Data-intensive science offers provenance some new challenges in terms of the need for richer adaptation to the sciences as well as scalability to meet exascale computing challenges. Conversely, some challenging aspects of data-intensive science can potentially be answered through provenance solutions. This presents many opportunities for the science and provenance communities to contribute to each other's future successes.

12.5.1 Science Provenance Challenges

Many opportunities exist for the scientific community to extend or develop new provenance capabilities, including extending provenance languages with scientific knowledge, contributing new approaches for organizing provenance information, and introducing compressed formats to support extreme scales.

12.5.1.1 Linking Scientific Knowledge to Provenance

The provenance community offers the scientific community tremendous capabilities to bridge the knowledge gap between producers and consumers. Currently, PML, OPM, and PROV provenance languages offer their own conventions for allowing users to add their own specializations of provenance vocabulary attributes and classes. Inference Web identified a layer called *knowledge provenance* [MCG2009] that tied domain-specific semantic knowledge to PML. Within Inference Web, The University of Texas El Paso has developed a tool, WDO-IT [TRU]. Based on the knowledge provenance concept, it allows scientists to define their own customized provenance vocabulary as a layer on PML where provenance is recorded in a representation that is recognizable to the end user. WDO-IT provides the Workflow-Driven Ontology (WDO) and Semantic Abstraction Workflow ontology layers [PIN2010] allowing scientists to define workflow components using their own customized vocabulary. Once the specification is complete, the WDO-IT provides users with support to export APIs called Data Annotators that record the user's provenance using their customized vocabulary.

The DCMI Provenance Task Group [DCP] is also investigating ways to map their DCM vocabulary to PROV. The Mapping Primer draft [DCPP] provides methodologies for simple vocabulary term mappings and more complex language mappings between Dublin Core to PROV. As the Primer matures, it could potentially serve as a template and methodology for future mappings between scientific ontologies and provenance languages.

A technique used by some scientific communities is to embed provenance within an existing scientific vocabulary or ontology. ISO 19115 [ISO2003] geographic metadata standard provides some provenance support within its vocabulary. The Chemical Information Ontology [HAS2011] introduces information entities to ensure the accurate exchange of biomolecular chemical information.

12.5.1.2 Innovative Organizational Structures

Another science provenance challenge is to innovate provenance representation in extreme scale computing and large-scale data repositories efficiently to represent and communicate provenance. An example approach successfully demonstrated was designing provenance in different tiers called multi-tier provenance inspired by Bruno Courcelle's ordered sequence of unranked subtrees [COU] and tree automata theory [COM1997]. Multi-tier provenance [STE2010] can simply be thought of as a way to divide up the provenance model into different submodels, or as in the case of the Atmospheric Radiation Measurement [ARM] program, organize the provenance to reflect different levels of resolution of detail. The more detailed layers can infer context from their parent tiers providing summary levels of metadata throughout the hierarchy.

12.5.1.3 Efficient Formats

To support exascale, computing provenance also needs to be recorded and stored in more efficient formats. Currently, the provenance community is relying on tagged languages such as XML, RDF, and OWL. While rich in their expressiveness, they can be a hindrance, especially if used as text representations at runtime as earlier described. Within HPC communities and large-scale data repositories, a common practice has been to use byte arrays to encode flags or other simple indicators. As part of our previous research on multi-tier provenance, we investigated the use of creating a reduced form of the provenance language and then created two-dimensional matrices by making the nodes "things" and any relationship between the nodes "edges." Once this is done, adjacency matrices or perhaps other combinatory analysis techniques could be used to monitor, assess, and report behavior. Another expansion on this idea is the use of Javascript Object Notation (JSON) or a more compressed Binary JSON as a way to more efficiently convey provenance.

12.5.2 Future Science Solutions

Due to either the lack of awareness of provenance or the perception that provenance representation is inefficient, the provenance community has not made a large impact on the data-intensive community. However, the use of provenance could help data-intensive science overcome current barriers experienced by the HPC community.

12.5.2.1 Intermediate Result Reproducibility

Scientists could use provenance knowledge that enables consumers to reproduce a portion of intermediate results from experiments and simulations. Scientists conducting large-scale studies, such as sensitivity analysis (SA) and UQ, are increasingly relying on archives or data grids to save results so fellow researchers can later access, analyze, and, if needed, reproduce the results. We recognize that most documented provenance knowledge including scientific publications lacks a full disclosure of hidden parameters used at run time, a list of files that can be easily reproduced, and methods used to recreate those files.

To do this, scientists need to identify which processes are responsible for generating intermediate and final files within each simulation run, the estimated work that it takes to regenerate these files, and what needs to be captured that describes the steps it takes to reproduce the results.

In a real-world example, we have seen evidence that at times, depending on the model, even small studies are capable of overwhelming the inode table of petascale storage systems because the studies rely on numerous intermediate small files. In a recent example, a 1 PB Lustre data system at a U.S. university was temporarily disabled when the 250 million inode table was overwhelmed by the many small files generated by a land use model. This is not an isolated incident. Standard archival procedures on the PNNL Environmental Molecular Science Laboratory require datasets to be archived in compressed tar and zip files.

Sharing and accessing large-scale UQ and SA resources becomes restricted depending on the network bandwidth of collaborators. To resolve issues like the one just described, administrators have to log in as users to remove temporary files or tar bundle the user's data. Although this solves the resource issues on the archival system, it creates one more obstacle for researchers wanting to later examine the results. This is because the results need to be extracted and it may be currently impractical to analyze some large SA or UQ studies if all results extracted overwhelm a given system. Currently, no facilities exist to reproduce such files. Network bandwidth is also a concern. The Climate Science for a Sustainable Energy Future (CSSEF) project [CSS] is generating relatively small 15 TB UQ datasets and sharing these on the ESGF grid. Transmitting this size unfortunately restricts the user community to those with high bandwidth.

12.5.2.2 Resiliency Support

Scientists need provenance tools to better support resiliency. Fault detection, fault information propagation, fault tolerance and resilience, and fault prediction are critical to the successful and effective utilization of exascale systems. The CSSEF project investigated the required infrastructure that uses a new approach to resilience, selecting appropriate strategies for fault tolerance and mitigation, based on real-time information on application execution and fault detection status, and captured in provenance records available throughout the software stack.

Currently, a number of resilience and fault mitigation strategies are under development at PNNL, including NWChem [VAL2010] molecular dynamics parallel codes. As an example an NWChem simulation might decide to terminate if too many errors have occurred that have affected its accuracy. Moreover, the system scheduler might decide to flag a node or core for maintenance if too many faults have been detected. In this case, the system would no longer schedule any tasks to the affected system region until the problem has been resolved.

12.5.2.3 Provenance Life Span

Extreme scale computers, emerging global cloud and grid services, and high resource demand simulations, such as sensitivity analysis and UQ; create challenges for the provenance research community. Ideally, any provenance connected to a digital object, including auxiliary knowledge, would be captured and made available in perpetuity to enable maximum usage of the resulting data and data products. However, the amount of effort and resources required would be prohibitive, and so it is necessary to carefully consider the trade-offs between the value of the knowledge, its usage, and its life span.

Unlike current practices, provenance for data-intensive computing is recorded for different purposes based on different kinds of client needs. Clients may include code developers, data producers, data curators, and data consumers (domain scientists or software services) to name a few. For HPC and data-intensive applications, a growing interest will be to use provenance at runtime and as a result will never be archived. Other provenance captured in the run used for post-processing and analysis may need to be kept indefinitely.

12.6 CONCLUSIONS

Despite the methodical and painstaking efforts made by scientists to record their scientific findings and protocols, a knowledge gap problem continues to persist today between producers of scientific results and consumers because technology is performing the exchange of data as opposed to scientists making direct contact. Provenance is a means to formalize how this knowledge is transferred. However, for it to be meaningful to scientists, the provenance research community needs continued contributions from the scientific community to extend and leverage provenance-based vocabularies and technology from the provenance community. Going forward, the provenance community must also be vigilant to meet scalability needs of data-intensive science.

REFERENCES

[ARM] Stokes, Gerald M., and Stephen E. Schwartz. 1994. The Atmospheric Radiation Measurement (ARM) Program: Programmatic background and design of the cloud and radiation test bed. *Bulletin of the American Meteorological Society*; (United States) 75.7.

[BOW2006] Bowers S, B Ludascher, et al. 2006. Enabling scientific workflow reuse through structured composition of dataflow and control-flow. IEEE.

[BUN2001] Buneman P, S Khanna, WC Tan. 2001. Why and where: a characterization of data Provenance. In Proceedings of the 8th International Conference on Database Theory, pp. 316–330, January 04–06, 2001. London, U.K.

[CAL2006] Callahan SP, J Freire, et al. 2006. VisTrails: visualization meets data management. In *Proceedings of ACM SIGMOD International Conference on Management of Data*, pp. 745–747.

[CAM] Collins, William D., and Coauthors. 2006. The Formulation and Atmospheric Simulation of the Community Atmosphere Model Version 3 (CAM3). *J. Climate*, 19, 2144–2161. Community Atmosphere Model, http://www.cesm.ucar.edu/models/atm-cam/

[COM1997] Comon H, M Dauchet, et al. 1997. Tree automata techniques and applications. Available on: http://www. grappa. univ-lille3. fr/[SIM2005] 10.

[COU] Courcelle B. 1989, On recognizable sets and tree automata. In ed. Maurice Nivat and Hassan Ait-Kaci. *Resolution of Equations in Alegebraic Structures*. Academic Press, New York.

[CSS2012] U.S. Department of Energy. 2012. CSSEF: Climate Science for a Sustainable Energy Future. http://climatemodeling.science.energy.gov/projects/cssef-climate-science-sustainable-energy-future.

[CTF2011] WC3. 2011. Connection Task Force Informal Report. Connection Task Force, http://www.w3.org/2011/prov/wiki/Connection_Task_Force_Informal_Report.

[DCMI] Dublin Core Metadata Initiative. 1999. Dublin Core Metadata Element Set, version 1.1: Reference description [Online]. Available at: http://purl.org/dc/documents/rec-dces-19990702.htm.

[DCP2012] Dublin Core Provenance Task Force, http://dublincore.org/groups/provenance/.

[DCPP2012] WC3. 2012. Dublin Core—PROV Mapping Primer, https://github.com/dcmi /DC-PROV-Mapping/wiki/Mapping-Primer#wiki-References.

[GOR2008] Gorton I, AS Wynne, JP Almquist, J Chatterton. 2008. The MeDICi Integration Framework: A Platform for High Performance Data Streaming Applications. In WICSA 2008. 7th IEEE/IFIP Working Conference on Software Architecture, February 18–22, 2008, Vancouver, Canada, pp. 95–104. Los Alamitos, CA: IEEE Computer Society. doi:10.1109/WICSA.2008.21.

[GRA1988] Grady, RK 1988. *Data Lineage in Land and Geographic Information Systems*. In Proceedings of GIS/LIS'88, Vol. 2, pp. 722–730. Falls Church, VA: American Congress on Surveying and Mapping.

[HAS2011] Hastings J, L Chepelev, E Willighagen, N Adams, C Steinbeck, et al. 2011. The Chemical Information Ontology: Provenance and disambiguation for chemical data on the biological semantic web. *PLoS ONE*, 6(10):e25513.

[INC2010] Y Gil, J Cheney, P Groth, O Hartig, S Miles, L Moreau et al. 2010. W3C Provenance Incubator Group Final Report, http://www.w3.org/2005 /Incubator/prov/XGR-prov-20101214/.

[INF] Inference Web, http://inference-web.org/wiki/Main_Page.

[ISO2003] ISO/TC211. 2003. ISO 19115: 2003 Geographic Information-Metadata. www.iso.org.

[MCG2004] McGuinness DL, P Pinheiro da Silva. 2004. Explaining answers from the semantic web: the inference web approach. *Journal of Web Semantics*, 1(4):397–413.

[MCG2007] McGuinness DL, L Ding, P Pinheiro da Silva, C Chang. 2007. PML 2: A modular explanation interlingua. In Proceedings of the 2007 Workshop on Explanation-aware Computing. Montpellier, France.

[MCG2009] McGuinness DL, P Fox, B Brodaric, EF Kendall. 2009. The emerging field of semantic scientific knowledge integration. *IEEE Intelligent Systems*, 24(1):25–26.

[MIL2007] Miles S, P Groth, M Branco, L Moreau. 2007. The requirements of using provenance in e-science experiments. *Journal of Grid Computing*, 5(1):1–25.

[MOR2008] Moreau L, BT Ludaescher, et al. 2008. Special issue: The first provenance challenge. *Concurrency and Computation. Practice & Experience*, 20(5):409–418.

[MOR2011] Moreau L, B Clifford, et al. 2011. The open provenance model core specification (v1.1). *Future Generations Computer Systems*, 27(6):743–756.

[MYE2005] Myers JD, TC Allison, et al. 2005. A collaborative informatics infrastructure for multi-scale science. *Cluster Computing*, 8(4):243–253.

[PAU2008] Paulson PR, TD Gibson, KL Schuchardt, EG Stephan. 2008. Provenance Store Evaluation. In PNNL-17237. Richland, WA: Pacific Northwest National Laboratory.

[PIN2006] Pinheiro da Silva P, DL McGuinness, R Fikes. 2006. A proof markup language for semantic web services. *Information Systems*, 31(4–5):381–395.

[PIN2010] Pinheiro da Silva P, L Salayandia, N Del Rio, AQ Gates. 2010. On the use of abstract workflows to capture scientific process provenance. In Proceedings of the Second Workshop on the Theory and Practice of Provenance (TaPP'10) February 22, 2010. San Jose, CA.

[PRI2012] K. Belhajjame, H. Deus, D. Garijo, G. Klyne, P. Missier, S. Soiland-Reyes, and S. Zednik. 2012. PROV Model Primer. Technical report. http://www .w3.org/TR/2012/WD-prov-primer-20120110/, Accessed February 14, 2012.

[PROV2012] The W3C Provenance Working Group (P Groth, L Moreau Chairs). 2012. http://www.w3.org/TR/prov-overview/.

[SIM2005] Simmhan YL, B Plale, D. Gannon. 2005. A survey of data provenance in e-science. *SIGMOD*, 34:31–36.

[SMI2004] Smith, MK, C Welty, DL McGuinness. 2004. OWL Web Ontology Language Guide. World Wide Web Consortium (W3C) Recommendation. February 10.

[STE2010] Stephan EG, TD Halter, BD Ermold. 2010. Leveraging the open provenance model as a multi-tier model for global climate research. In Provenance and Annotation of Data and Processes—Third International Provenance and Annotation Workshop, IPAW 2010, June 15–16, 2010, Troy, New York. Lecture Notes in Computer Science, vol. 6378, ed. DL McGuinness et al. pp. 34–41. Springer, Berlin, Germany.

[TRU2006] Salayandia, Leonardo, et al. 2006. Workflow-driven ontologies: An earth sciences case study. *e-Science and Grid Computing*, 2006. e-Science'06. Second IEEE International Conference on. IEEE.

[VAL2010] NWChem M, EJ Valiev, et al. 2010. NWChem: A comprehensive and scalable open-source solution for large scale molecular simulations. *Computer Physics Communications*, 181:1477.

CHAPTER 13

In Situ Exploratory Data Analysis for Scientific Discovery

Kanchana Padmanabhan, Sriram Lakshminarasimhan, Zhenhuan Gong, John Jenkins, Neil Shah, Eric Schendel, Isha Arkatkar, Rob Ross, Scott Klasky, and Nagiza F. Samatova

CONTENTS

13.1 INTRODUCTION

Petascale simulations utilize the massive computational power available from supercomputers to simulate scientific phenomena at previously unseen levels of detail. Resulting from simulations being performed at such a scale is the ability to easily generate several terabytes of a data in a single day. Unfortunately, the rate at which data are generated across simulations exceeds the bandwidth available to ingest data into external storage devices. This disparity oftentimes leads to simulations being delayed while waiting for input/output (I/O) operations to complete.

For example, the computational capability of leadership computing facilities grew from 14 TFlops to 1.8 PFlops (the Cray Jaguar at Oak Ridge National Laboratory) from 2004 to 2009, an increase of over 100× [27]. This trend is likely to continue into the exascale, and in 2011, the K computer in Japan has reached 10.51 PFlops. On the other hand, from 2004 to 2009, the aggregate I/O rate on leadership computing facilities has only increased from 140 GB/s (the ASC Purple at LLNL) to 200 GB/s (the Cray Jaguar) [27], through the support of high-performance parallel file systems, such as IBM general parallel file system [57], parallel virtual file system [9], and Lustre [58]. At these performance trajectories, leadership-class facilities will quickly hit a performance ceiling, where increases in computational capability will go unutilized due to the much slower increases in I/O. Thus, it is imperative that we remove or at least lessen the severity of I/O limitations.

There are many ways to help address I/O performance limitations, including developing new storage technologies such as solid-state drives (SSDs), optimizing parallel file system implementations to provide peak disk bandwidth, and overlapping I/O and computation. However, more to the point of this chapter, we can rather exploit this gap in performance by using the otherwise idle CPU cycles on other meaningful computation. This allows us to not only make use of the computational resources available, but also perform operations on the data previously considered to be post-processing, while the original data are still in memory.

This process of operating on data in memory at simulation time is called in situ processing. Although technically any parallel algorithm that can be inserted into a simulation code can be considered in situ, there are a number of optimization goals particular to the high-performance computing field.

1. First and foremost, in situ algorithms must minimize the time added to the simulation as a whole. Project budgets are typically strict about computational time allocated.

2. Related to the first point, in situ algorithms should minimize or avoid global communication, opting for local communication or even being communication free. Requiring idling while waiting for network operations to complete (e.g., global communication) defeats the purpose of performing in situ computations.

3. In situ algorithms should also leave a minimal memory footprint. In systems with hundreds of thousands of cores, available memory per core is limited, with simulation codes using the majority of it.

There are a number of ways that in situ methods can satisfy these goals, but they are, of course, application dependent. This brings us to another question: what kinds of computation are applicable for performing in situ?

To answer this question, it is useful to review the categories of data produced by applications:

Checkpoint and restart (C & R) data: These data are written out by simulations at routine intervals (e.g., every hour) so that simulations can avoid having to restart from scratch in the case of any execution failure. These data, which are usually voluminous, can benefit from the application of lossless compression techniques to reduce the burden on I/O.

Analysis and visualization data (A): These data are often written in more frequent intervals than C & R data and are repeatedly accessed by visualization and analysis routines. Unlike C & R data, the data used for analysis are inherently lossy as scientists typically sample data on some fixed timestep interval to keep the data size within manageable proportions. Although lossless compression techniques can be applied to these data, it usually requires repeated access and could benefit from lossy compression algorithms that can provide higher compression ratios, as long as the loss is within an acceptable bound. These data are mined for exploration and analysis, leading to scientific discovery. Some simulations treat the checkpointed and analysis data the same.

Verification and validation data (V): These data are written out every few timesteps to check the sanity of the running simulation. These data are usually small, amounting to a few megabytes every timestep.

Table 13.1 shows example statistics on each type of data generated by the GTS [70] simulation, a particle-based simulation for studying plasma microturbulence in the core of magnetically confined fusion plasmas of toroidal devices simulation.

From these types of data, there are two broad types of computations applicable for in situ computation, which we focus on in this chapter. One is *data*

TABLE 13.1   Summary of GTS Output Data by Different Categories

Category	Write Freq.	Read Access	Total Size	In Situ Algorithms
C & R	1–2 hours	Once or never	≈TBs	Data reduction
A	10th timestep	Many times	≈TBs	Data reduction
				Transformation, analysis, and visualization
V&V	2nd timestep	A few times	≈GBs	Analysis

analysis, performed on analysis and visualization data. There are two sides to this coin: one involves analyzing the data for specific phenomena without user interaction, computable through data mining algorithms (see Section 13.2). For example, *feature extraction*, where the definition of a feature is dependent on the application domain (such as vortices in a fluid-flow simulation), can be performed in situ and without user guidance. The other side of the coin instead requires user interaction and resides in the space of database optimization. That is, the data can be prepared for user-defined *query processing*–the retrieval of data given constraints in the spatio-temporal domain as well as on variable values. Query processing on the data in situ is a complex problem, one that would require running queries on data that might or might not exist in the simulation working set and one that produces a number of interesting research directions, such as simulation steering, validation, and diagnostics at simulation time. A more clear-cut goal is to build the global database in situ, performing data reorganization, indexing, reduction, and so on to optimize post-simulation query processing, rather than doing all these data-intensive operations from disk. This is described in Section 13.5.

The second type of computation applicable is *data transformation*, which overlaps with the (future) analysis portion through *data reorganization*, which prepares the data for efficient future access, and *data reduction*, through saving only the analysis results, such as features obtained from feature extraction. However, there is another important class of reorganization that is not addressed in the analysis space: compression. Compression not only has the potential of reducing the overall data footprint and thus the overall I/O cost, but it also exploits the growing compute-I/O performance gap. Therefore, with some appropriate level of overlap, compression (lossy or lossless), obviously in situ-capable, has the potential of improving overall application performance. This is discussed in Section 13.3.

Finally, an important question for system designers to answer is: where in the software stack should in situ computation be enabled? Scientific codes can reach into the hundreds of thousands of lines: a naïve inclusion of in situ computation would require an enormous programming effort from the application designers to include the computation and efficiently handle compute, network, and I/O resource utilization to allow efficiency. Obviously, this is unacceptable. Furthermore, without the ability to decouple the computation of simulations and their corresponding I/O, in situ algorithms will only add to the overall simulation time, rather than be properly overlapped. Therefore, we must enable in situ computations at varying subsystems of the scientific code, so that we can allow the scientific code itself to incur minimal changes, while allowing for a large degree of flexibility and resource utilization. The following subsystems need to be looked at in greater detail to allow this enabling to occur:

High-level I/O APIs: APIs such as adaptive I/O system (ADIOS) [37] are highly extensible, allowing for the definition of custom operations to occur while the data are "in-flight," or on their way to disk. Libraries such as these represent the entry point of in situ codes, plugging in libraries and frameworks that can include in situ algorithms.

I/O optimizing middleware: A popular I/O optimization is to dedicate a subset of nodes in a compute cluster to be responsible for all I/O in the system and even an additional layer of nodes to control resource scheduling; this way, compute nodes send data, often asynchronously, across a higher-bandwidth network rather than directly to disk, reducing idle time and enabling efficient overlap. In situ code can be enabled in these middlewares, hooked into the high-level I/O APIs, so that the compute nodes need not incorporate the additional code nor perform the actual in situ computation, leading to the low-latency, transparent integration of in situ algorithms.

Section 13.4 discusses both the high-level I/O APIs that many scientific applications use and the enabling middleware technologies.

The benefits of answering these questions and providing robust, efficient in situ computation will not only benefit high-performance computing, but also find much benefit for other "big data" application domains and online/streaming analysis scenarios. For example, the ever-increasing sample rate of sensor data in both the climate and physics domains necessitates the usage of analysis methods that can analyze and process data in real time [18]. Although the conditions by which the data are generated may be different, the components necessary to work with distributed data before outputting to storage share numerous similarities with the in situ processes and frameworks discussed in this chapter.

13.2 IN SITU DATA MINING

In any scientific simulation, the application scientist is looking to understand certain phenomena (or patterns) that may not be easy to analyze in the physical world. An example of such a pattern is the vortex of a fluid flow, including fluids such as liquid, gas, and plasma. *Vortices* are spinning flows of fluids that are typically turbulent. Vortices have certain physical properties; for instance, their speed and rate of rotation is the highest in the center and gradually decreases when moving away from the center and they exhibit a minimum fluid pressure in the center. Scientists may be interested in the vortices themselves, which in climatology signify cyclones or hurricanes, or in more generic patterns such as regions with abnormally high temperature or low pressure, where low pressure systems may signify a coming storm.

Scientific simulations not only help to understand a physical phenomenon but also provide information that will help replicate the phenomenon under controlled settings. One example is thermo-nuclear fusion, widely considered the "Holy Grail" of renewable energy generation. Fusion energy production could provide an alternative, environmentally friendly, and renewable energy source for our planet, but it is an extremely complex problem in the realm of physics and chemistry. The technical challenge for fusion energy production is stabilizing the flow of heated plasma in magnetic fields inside a fusion energy reactor—scientifically, this problem is oriented towards controlling plasma instability or turbulence. Scientists have conducted simulations for several years now to model fusion reactions, but lacked appropriate technology to find the onsets of turbulence (called *fronts*) from analysis data [59].

Using some known properties of the physical patterns, analysis and visualization data produced by a simulation can be mined for regions or volumes that satisfy these properties, known as *features*. The process of identifying these features is called *feature extraction*. In simple terms, feature extraction is the process of separating the *important* regions from the *background* [63]. Traditionally, this analysis is performed as a postprocessing step. However, the ability to perform these in situ would have numerous benefits. First, features in many cases are the only important data scientists wish to analyze, so storing only the features results in huge data reduction [38,79]. Second, feature extraction done at simulation time allows *real-time* analysis of features, that is, analysis of the data while the simulation is running. This is especially important for simulations with running times in the order of days and weeks or for future applications that allow scientists to steer or verify the simulation as the simulation is running.

Scientific simulations deal with dynamic phenomena, that is, phenomena that change over time. Hence, performing feature extraction alone is not sufficient. We also need to be able to study the changes the features undergo over time. This process is called *feature tracking*. Typically both feature extraction and tracking are performed to better understand the simulation.

Feature tracking involves looking for the following events [54,63] illustrated in Figure 13.1:

Continuation: A feature identified in timestep $t - 1$ exists in timestep t.

Creation: A new feature appears in timestep t.

Dissipation: A feature from timestep $t - 1$ is no longer present in timestep t.

Bifurcation: A feature from timestep $t - 1$ separates into two more features in t.

Amalgamation: Two or more features from timestep $t - 1$ merge into one feature in timestep t.

13.2.1 Property-Based Methods

Features could be defined as volumes in the data that satisfy a threshold for certain attributes (e.g., pressure, temperature). The region of interest may be constrained by a single attribute or a combination of several attributes. There are two strategies that are typically deployed for feature extraction, similar to density-based clustering techniques that separate the useful regions from noise.

Region growing (or filling) [63] is an iterative method that uses certain data points (data cells, voxels, pixels, or particles) in the current timestep

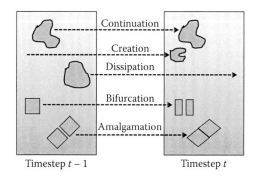

FIGURE 13.1 Feature tracking events.

as seeds, and the feature is "grown" by utilizing the neighboring points until the value of the attribute falls below (or rises above) a certain threshold. *Seed points* are typically defined as points with simulation-specific extreme values. For example, in combustion simulations to identify high-temperature regions, the simulation points with the highest temperatures are chosen as seeds and the feature is grown using the neighbors of the seeds, stopping when the temperature falls below a certain user-defined threshold.

Connected Components [76] is a method where all data points that satisfy certain criteria (e.g., pressure below a threshold) are identified, followed by grouping neighboring points into one feature. Neighboring points here are defined as those not separated by other data that do not satisfy the criteria.

Once the features are identified in some timestep, they need to be tracked in the following timesteps. As discussed, this involves looking for any of the five events defined. Feature extraction takes place at each timestep and then the lists from the current and previous timesteps are correlated. This is called the *correspondence problem*. A brute-force method would be to compare each feature identified in the previous timestep with all features and all combinations of features (*bifurcation* event) in the current timestep [63]. The comparison is based on certain domain-specific attributes. Because the brute-force method must perform comparisons on all subsets of the feature set, the number of comparisons is exponential, which is certainly not applicable for in situ computation. A more reasonable method is to split the correspondence into two steps. The first step is to look for those feature pairs that overlap [54,63], as shown in Figure 13.2. An efficient way to perform this overlap is to sort the feature lists into

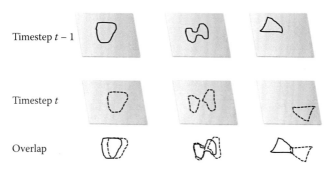

FIGURE 13.2   Feature tracking using overlap.

two consecutive timesteps and then merge the two lists [10–12]. Using this method, the amount of overlap between a pair of features can more easily be determined. This gives us candidates for corresponding features. The second step is to identify for each feature the *best-matched* feature pair. This can be done by calculating the *normalized correspondence metric (NCM)* [63,64], originally defined for overlapping volumes. Let f_i and f_j be features occupying a volume in space. The NCM for the pair is given by

$$\text{NCM}(f_i, f_j) = \frac{\text{Volume}(f_i \cap f_j)}{\sqrt{\text{Volume}(f_i) \cdot \text{Volume}(f_j)}} \tag{13.1}$$

In the sorting-based strategy, the entire list of features from each timestep has to be calculated before the feature is tracked by solving the correspondence problem. This requires us to analyze all the data produced in every timestep. In reality, only a few features out of the entire list will be of interest. Therefore, instead of segmenting the entire data in the current timestep, a more efficient method would be to take advantage of the feature information from the previous timestep(s) to predict the position of the feature in the current timestep [43].

This problem can be mathematically modeled using one of three variations (Figure 13.3). The *direct method* merely performs direct projection of the same feature in the previous timestep to obtain the feature in the

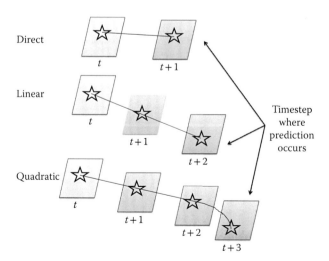

FIGURE 13.3 Feature prediction in the current timestep utilizing information from previous timesteps.

current timestep. However, this does not account for features migrating during simulations. To predict the current position of a feature more accurately, two additional methods can be used. The *linear method* uses the projection of the feature in the previous frame, but the position is offset using the difference in centers of the feature's region in the previous two timesteps using a linear model. The *quadratic method* also uses the projection of the feature in the previous frame, but the position is offset by modeling a quadratic curve that passes through the centers of the same feature in the previous three timesteps. From the description, it is fairly clear that feature tracking using the direct method can begin in the second timestep, the linear can begin only in the third timestep, and the quadratic can begin only in the fourth timestep.

A drawback for these kinds of predictions is that they only consider motion with respect to the feature centers, ignoring other parameters such as angular motion, change in volume, or the merge or split that can occur in subsequent timesteps. Furthermore, these projection-based methods cannot identify new features in the current timestep.

The prediction made is only an estimate; to identify the actual feature, an additional step called *region morphing* is performed. Region morphing is similar to region growing but includes both boundary growing and shrinking. It uses a simple breadth-first search strategy to include neighboring data points that satisfy the attribute thresholds and excludes those data points from the feature that do not satisfy the attribute thresholds. The search begins with a seed point in the feature region.

The algorithms discussed in this section so far deal with identifying features based on some attribute value. There are also some application-specific features that may not be dependent on the attribute value but only on the distance or location with respect to other particles. In dark matter cosmological simulations, one particular phenomenon of interest is the evolution of *dark matter halos*. These halos are of great importance because it is said that almost all the mass in the universe ends up in these halos, which are objects with dynamical equilibrium. These also have the property that smaller halos over time merge to form larger halos. Objects like the galaxies are said to form and evolve in these halos [32]. However, not much is known or understood about these, except that they are clusters of particles in space where all particles are within a certain threshold distance from all the other particles in the halo. This threshold is known as the *linking length*.

The implementation to identify these features uses a *friend-of-friend (FOF)* algorithm, which employs these linking lengths to identify halos.

A *friend* of a particle is a particle within the linking length. It is sufficient to identify the halos by then comparing friends of the friends and clustering them together. This lends itself to an efficient parallel implementation [3,4,74], where the overall space is partitioned among the processes (when in situ, this is predetermined), and spatial overlap is utilized to minimize local communication, only needing to update particle locations after each timestep. This is similar to *stencil computation*, which defines a set of *ghost cells* on array boundaries, which are updated each step in the computation. In addition to the identification, the halos are also further classified based on the number of particles they contain. The *light halo* has about 10–40 particles, the *medium halo* has about 41–300 particles, the *heavy halo* has about 301–2500 particles, and the *extra heavy halo* has more than 2500 particles [3]. The data reduction by storing only the halos is significant: up to 50% of the nonhalo data can be thrown out.

13.2.2 State Change Detection

In the previous section, we identified *salient* features by analyzing some specific attribute values. There are other feature extraction algorithms where the features represent specific significant events that take place within the system and, in some cases, it represents the system moving from one state to another. For example, in the lifted ethylene jet flame simulation, the extinction and reignition events are important and complementary. The flame may get extinguished because the heat release rate cannot keep up with the heat losses, and if the extinguished regions continue obtaining a supply of fuel and oxygen, the flame could reignite. These events are important and require further analysis [33].

The regions in each timestep in the Lifted Ethylene Jet Flame Simulation can be characterized by some subset of chemical species that is present in the input. Each timestep is divided into data blocks and *principle component analysis (PCA)* is performed on each block [69]. The eigenvalues and eigenvectors provide enough information to understand the chemical species that the block correlates with. This information, when taken over the entire simulation, can characterize the various possible events that have taken place during the simulation. However, the number of such states may be very large; hence, some local clustering using the chemical species as attributes is used to provide a more abstracted view. This view can also help understand the timesteps where the ignition or extinction events took place. To reduce the dimensionality of the data block, PCA is used to project the data onto a lower dimensional space by removing highly correlated dimensions. However, instead of performing it per block, the global (per timestep) PCA is calculated using the local covariance matrices of each block combined using an update function [51]. The global projection is then applied to each data block. This reduces the problem to performing one single PCA.

Another example of a feature representing the underlying system's state change is a *front*. Fronts, as discussed in the introduction, signify the beginning of a turbulence. Thus, there is a change in the state of the system, which needs to be captured. An algorithm has been developed that enables automatic turbulent front detection and is able to track and quantify front propagation over spatiotemporal regions [59]. Mathematically, fronts tend to occur at points during the simulation where the potential energy function reaches maximum curvature. However, calculating curvature numerically (via second derivative) is unreliable and ineffective, as it produces noisy patterns that challenge finding the points of interest. Thus, because of the inherent complexity of the simulation data, automated detection of fronts calls for a statistically robust and productive method.

The front detection method is shown in Figure 13.4. The method first smooths raw simulation data to reduce noise and make analysis more

Data preprocessing steps

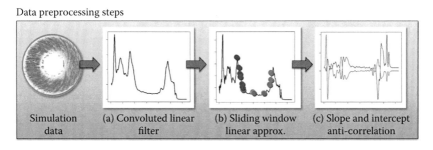

Front detection and tracking steps

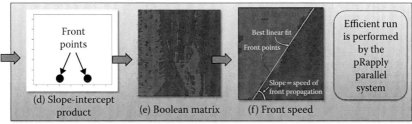

FIGURE 13.4 **(See color insert.)** An end-to-end, multistep analytical pipeline for spatiotemporal turbulent front detection and tracking.

manageable (Figure 13.4a). The smoothing technique uses a convoluted linear filtering algorithm, where values of the smoothed function are dependent on the weighted average of surrounding values. Next, the data are traversed in a sliding fashion and linear regression is applied on each window of points (Figure 13.4b). The slope and intercept values are collected and normalized to the $[-1,1]$ range because of their varying magnitudes. The collected values suggest that slope and intercept values over the windows will be anticorrelated—a positive slope likely results in a negative intercept and vice versa (Figure 13.4c). This strongly anticorrelated pattern suggests that multiplying these slope and intercept values to create a SIP (slope–intercept–product) "signal" could effectively enable identification of the curvature points of interest. This signal is made entirely positive to further amplify the differences in the data's direction and magnitude. Next, a low ($T \approx 0.01$) threshold is applied for all such signals over varying timesteps (Figure 13.4d). The theoretical justification behind this process is that points of curvature will be found where the data change from having a slope parallel to the y-axis to the x-axis and vice versa. At these points, the slope will be near 0, and thus the product will also be near 0. The thresholding operation converts all the SIP signals to matrices of Boolean TRUE/FALSE values, where TRUE corresponds to signal values v, $v \geq T$, and FALSE represents values for which $v < T$. Finally, a two-color heatmap is constructed, representing TRUE and FALSE, respectively (Figure 13.4e). The heatmap clearly depicts propagations of turbulent fronts over data. This analysis technique also enables users to specify regions of interest and accordingly calculates the direction, duration, and speed of propagation of fronts located within that region. This entire process is simply parallelized over different timesteps using the pRapply software.

13.2.3 Importance-Driven Feature Extraction

A strategy for feature extraction that is different from those previously defined is to quantify the various regions in the data based on the amount of information they convey. An information-theoretic method quantifies the significance of a feature based on the amount of information it conveys by itself and the amount of information it conveys in comparison to the same feature in a previous timestep [71]. Each timestep is divided into spatial data blocks corresponding to processors at simulation time. Each data block is characterized by a set of attributes, for example, pressure, temperature, and so on, which is used to build a *multidimensional histogram* for each block. Each bin in the histogram contains the data (voxels, pixels, particles) that satisfy a certain combination of attribute values. This histogram is used to calculate the *entropy* that will quantify the amount of information a data block holds on its own. A high entropy value signifies a more important block. The *mutual information* is used to quantify the amount of information a data block has with respect to the block in the same spatial location but in the previous timestep. This is calculated by building a joint *feature temporal* histogram. A low mutual information value signifies a more important block. The normalized heights of each bin in the histogram provide the required probabilities to calculate both mutual information and entropy. The importance of a timestep is calculated as a function of the importance of each data block it contains.

This method helps separate out regions that offer significant information. Additionally, this method helps bring to attention the abnormal events that took place in the spatiotemporal simulation. It helps the questions of *when* and *where* the abnormal (or interesting) events happened in the simulation. The timesteps with the highest importance scores answer the *when* question. Within this timestep, the data blocks with the highest significance values answer the *where* question.

13.2.4 Communication Strategies

An important optimization goal of in situ analysis techniques is to minimize the amount of interprocess communication at simulation time. Interprocess communication is required to analyze features spanning multiple processors. In the worst case, these features could span every processor. There are three general interprocess communication strategies found in the literature. In the *complete merge* [10–12] strategy (Figure 13.5), each

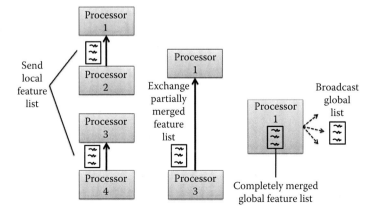

FIGURE 13.5   Complete merge strategy.

processor identifies its local set of features and then communicates the information to all the other processors, using *all-to-all* communication methods. Using tree-based communication algorithms, logarithmic communication steps are needed for the all-to-all communication, relative to the processor count. At the end of this communication, all processors have the *global* feature table with correlated information. Obviously, this is highly communication-intensive, which is undesirable for in situ computations.

In the *partial merge* [10–12] strategy, each processor communicates only the feature information with its immediate neighbors (Figure 13.6). At this stage, the partially merged feature tables are sent to another server (called *viz-accumulator*) that takes care of the merging. This server may then pass on the globally merged feature table back to each processor or to a visualization system. The communication is much less intensive, though it still requires global communication. Both the complete and the partial merge strategies have been deployed by the algorithms [10–12] discussed in Section 13.2.1.

The third strategy is called *no merge*. In this strategy, each processor identifies the features at each timestep but no merging takes place during the simulation run. As a post-processing step, the entire simulation feature set is analyzed and merging takes place. This maintains simulation speed, but as a result moves the problem of merging elsewhere, so using no merge will be application-dependent.

In conclusion, we find that in situ data mining is an intelligent way of extracting useful information from the raw simulation data. The features identified characterize important physical patterns and provide a way to analyze and understand them better. Apart from the techniques discussed

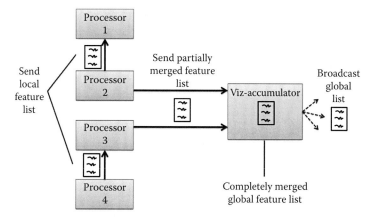

FIGURE 13.6 Partial merge strategy.

in this section, parallel implementations for several existing data mining algorithms [13,31,36,40,44–46,48,60–62,72,73] could potentially be translated for in situ data analysis.

13.3 IN SITU COMPRESSION OF SCIENTIFIC DATA

The last decade in the high performance computing realm has witnessed an increasing imbalance between computational power and file system bandwidth. As this imbalance is expected to grow even further in the exascale era, data compression becomes a necessity rather than an option. With compression being co-located with the simulation process, idle CPU cycles can be utilized to bridge the gap between data generation and ingestion.

However, compression of single and double-precision spatiotemporal scientific data has proven to be a challenge due to its inherent complexity and high entropy associated with the values. General lossless compression libraries, such as BZIP2 and ZLIB, tend to fare poorly on such scientific datasets, offering less than 10% reduction in data size in majority of the simulation data [56]. To alleviate this problem, several compression techniques have been proposed, both in the lossless and lossy front, specifically designed to work on data from scientific simulations. Although suitable for compressing in the post-processing phase for archival and storage, some of the techniques are inefficient for in situ application. In this section, we evaluate the effectiveness of some of the existing lossy and lossless compression techniques and analyze the applicability of compression algorithms in in situ processing environments.

13.3.1 Lossless Compression

Lossless compression of floating-point data is the only applicable type of compression in C & R data, since the copy of the data must be exact for reproducibility. Lossless compression is also applicable to high-fidelity simulation data or for analysis methods sensitive to noise, such as Fourier analysis. However, lossless compression requires a more complex processing scheme that general purpose compressors traditionally do not exploit. To understand what makes floating-point compression hard, one can look at the bit-level probability distribution values on scientific datasets. An ideal distribution on a bit for compression would be every bit position in the data holding the same value (0 or 1), while the least ideal distribution would be equal occurrences of each bit. Highly repeating values, represented by higher probability distribution values closer to 1, can be easily predicted and compressed. However, an equal probability distribution (closer to 0.5) makes the bits unpredictable and hence hard-to-compress (HTC). Of course, other properties, such as spatiotemporal correlations in bit probabilities, may be present. Compressors can take advantage of these properties, though property discovery for general-purpose compression is a nontrivial problem, and least significant bits in double-precision variables tend to have nearly no correlation of any kind with each other.

Figure 13.7 shows the bit-level probability distribution patterns for a typical floating-point variable in scientific datasets (the one given is from FLASH astrophysics simulation code [5]). Widely used general purpose compressors like ZLIB and BZIP2 are oblivious to this fact and perform poorly as a result. In this section, we study some of the state-of-the-art compression utilities, namely FPC [7], FPzip [34], and ISOBAR [56], and their applicability in in situ processing environments. A summary of the strategies behind these algorithms is given in Table 13.2.

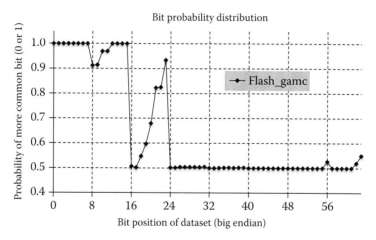

FIGURE 13.7   Plot of probability distribution of each bit position from the gamc variable in FLASH simulation data.

TABLE 13.2   Summary of Different Compression Algorithms

Compression	Lossless?	Strategy
ISOBAR	Yes	Frequency analyzer, black-box compressor
FPC	Yes	FCM/DFCM predictor
FPzip	Yes	Lorenzo predictor, arithmetic encoder
BZIP2	Yes	Burrows-Wheeler transform
ZLIB	Yes	LZ77 and Huffman encoder
Wavelets	No	Wavelet transform
ISABELA	No	Sort preconditioner, B-Spline reduction

13.3.1.1 ISOBAR

The in situ orthogonal byte aggregate reduction (ISOBAR) compression utilizes a preconditioner to extricate compressible data from HTC datasets. ISOBAR is composed of two main components to provide improved compression efficiency:

1. The ISOBAR-analyzer, which identifies the portions of the data that can be compressed by an entropy-based encoder versus the portions that cannot.

2. The ISOBAR-partitioner, which reorganizes the compressible and incompressible portions of the input data identified by the analyzer.

ISOBAR-compress is highly flexible in that varying compression libraries such as ZLIB or BZIP2 or even FPC and FPzip can be utilized to encode compressible bytes, based on user preferences (compression ratio vs. speed). The overall workflow of ISOBAR compression is shown in Figure 13.8.

ISOBAR-analyzer: The objective of ISOBAR-analyzer is to identify byte column clusters in the floating-point dataset that are ineffective when compressed. That is, in a double precision dataset, each significant byte in all doubles in the stream is considered for compression together, rather than compressing on a double-by-double basis. ISOBAR-analyzer works on a byte-level granularity to efficiently identify high-entropic byte columns that cannot be efficiently compressed. The ensuing ISOBAR-compress operation reorders the compressible and incompressible portions and pipelines them to be used by general purpose compressors. This byte-level ordering ensures better compression ratios with entropy encoding.

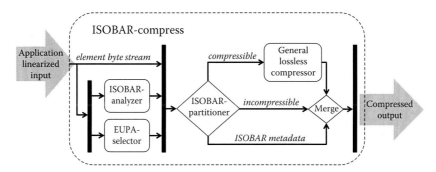

FIGURE 13.8   ISOBAR-compress preconditioner workflow.

To ascertain which bytes are compressible, ISOBAR-analyzer first transforms an n-dimensional input data into a linearized stream of floats and then generates a frequency distribution for each byte column in the set of floats. For example, consider the case when ISOBAR-analyzer works on a one-dimensional (1D) dataset of N elements of w bytes each ($w = 8$ for double-precision, $w = 4$ for single-precision float). The analyzer starts with the initialization of a frequency counter array of length w, with zeroes. The data are scanned through, and the frequency counter is updated to calculate the distribution of each of the possible $2^8 = 256$ unique bit patterns for every byte column. When this value is less than $N/256$, ISOBAR considers the column-based dataset to negatively affect the overall compressibility. Removing this set of bytes in the compression phase would invariably result in both better compression ratios and throughput.

ISOBAR-partitioner: Once the compressible byte columns are identified by ISOBAR-analyzer, the ISOBAR-partitioner determines whether the column would contribute to improved compression ratios. Columns identified as *improvable* are subsequently passed to the compress phase after repartitioning. In some cases, it is possible that the analyzer concludes that either none or all the columns in the dataset are compressible. When this happens, during the analyzer phase, the partitioner classifies the compressibility type of the dataset to be *undetermined* and passes the entire dataset to the compressor.

The ISOBAR-partitioner essentially converts the input data into two segments. One is the compressible byte columns that are packed together and fed to the underlying compression library. The other is the set of incompressible bytes that are stored as-is. In this phase, different linearization strategies can be employed like row-order or Hilbert or Z-order mapping (see Section 13.5.1.3) before passing the data on to the compressors.

The partitioner realigns the data, as determined by the linearization strategy. For example, suppose ISOBAR-analyzer returns that columns 1, 2, and 4 are compressible, and row-major linearization order is used. The ISOBAR-partitioner would cluster all 1st byte values belonging to the N input elements and reorganize them in row-major order. These data would then be appended with the 2nd column, followed by the 4th. So, in this case, only $3 \times N$ bytes are passed to the compression process (a ratio of $3/w$), guaranteeing an improvement in both throughput by compressing less, and compression performance by keeping high-entropy content out of the compression algorithm.

13.3.1.2 FPC

FPC is a fast lossless floating-point compression algorithm for 64-bit double precision data. Unlike other compression algorithms like ZLIB and BZIP2, FPC is a single-pass linear-time algorithm that is designed to deliver high-throughput on both compression and decompression. One important characteristic of the FPC algorithm is that it does not depend on the underlying structure of the data. FPC compresses a linear sequence of doubles by predicting each encountered value and storing the XOR difference between the predicted and the actual value. The leading zeros in the difference are then encoded to achieve compression.

The prediction algorithm used in FPC is a variant of the FCM [78] and DFCM [19] predictors, using finite context models of order n to predict the next occurrence in a stream of values based on the preceding n occurrences. For example, a five-order FCM would predict an element using the previous five values. In the case of a differential model, the context is modeled as the difference between successive occurrences. The FCM and DFCM predictors have shown accurate performance in predicting and prefetching instructions, thus eliminating dependencies and improving parallel execution. With FPC, this concept is extended to predict bit patterns of the first two significant bytes of floating-point numbers containing the sign bit, exponent bits, and top 4 mantissa bits in an efficient manner. FPC, for each prediction, uses both FCM and DFCM and selects the best predicting one, using an additional bit per prediction to store the choice. The primary feature of FPC is its efficient implementation that results in high compression throughputs. To accomplish this, FPC is *cache-aware*, bounding the memory needed by the hash functions of FCM and DFCM to store the recent m contexts (where $m < 26$ and user specified) to fit within the cache. Because difference coding is used, FPC can fail to provide adequate compression performance on data from petascale simulation application that contain little or no repeating values or those without much point-to-point correlation.

13.3.1.3 FPzip

The FPzip compression utility was designed to compress values from 2D- and 3D-structured grids, fields, and unstructured meshes. FPzip compression processes data in a coherent fashion and employs a Lorenzo predictor to predict values, and the difference between the predicted and the actual values is encoded using a high speed entropy encoder.

The *Lorenzo predictor* [24], used in FPzip, is a generalization of parallelogram prediction algorithms to an arbitrary number of dimensions. The prediction algorithm relies on the values of its neighbors in a grid to predict the current value. This is in contrast to FPC, which implicitly uses only spatial correlation in its prediction, if the linearization scheme captures it. For example, with a 2D Lorenzo predictor, a point $\langle x,y \rangle$ whose value is given by $f(x,y)$ is predicted as

$$f'(x,y) = f(x,y-1) - f(x-1,y) + f(x-1,y-1) \qquad (13.2)$$

For an n-dimensional grid, an $(n - 1)$-dimension slice must be maintained in memory. Since the addition and subtraction of a large number of floating-point numbers might lead to overflow and underflow, the floating-point values are monotonically mapped to unsigned integers and extra care is taken to identify the propagation of the carry bit. In the case a lossy compression is desired, FPzip allows some of the least significant bits to be discarded during the mapping phase. FPzip exploits an inherent characteristic of data where exponent values can be correlated with neighboring values, and with a light-weight metadata for the predictor, FPzip performs effectively on several datasets.

13.3.2 Lossy Compression

Lossy compression techniques are irreversible operations that trade some form of loss in information or precision to achieve significant reduction in data sizes. Some of the common forms of reduction include quantization, histogram binning, subsampling, and transform-based techniques like wavelet transform and discrete cosine transform (DCT) [39]. The advantage of lossy compression techniques over lossless ones is the ability to provide multifold reduction even on HTC datasets. In cases where domain knowledge about the generated data can be utilized, reduction by over an order of magnitude space is not uncommon. Additionally, space savings can be incorporated when it is known apriori that the data will be used for visualization and analysis routines that do not require high fidelity or full precision data. This amounts to considerable savings in data storage and time otherwise spent in expensive data movement.

13.3.2.1 Subsampling

Subsampling in the spatial or temporal resolution is a common method employed by application scientists to achieve data reduction. For example, in the case of fusion simulations such as GTS [70], the data that are used for analysis are saved to storage only every 10th timestep. In the case of astronomical simulations like Supernova, this number can be as high as 100. Although skipping timesteps help keep the output data within manageable proportions and alleviate the bottleneck on I/O, this technique is not scalable. Skipping a large number of timesteps can result in extreme events or features being missed from identification during the analysis phase.

13.3.2.2 Quantization

Quantization in its simplest form can be defined as a process that maps a set of values representing a large, possibly infinite, range into a smaller range of values. The smaller range of values is called a *codebook vector*, and each individual value is called a *code word*. Although converting the values into codewords results in loss of information, its subsequent encoding results in data compression. The design of the quantizer function determines the amount of compression achieved and information loss that is incurred. The simplest form of quantization is the scalar quantization, which applies a quantizer function to each input value, individually. An example scalar quantizer is a function *round()* that maps values in the real domain to integral domain. Encoding the values in the integral domain provides better compression ratios than its original form.

Scalar quantization techniques are usually easy to apply, and its fast operation makes it an attractive technique for in situ processing. The quantizer function can be further classified as either uniform or nonuniform. Uniform quantization uses the same number of bits to encode every value in the input. This can lead to varying error rates, with regions in the input value that have low entropy incurring a higher penalty. Therefore, a nonuniform quantization can be applied that can adapt to different levels of compression in regions with varying information content, thus leading to more accurate approximation. While nonuniform quantization requires additional metadata to reflect the level of compression, superior quantization performance usually justifies the trade-off. Methods like the Lloyd-Max method [22], which ensures the quantizer boundaries are changed to match the data statistics, or adaptive scalar quantization that selects quantization boundaries "on the fly" to reflect data statistics in local context are effective examples of nonuniform quantization.

An extension of scalar quantization in higher dimensions is vector quantization, which has been shown to be a promising approach for compression. The principle behind vector quantization (VQ) is the fact that coding blocks of spatially correlated values provide more compression than when coding individual values. The overhead involved in sampling and training the input data to generate the codebook vector is nontrivial. To overcome this overhead, some variants of this technique employ no codebook training and adapt the codebook based on local statistics. In either case, while compression rates provided by VQ are usually high, the slow convergence nature of the algorithm, along with issues in precision of computation, complicates its deployment as an in situ processing method.

13.3.2.3 Discrete Transforms: Cosine and Wavelet

The DCT and discrete wavelet transform (DWT), namely Haar and Daubechies [14], are two extensively used techniques in visualization and multimedia routines to achieve high compression ratios. DCT works by subdividing the input data into blocks and traversing the values in a coherent order, such as zigzag (to take advantage of spatial correlation), and then converting the values in each block in its spatial domain to its frequency domain. The resulting compact representation stores a large amount of information in a small number of coefficients. Coefficients with lower values can be eliminated by thresholding, and the remaining few coefficients are optionally quantized and encoded, thus achieving high compression rates. The decoder part reads the frequency values and applies the inverse transform to obtain the data in its original form. Since the majority of the information is contained in a smaller number of coefficients, the decompression results in data that exhibit good visual fidelity compared to the original.

DWT differs from DCT, in which the transformation occurs from the spatial domain to the time–frequency domain. With DWT, the transform allows good localization in both the time and spatial frequency domains, which enables DWT to achieve high compression rates. Wavelets are predominantly used for multiresolution analysis where wavelet transformation can be applied recursively within each block. The compression properties still hold at higher resolutions, and visualization functions take advantage of these properties by allowing different levels of detail to be selected during runtime. Both transform-based compression techniques are ideally suited for in situ data reduction since they offer high compression throughput.

13.3.2.4 Transform Preconditioners: ISABELA

The core idea behind applying DWT and DCT is that the transformation from the spatial domain to a more natural frequency-based domain yields values that are highly compact in its representation. The coefficients (values in the transformed domain) are clustered together, leading to an ideally optimal encoding. Unfortunately, because of inherent high-entropy content in many scientific datasets, the transformations generate a scattered representation, which lowers the compression rate that can be achieved. This means that existing transform-based reduction methods are effective in their natural form only when they can exploit spatial and temporal correlation in the underlying data. To circumvent this problem, new methods have emerged that apply a *preconditioner* to the data before domain transformation, leading to a more natural, high compressible representation upon doing so.

Preconditioners are known to be effective tools to tackle a large number of problems in linear algebra. For example, to calculate the determinant of a large matrix, a transform may be applied to convert it to a lower or upper-diagonal form, thus simplifying the problem into an easier task of multiplying diagonal elements.

In the case of data reduction, what would be an ideal preconditioner that transforms the high-entropy spatiotemporal data into a function of high global regularity? One example of a state-of-the-art lossy compressor using preconditioners is ISABELA [28], which is capable of providing high-accuracy compression with multifold reduction of scientific data. ISABELA uses a sorting preconditioner that sorts the noisy original input signal into a monotonically increasing curve. The exponent values in scientific floating-point datasets exhibit low entropy, as seen in Figure 13.7, thereby producing a gradually increasing smooth curve of high regularity upon sorting. This global regularity leads to rapidly vanishing moments in the transformed space, and compared to the data approximation without the above preconditioner, the transformed signal can now be modeled with high precision using a smaller number of coefficients.

ISABELA Methodology: The core methodology behind ISABELA is quite similar to DWT- and DCT-based encoding. The data from the input signal are divided into smaller chunks, transformed, and subsequently quantized and encoded. With ISABELA, a sorting preconditioner is applied to the input data, which produces a curve whose rate of change in values is the slowest. A smooth curve of this sort can be approximated efficiently using curve fitting techniques. ISABELA compression utilizes cubic *B*-splines, which are piecewise-polynomial functions, to approximate the sorted curve. In contrast to higher-order nonlinear polynomial functions, B-splines model complex curves by dividing the curve into piecewise parametric curves of lower order, and hence, the shape of the curve can be controlled locally without affecting the other parts of a curve. Since cubic B-splines use only a third-order polynomial function, they are fairly efficient in the time taken for both curve approximation and interpolation.

Compared to DCT-based reduction, which performs well on smaller chunks that exhibit a large degree of spatial correlation (neighboring pixels in an image usually have little variation in intensity values), preconditioning a larger chunk ensures a larger number of spatially clustered values. This results in the need to store only a few B-spline coefficients to model the entire data accurately. Thus, the ideal scenario for accurate modeling is when the entire data are sorted. However, the overhead introduced to maintain a large index that tracks the position from the original to the sorted signal eliminates any gain in compression achieved by using a smaller number of coefficients. To balance between the compression ratio and the accuracy levels, ISABELA divides the data into smaller chunks, or *windows*, of fixed sizes. Within each window, a sorting preconditioner is applied, and the curve is then reduced to a set of B-spline coefficients.

The total storage used by ISABELA is the sum of a heavy-weight index (I) and a significantly reduced number of constant coefficients (C), which is light-weight. The index $I = \{i_1, i_2, \ldots, i_N\}$ is a bijective mapping that translates the position in the input array to the new location in the sorted array, where N is the number of elements. Since the range of values in I is a permutation of numbers from 1 to N, each value in the index can be represented using $\log_2(N)$ bits. Clearly, the storage taken up by I is determined by the number of elements in a compression window W. Using the compression ratio metric of original data size divided by compressed data size, the compression ratio of a 64-bit double precision dataset in each window (and hence the entire data) is given by

$$CR = \frac{W \times 64}{W \times \log_2(W) + C \times 64} \tag{13.3}$$

An optimal choice of window size is one that balances the storage taken up by I and incorporates a sufficient number of elements to generate a smooth curve when conditioned. Therefore, to balance the compression ratio and accuracy, the ideal strategy for choice of the input parameters would be to choose the smallest window size.

13.4 MIDDLEWARE FOR IN SITU PROCESSING

As noted, in situ processing is best performed below the application level of the scientific software stack for numerous reasons. Efficiently adding in situ capabilities to scientific codes relies on a few I/O-related optimizations:

- First, compute nodes in current HPC architectures perform *I/O Forwarding* [47]; rather than directly interacting with parallel file systems and causing costly disk and metadata server contention, the data are sent to a smaller number of dedicated I/O nodes across a much faster network interconnect. This not only simplifies super-computer architectures by cutting down on the number of nodes connected to the I/O hardware, but opportunities exist for allowing the I/O nodes to coordinate and optimize file reading/writing. Compute nodes can then initiate I/O requests which, from the compute node point of view, finish much more quickly compared to direct disk interaction. Using an asynchronous interface, the compute nodes may also move on with their computation, providing a higher degree of autonomy.

- Although I/O forwarding provides numerous advantages, it still cannot increase the aggregate throughput restricted by the I/O architecture. Furthermore, I/O patterns of the applications may still cause problems in the I/O forwarding pipeline, such as resource exhaustion on the I/O nodes. Therefore, while it mitigates much of the problems in allowing thousands or more nodes unrestricted access to the file system, the sheer size and scope of the data make I/O prohibitively expensive and the use of a smaller number of I/O nodes introduces unintended consequences. Therefore, large gains in efficiency can be made by adding a layer of *staging nodes* [1] between the compute and I/O nodes. The benefits here are twofold. First, more advanced resource management protocols can be used to minimize the effect of undesirable I/O patterns on compute node stalling. Second, these nodes can additionally perform analytics and visualization "in-flight," while the data are still in memory, reducing the I/O pressure on post-processing scenarios on the data, an espe-cially important task when working with petabyte datasets. A large degree of operators on the data can be performed here, such as those discussed in the previous sections and can be pipelined with applica-tion computation and I/O.

To take advantage of these hardware capabilities, efficient abstractions must be provided that both allow *transparency* of the particular architecture to the user and allow *flexibility* in allowing users to use the underlying hardware in interesting and efficient ways and not restricting them to an overly concise set of functionalities. For instance, it would be unacceptable to force application designers to perform manual bookkeeping across each layer of nodes; instead, an I/O operation should look like an I/O operation from the compute node point of view. At the same time, users who have prior knowledge about their application demands should be able to use the same libraries in a fashion optimal for their application. These software abstractions are called *middleware*, and this section describes numerous aspects of middleware, such as system design, interface to the application designer, challenges in bridging the gap between simplicity and optimality, and so on. These technologies allow application designers to tap into current-and-future generation hardware capabilities, while avoiding the high development costs of writing optimal code directly targeting the hardware.

Before a discussion on middleware ensues, it is important to describe some *enabling technologies*, technologies that the middlewares rely on to provide their functionality to application designers. These are described in the next section. Once we have a brief familiarity with these software codes, we move on to middleware implementing the I/O forwarding and built on top of these enabling technologies and finally discuss staging middleware. These are described in Sections 13.4.2 and 13.4.3, respectively.

13.4.1 Enabling Technologies

13.4.1.1 Collaborative I/O with MPI-IO

MPI-IO [23] is a set of portable I/O APIs that are specifically designed for high-performance parallel MPI programs. It incorporates MPI features, such as interprocess communication and synchronization, into its I/O routines for cooperative reading and writing among a large number of processes. It applies collective I/O to group small noncontiguous accesses to improve data access performance. It is also optimized for parallel file systems to achieve high-throughput parallel data access. MPI-IO usually refers to the interface; there are a number of implementations. One in particular, ROMIO [68], is a high-performance and portable implementation of MPI-IO. It is optimized for noncontiguous data access patterns by data sieving and collective I/O [67]. For multiple read requests, instead of reading each piece separately, ROMIO reads a contiguous chunk of data, which starts from the first requested byte until the last requested byte into a temporary memory buffer. It then extracts the portions the user requested from the temporary buffer and copies them to the user's buffer. ROMIO has been included as part of MPICH2 and MPICH-1 MPI implementations.

13.4.1.2 Hardware-Agnostic I/O Optimization with ADIOS

Existing I/O routines used in scientific codes vary from the standard POSIX I/O and MPI-IO interfaces to higher-level I/O libraries, providing advanced data layout functionality. Two examples of advanced data formatting are HDF-5 [17] and parallel netCDF [30], which provide a set of tools and libraries to help users create, organize, and manage high-volume complex datasets using their data models and formats. They support and optimize for various I/O and storage systems and are portable across different platforms. Different optimization techniques are applied based on the I/O libraries and file systems used. When running the same scientific code in different environments, different I/O routines and configurations need to be applied to achieve optimal performance, which could require significant changes to the application code.

The ADIOS [37] allows us to handle all these differing architecture and configuration environments gracefully. Figure 13.9 shows the architecture of ADIOS and its interaction with scientific codes and I/O libraries. ADIOS provides a set of standard I/O routines, which can be dynamically configured by XML configuration files, without the need to rewrite and recompile the code. Through this design, ADIOS provides extreme flexibility in allowing application scientists to tailor their code to a particular I/O optimization technique without modifying their source code, as well as achieve optimal performance on different platforms through dynamic configurations.

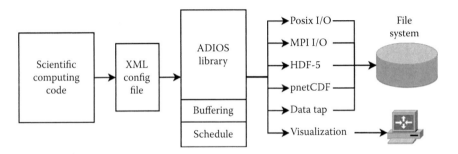

FIGURE 13.9   ADIOS architecture.

More importantly, for staging and in situ analytics middleware, ADIOS is extensible to include various auxiliary tools, such as analytic and monitoring codes, which provide a framework for in situ data analysis and processing. These analytics functions can be embedded within the I/O calls themselves to provide transparent in situ analytics and metadata generation to applications, again without requiring source code changes.

ADIOS accomplishes this by describing data in the BP file format, a portable format compatible with popular scientific file formats such as HDF-5 and netCDF. The BP file format is designed to support *delayed consistency* (performing synchronization operations only at the start and end of each I/O operation to reduce synchronization delay and improve I/O throughput), lightweight in situ data characterization, and resilience. Each process writes its own output into a process group slot. These slots are variably sized based on the amount of data required by each process. Data characteristics for the variables, such as user-defined analytics, are included in process output. These slots can be padded to parallel file system stripe sizes to improve access performance.

13.4.1.3 Resource Overlap with Asynchronous Operations

Regardless of I/O forwarding or data staging, synchronizing I/O requests, even just for transfers from compute to I/O nodes, produces idle times for the compute nodes, something which must be avoided in peta- and exascale applications. Therefore, asynchronous communication and I/O are of paramount importance for efficient large-scale applications. Underlying the need for asynchronous I/O and communication requests is the desire to utilize all resources concurrently and minimize resource-to-resource dependencies. In the context of computing, this means to overlap transfers on the network and to disk with computation that further advances the application.

In hardware, the mechanism of separating the requirement of CPU involvement in memory transactions is called *direct memory access (DMA)*. DMA-enabled hardware subsystems (such as disk controllers and network cards) can access memory independently of the CPU state, though cache coherence on DMA-enabled systems becomes more complicated. Expanding the concept of DMA across computational units, *remote direct memory access (RDMA)*, enables devices on one node (in particular, the network adapter) to access the memory on another node, without CPU involvement. RDMA enables zero-copy, high-throughput, and low-latency networking for high-performance parallel compute clusters. Examples of modern RDMA infrastructures in supercomputing environments include InfiniBand, iWarp, Quadrics Elan, IBM BlueGene Interconnect, and Cray SeaStar [1].

To test the benefits of asynchronous operations in empirical terms, a detailed evaluation was conducted to study strategies of resource overlap in MPI-IO [50]. Different strategies are considered and evaluated for the overlap of I/O with computation and communication to show the performance benefits of asynchronous operations in a distributed context. Specifically, the following overlap strategies are presented and experimentally evaluated: (1) overlapping I/O and communication; (2) overlapping I/O and computation; (3) overlapping computation and communication; and (4) overlapping I/O, communication, and computation. Experiments show that all these techniques are effective and bring performance improvements averaging 25% or greater on a number of applications and maxing out at 85% or greater, parameterized by buffer sizes and number of processors.

13.4.2 I/O Forwarding Middleware

As mentioned, I/O forwarding replaces direct disk access from compute nodes with transfers over the network to dedicated I/O nodes, where the data are subsequently read/written. Typically, as in the Blue-Gene architectures, there is a static compute-to-I/O node mapping. Thus, the primary goal of middleware implementing I/O forwarding is to make the operations as transparent as possible, as well as to supply users with common optimizations such as asynchronous operators. In other words, the user should be able to issue I/O commands from compute nodes as if the nodes had direct connections to the file system and without knowledge of where the data go en-route to/from the disk. A number of questions arise from these relatively simple goals:

1. Given the static compute-to-I/O node mapping, what is the best way to schedule the transfer of data to I/O nodes in collective I/O requests?

2. What opportunities exist for I/O nodes to optimize collective I/O from an arbitrary collection of compute nodes to arbitrary locations in disk?

3. What are some I/O specific optimizations (such as prefetching) that I/O nodes can perform to minimize I/O request completion time?

4. What application-specific phenomena, such as bursty I/O patterns that typically occur in scientific applications [41], can lead to problems such as exhausting memory resources in the pipeline? How does I/O forwarding handle these phenomena?

One example of a software implementation of I/O forwarding is the I/O delegate cache system (IODC) [47]. Figure 13.10 shows the architecture of the IODC system. The IODC divides system cores into two categories, application nodes and I/O delegate nodes (IOD nodes), using ROMIO [68] as its backend. IODC intercepts all I/O requests initiated by applications running on application nodes and redirects them to IOD nodes. These nodes store data in a collective cache system that aggregates small redirects into larger ones for higher throughput. The IOD nodes run a *request detector*, intercepting I/O requests from application nodes. Upon interception, the IODs update their in-memory cache using MPI collective communication between themselves and perform the redirect operation. To simplify cache coherence, the caching system partitions files into pages corresponding to the file system block size and only holds a single cached copy of files among the nodes. The IODC system is embedded in MPI-IO, allowing an abstraction of the underlying parallel file system and

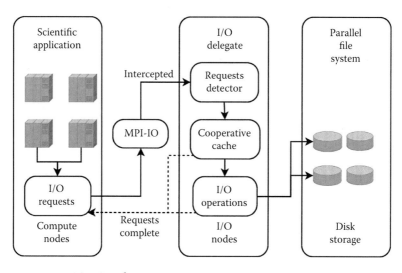

FIGURE 13.10   IODC architecture.

all redirection/cache operations to the application nodes. Therefore, any application using MPI-IO requires no changes to the source code.

13.4.3 Data Staging Middleware

The I/O forwarding mechanisms provide an efficient method of handling data coming to or from disk, while minimizing idle time for compute nodes. In this sense, the addition of data staging nodes represents an evolution of I/O forwarding into a model with a higher degree of flexibility, allowing more complex operations on the data as it makes its way to disk. Part of the expressive power of data staging includes the ability to allow users to define functionality on the staging nodes, opting for in-flight operations such as data analytics, reduction, and reorganization.

However, with this added flexibility comes an added responsibility for an application designer, as well as concerns for library developers. Note that many characteristics of data staging coincide with I/O forwarding, so the questions presented in Section 13.4.2 apply here as well. A sampling of concerns specific to data staging includes the following:

1. How much additional effort must application designers expend to make efficient use of the staging nodes, both on compute nodes (data packaging or preprocessing for staging operators) and staging nodes (custom staging operators)?

2. What are optimal ways of taking arbitrary input from compute nodes, load-balancing the staging operator workload among multiple staging nodes, and utilizing I/O forwarding efficiently and in a pipelined manner?

The simplest staging middlewares exist to combat issues that come up in scientific application I/O patterns such as bursty I/O behavior. Decoupled and asynchronous remote transfers (DART) [15] is an example framework to improve I/O through staging that consists of three components: the client layer, streaming server, and receiver. The client layer runs on compute nodes and links with the application. It notifies the streaming server when the application has data ready to transfer. The streaming server runs on service nodes and extracts data from compute nodes after receiving notification. It then transfers the data to the receiver running on remote nodes. It can also directly write data to local storage if the receiver is not specified. Using an additional layer between compute and I/O nodes allows DART to control in a finer-grain manner how to transfer data from the compute nodes to the I/O nodes, leading to smoother access patterns and resource utilization. A similar system is PDIO (portals direct I/O) [66], which adopts the three-tier architecture and supports wide-area network (WAN) data transfer. PDIO runs daemons on externally connected I/O nodes. The daemons receive messages and data from the clients on the compute nodes, aggregate them into memory buffers with optimal size, and asynchronously send them to remote receivers over the WAN via parallel TCP/IP data streams.

DataStager [1] provides an extensible framework for staging and I/O forwarding, providing mechanisms for scheduling and load-balancing transfers as well as allowing user-defined computation. The staging nodes serve as another layer to forward I/O for compute nodes and also perform in situ data capture and analysis for visualization [16] and metadata generation. DataStager leverages RDMA-based network infrastructures and enables the staging area to take charge of data processing jobs that require high throughput, including synchronization, aggregation, collective processing, and data validation. The processing results can be used as input to a variety of extensible service pipelines such as data storage on disk, visualization, and data mining.

DataStager has two components: a library called DataTap and a parallel staging service called DataStager. DataTap is a client-side library co-located with the compute application. It is implemented with the ADIOS API to support asynchronous I/O and keeps application code changes to a minimum. DataTap uses an efficient, self-describing binary format, FFS [8], to mark up its binary data. The FFS format makes it possible for binary data to be analyzed and modified in transit and enables the graph-structured data processing overlays, termed I/OGraphs [2]. The overlays can be customized for a rich set of backend uses, including in situ processing, online data visualization, data storage, and transfer to remote sites.

The staging service, DataStager, is composed of server-side processes to actively read data from DataTap clients using RDMA techniques. It applies server-directed I/O for asynchronous communication to fetch data from compute nodes. Server-directed I/O lets the I/O nodes control the data transfers and hence the resources based on their capacity, which allows for smoother access patterns. In addition, the server-controlled data transfer causes minimal runtime impact by allowing the application to progress without actively pushing the data out. The server-directed I/O is particularly useful in HPC environments where a small partition of I/O nodes serves a large number of compute nodes. The disparity in the sizes of the partitions, accompanied with the bursty behavior of most scientific application I/O [41], can exhaust memory and I/O resources on the I/O nodes. DataStager uses resource-aware schedulers to select I/O operations to carry out. The selection of an operation is based on the memory space on I/O nodes and the status of the application nodes and schedulers (e.g., idle vs. nonidle). If all schedulers agree to issue a transfer request, an RDMA read request is issued to the originating application node. Multiple requests may be serviced simultaneously if resources permit. Once the RDMA read is completed, the staging handler is notified and will then process the message according to the configuration. It can direct write data to disk, forward data to network, query and analyze data online, and so on.

Based on DataStager, an in situ data processing middleware called preparatory data analytics (PreDatA) [80] has been built. PreDatA is an approach for preparing and characterizing data produced by the large-scale simulations running on petascale machines in an in situ manner. It exploits the computational power of staging nodes and provides a pluggable framework for executing user-defined operations such as data reorganization, real-time data characterization, filtering and reduction, and select analysis. Data can be treated as streams and the operations can be specified in ways natural to the "streaming" context, so that streaming data processing techniques can be applied in PreDatA with little additional porting effort.

PreDatA middleware can be placed in both the compute nodes and the staging nodes. Data analysis and operations can be plugged and hosted in either location. When application performs I/O operations, PreDatA acquires output data through ADIOS and stages data from compute nodes to staging nodes. In-transit data processing is performed along the data flow. PreDatA schedules asynchronous data movement from compute nodes to staging nodes to minimize interference with the simulation. PreDatA supports user-defined data operations such as buffer management, scheduling, data indexing and query, and data exchange and synchronization across staging nodes. It also provides a pluggable framework for end users to specify, deploy, and debug data processing and analysis functions.

13.5 IN SITU DATA LAYOUT OPTIMIZATION

The data staging model provides opportunities for in situ data processing and analysis, which provides huge benefits for understanding the data without performing costly disk reads over the raw data for analytics. However, in situ data processing has its limits. For instance, there simply is not enough memory available for analysis in a *global context*, which is essential for *exploratory data analytics*. Therefore, it is infeasible to perform all data analysis at simulation time, and the data may have to be read multiple times for post-simulation analytics in a global context.

With this requirement in mind, there are a number of ways we can accelerate future analytics operations. Preferably in the staging process, applications can use data reorganization to optimize future post-processing. One scenario for this occurs in scientific databases, emphasizing query processing with heterogeneous constraints.

Scientific database technologies, especially when considering in situ computation, involve a number of considerations. First, what types of queries and data access patterns are these database systems optimized for (Section 13.5.1.1)? Second, what optimization techniques do databases currently apply for different access patterns (Sections 13.5.1.2 and 13.5.1.3)?

13.5.1 Basic Layout Optimization Techniques

Scientific simulation codes usually generate multidimensional spatio-temporal data, which is quite different from structured data stored in traditional relational databases, and thus requires different optimization techniques. Data accesses are usually accompanied by value and spatial constraints. For example, for a climate dataset, the user might want to know what regions within certain latitudinal and longitudinal ranges have abnormally high temperature values.

Different query types lead to different data access patterns, but contiguous access patterns are the most efficient. Thus, the goal of data layout optimization is to store data based on potential access patterns to achieve more contiguous access patterns in queries. But first, a discussion of the access patterns themselves, defined by query semantics, must be discussed.

13.5.1.1 Query Types and Data Access Patterns

We summarize common query types and data access patterns on scientific spatiotemporal data as follows:

Value-constrained: Queries that request spatial regions and/or their corresponding variable values, subject to constraints on those or other variable values. For example, what (latitude, longitude) pairs at some time in a simulation have an abnormally high temperature? What are those temperature values? These queries are also known as range queries.

Region-constrained: Queries that request spatial regions and/or their corresponding variable values, subject to regional constraints. For example, what are the temperature values within North Carolina at some time?

Value- and region-constrained: Queries that request spatial regions and/or their corresponding variable values, subjects to constraints on both the regions as well as the variable values. For example, what regions within North Carolina have an abnormally high temperature?

Besides the query types listed above, multivariate and multiresolution data access should also be considered. In multivariate data access, multiple variables may be accessed with constraints on different variables. In multiresolution data access, only part of the data in certain resolution is accessed to satisfy coarse-grained analytic requirements. These are also common access patterns for scientific spatiotemporal data and require different layout optimization techniques to achieve best access performance.

13.5.1.2 Optimization for Value-Constrained Queries

To optimize the data layout for value-constrained access, points with similar values should be stored together to achieve more contiguous data access. Data can be partitioned into *bins*, where each bin represents a set of similar values. *Value-constrained binning* is applied to assign data points in different bins based on their values, and data points within the same bins are stored together on the storage space to achieve contiguous data access patterns for value-constrained access.

A database popularized by fast value-constrained query performance is FastBit [75], a state-of-the-art bitmap indexing scheme, which applies different binning techniques to optimize for range queries. Bitmap indexing techniques traditionally employ any combination of three tasks: binning, encoding, and index compression. FastBit binning maps variables to bins so that variables of a similar property or value are co-located for quick lookup. For example, given a bin B that represents a range $[0,5)$, for each record, a bit is used to represent whether the record falls into a bin or not. A bitmap vector is encoded using this technique for each defined bin. Since the space taken up by the bitmap vectors becomes unmanageable for large datasets, FastBit employs a word-aligned-hybrid compression scheme [77], based on run-length encoding, to reduce the index size. To support and optimize for multivariate queries, FastBit applies fast bitmap operations such as AND and OR among bitmaps of different variables to achieve fast selection. However, building the bitmap indexes is an expensive operation, lowering the capability for performing it in situ. Furthermore, the storage requirements for the indexes are on the order of the dataset size, adding significantly to I/O and storage costs.

ISABELA-QA [52] is a query engine based on ISABELA-compressed data (see Section 13.3.2.4), designed to occupy far less space than existing database technologies and indexes. Similar to FastBit, ISABELA-QA employs binning of values to cluster data with similar values. However, ISABELA-QA, rather than building bitmaps to identify particular values with bins, co-locates the ISABELA-generated permuted indices for each window on a per-bin basis, taking advantage of the sorted nature of the windows to minimize the ensuing metadata. When processing queries, bin metadata is fetched, compression windows are identified, and the B-spline coefficients and permuted indices are read. One particular advantage of this setup is that data points can be interpolated independently of one another, so only a subset of the window need to be decompressed. Furthermore, the storage footprint is nearly the same as a linear organization of the compression windows, occurring additional storage costs only for the metadata.

In a similar vein, ALACRI²TY [25] fuses lossless compression and database indexing to produce a lightweight data and index representation. The method relies on the representation of floating-point data in memory, consisting of a fractional mantissa component exponentiated by an exponent component and given an explicit sign bit. The most significant bytes, containing the sign, exponent, and most significant mantissa components, are removed from the dataset and unique value encoding is used to compress them. These unique values then form the bin edges, and data are reorganized into bins corresponding to their high-order bytes. An *inverted index* map bins to record IDs to maintain the original mapping.

13.5.1.3 Optimization for Region-Constrained Queries

Region-constrained queries usually access subvolumes of multidimensional data, in which the points are spatially contiguous. However, multidimensional data need to be linearized before being written to disks since storage space is 1D. If data are linearized as they are stored in memory (e.g., row-major order for 2D data), there will be high performance disparity when accessing data along different dimensions. Thus, the key challenge is how to linearize data to achieve better spatial locality.

One common technique is to divide multidimensional data into chunks. The chunking technique divides each dimensions into equal- or variate-sized partitions. The SciDB system [6] is a distributed spatial database system, which applies chunking techniques to optimize for spatial-constrained queries [65]. Chunks are distributed over multiple nodes to achieve parallel data access. SciDB includes special optimizations for sparse matrices and introduces overlapping areas to optimize for certain calculations.

A more fine-grained optimization is to use *space-filling curves (SFC)* [53] to linearize data on disks to improve spatial locality. Rather than storing data blocks in row-major or column-major order, data are reorganized in SFC order. The SFC order helps to ensure that spatially contiguous points in multidimensional space are also placed contiguously on disks, thus reducing potential seek operations in queries accessing certain spatial regions. Popular SFCs include Hilbert SFC, Z-order SFC, Moore SFC, and so on. An illustration of these is shown in Figure 13.11. Among all SFCs, Hilbert SFC has been demonstrated to have best preservation of spatial locality properties. A detailed analysis and comparisons of clustering achieved by Hilbert curves, Z-order curves, and Gray code curves are presented in Moon et al. [42] The Hilbert curve reduces the number of clusters by 50% compared with Z-order curve and Grey code curve for 2D square and sphere access. For 3D cube access, Hilbert curve reduces the number of clusters by 45%, and for sphere access, it reduces the number of clusters by 30% compared to the other space-filling curves.

Since space-filling curves are usually defined in a recursive manner, some SFCs can additionally be thought of as a hierarchical representation to support *multiresolution* data access. A hierarchical indexing scheme based on Z-order SFCs [49] has been introduced. For each level of resolution, a subset of points is stored together in Z-order, and lower levels of resolution contain fewer points than the higher resolutions. Higher

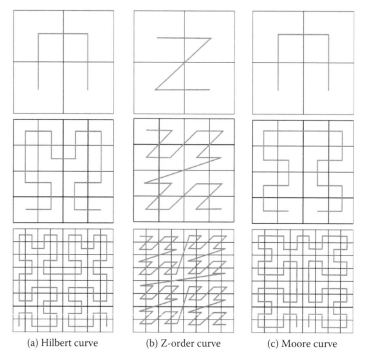

(a) Hilbert curve	(b) Z-order curve	(c) Moore curve

FIGURE 13.11 The first three iterations of Hilbert/Z-order/Moore space-filling curves.

resolution access can be always achieved by fetching data that are contiguously located on storage space.

13.5.2 Toward Heterogeneous Access Patterns

One major drawback of existing storage layout techniques is that they primarily focus on optimizations for a *particular* access pattern(s). For example, the SFCs mentioned in the previous section only improve performance for access patterns induced by spatial constraints on subplanes/subvolumes of the data space. For value-constrained access patterns, the entire dataset must be scanned to select qualified points. Similarly, FastBit and ISABELA-QA are optimized only for value-constrained accesses and cannot handle spatially constrained accesses efficiently. A naïve approach to support multiple access patterns would be creating copies of dataset that favor different access patterns. However, the ever-increasing sizes of simulation data make multiple replications infeasible at extreme scale.

One potential way to address these challenges is to utilize multiple value-constraint-optimizing and region-constraint-optimizing techniques. One example of this is in a configurable, layered layout optimization technique, which iteratively applies SFC and ISABELA-compression to provide query processing performance improvements for numerous access patterns [20]. The layout scheme combines Hilbert space-filling curves and value-constrained binning, through ISABELA-QA, together. It divides data into blocks and interleaves these blocks across value-constrained bins along Hilbert space-filling curves to optimize for both value and region queries with both value and spatial constraints. Based on this scheme, MLOC, a flexible multilevel framework, has been presented [21]. MLOC applies a flexible, hierarchical multilevel architecture. Multiple fine-grained optimization techniques can be flexibly placed within the framework to achieve optimizations for multiple access patterns in user-defined priority. Users can specify the order of the optimization techniques within the framework to generate a different layout on storage based on the frequency of access patterns induced by real queries.

13.6 LIMITATIONS AND FUTURE DIRECTIONS

Although there is a large amount of pioneering work being performed to bring robust, efficient in situ processing to reality, the problems necessitating in situ processing, namely I/O bottlenecks in large-scale systems, remains far from being solved. Specifically, there are a number of limitations with the current state of the art that must be improved upon in the coming years as compute capability continues to scale at a faster rate than I/O capability.

13.6.1 Data Analysis

In situ data analysis methods, some of which are identified in this chapter, have shown impressive progress in taking advantage of large-scale data resident in memory to accelerate knowledge discovery. In spite of this, it is infeasible to perform *all* data analytics at simulation write time, especially for interactive processes such as query processing. Furthermore, the act of aligning simulation code and staging architectures with scalable analytics kernels is a nontrivial problem.

For data analysis methods that are feasible to perform in situ, it is crucial to continue advancing the state of the art, developing and optimizing scalable, robust analysis algorithms, especially for increasingly complex simulation datasets and analytics scenarios.

13.6.2 Data Reduction

Recent compression algorithms, focused on scientific data, have advanced both compression speed and compression ratio for *hard-to-compress* double-precision datasets, which scientific simulations predominantly use for accurate computation. However, it has been observed that average lossless data reduction rates for these datasets, so far, fall in the range of 30% [56]. While this degree of reduction is significant for datasets in the terabytes, petabytes, and beyond, it does not come close to bridging the performance trends gap of compute and I/O. While lossy compression can be a solution to this, with far higher data reduction rates, the loss of precision may not be acceptable to application scientists, simulations requiring pristine data for checkpointing, and analysis functions sensitive to changes in precision.

To achieve higher compression ratios, it is necessary to exploit all possible relationships in the data, creating a *context-sensitive* approach to lossless compression. For example, FPzip takes advantage of spatial correlations between data points for prediction, and ISOBAR takes advantage of similarities in the double-precision data representation for compression. These techniques are a good start, but more research taking advantage of these and other characteristics need to be performed. For instance, is it possible to use temporal relationships in compression, and if so, how to maintain the necessary information between simulation timesteps within a staging framework? Are there correlations between grid nodes of a particular level of resolution in adaptive mesh refinement (AMR) simulations?

For lossy compression, it is crucial that the loss of precision be able to be quantified and bounded so that application scientists can reason about the trade-offs between data accuracy, analysis accuracy, and data reduction. For instance, ISABELA provides such a bound at compression time through error encoding, trading off additional storage cost for guaranteed accuracy bounds [29]. As with lossless compression, increasing the context-sensitivity of compression techniques is necessary to see huge reductions of data without suffering too much loss of precision.

Furthermore, compression methodologies must be compatible with staging architectures in order for the benefits of data reduction to translate into I/O cost reductions. Effective pipelining of compression and I/O operations is a necessity to hide compression costs in terms of CPU time. Initial work, for example, has been done pipelining ISOBAR-compression of byte columns with I/O operations using ADIOS [55], resulting in both reduced storage footprints and write times. The trick behind the performance seen in the ISOBAR pipelining method is the semantic division of tasks (e.g., multiple compression streams for both compressible and incompressible byte streams).

13.6.3 Architecture

The march of new hardware trends is a constant in the computing world, and heterogeneity is increasing in HPC systems as a result. Nearly all compute clusters are comprised of multicore processors, and alternative computing hardware such as GPUs and FPGAs are highly active in the research community as a method to achieve a high computation rate with lower relative power consumption. There are a number of implications for in situ computation as a result. First, how do we best utilize and integrate emerging hardware architectures, such as GPUs and SSDs, within a robust in situ framework? Second, access to CPU main memory is required for network and I/O operations, memory transfer to co-processors (such as GPUs, with discrete main memory), not to mention the simulation computations themselves. With so many competing resources for main memory, how does one additionally include in situ operations in such a way as to not be bottlenecked by memory bandwidth and to optimize multicore/memory hierarchy usage?

On an intranode level, the increasing complexity of multicore processors requires new ways of thinking about how computation maps to resources and how to best utilize the cache hierarchy, which can be performed at both the application and compiler levels [26,35]. In situ processing techniques must take advantage of these architectural advances and utilize richer forms of inter- and intranode parallelism (e.g., MPI and OpenMP).

On an off-chip level, new technologies such as SSDs, GPUs, and so on pave the way toward new and exciting in situ capabilities. SSDs, while being more expensive and having less data density than traditional HDDs, have low power consumption and far better random I/O rates, making them well suited for local disk caches on compute nodes. These would allow for in situ computations and analysis requiring too much memory to be performed in memory alongside simulation data. GPUs provide the possibility to accelerate compute-intensive in situ operations and allow CPUs to instead perform data management tasks.

CONCLUSION

It is highly unlikely that there will be a "silver bullet" solution to the growing disparity between computational capability and I/O bandwidth. This means that new programming paradigms must be adopted to continue the phenomenal increases in the ability to simulate, solve, and/or analyze larger and more complex problems into the exascale. In situ processing of data is a strong first step in this direction. Rather than cutting corners to allow scientific codes to complete in a reasonable amount of time, in situ algorithms can exploit the far superior compute performance to allow meaningful computation at simulation time, eliminating the need for costly post-processing.

There are many such cases where in situ computation makes sense in the data analysis realm, which have been presented here; the take-away point is that we get insightful analysis of data at minimal cost by performing the operations while waiting on I/O. Furthermore, we can prepare the data in situ for future analyses, which is essential for efficient database operations such as query processing. Rather than incurring the enormous cost of reorganizing the whole database from disk, the data can be shuffled at write time for a far more efficient database construction methodology. Finally, we can directly optimize the application on multiple fronts using in situ compression techniques: lossless (and especially lossy) compression reduces the I/O pressure and storage costs, while making use of the huge aggregate throughput of compute clusters.

Given the number of operations we can perform in situ, we can ensure that, when I/O becomes rate-limiting for scientific codes, there are ways in which we can enable maximum utilization of the underlying hardware and continue the relentless pace of scientific discovery provided by supercomputing.

ACKNOWLEDGMENTS

We would like to acknowledge and thank the Deparment of Energy, Office of Science, and National Science Foundation for their funding support.

REFERENCES

1. H. Abbasi, M. Wolf, G. Eisenhauer, S. Klasky, K. Schwan, and F. Zheng. DataStager: Scalable data staging services for petascale applications. In *Proceedings of the 18th ACM International Symposium on High Performance Distributed Computing*, HPDC '09, pages 39–48, New York, NY, USA, 2009. ACM.
2. H. Abbasi, M. Wolf, and K. Schwan. LIVE data workspace: A flexible, dynamic and extensible platform for petascale applications. *Cluster Computing, IEEE International Conference on*, Austin, Texas 0:341–348, 2007.

3. J. Ahrens, K. Heitmann, M. Petersen, J. Woodring, S. Williams, P. Fasel, C. Ahrens, H. Chung-Hsing, and B. Geveci. Verifying scientific simulations via comparative and quantitative visualization. *IEEE Computer Graphics and Applications*, 30:16–28, 2010.

4. J. Ahrens, J. Woodring, S. Williams, C. Brislawn, S. Mniszewski, P. Fasel, J. Patchett, and L. Lo. Visualization and data analysis challenges and solutions at extreme scales. In *Proceedings of the Scientific Discovery through Advanced Computing Conference*, 2011, Denver, Colorado.

5. B. Fryxell, K. Olson, P. Ricker, F. X. Timmes, M. Zingale, D. Q. Lamb, P. MacNeice, R. Rosner, J. W. Truran, and H. Tufo. FLASH: An adaptive mesh hydrodynamics code for modeling astrophysical thermonuclear flashes. *The Astrophysical Journal Supplement Series*, 131:273–334, 2000.

6. P. G. Brown. Overview of SciDB: Large scale array storage, processing and analysis. In *Proceedings of the 2010 International Conference on Management of Data*, SIGMOD '10, pages 963–968, New York, NY, USA, 2010. ACM.

7. M. Burtscher and P. Ratanaworabhan. High throughput compression of double-precision floating-point data. In *IEEE Data Compression Conference*, pages 293–302, 2007, Snowbird, Utah.

8. F. Bustamante, G. Eisenhauer, and K. S. Widener. Efficient wire formats for high performance computing. In *Proceedings of Supercomputing 2000*, pages 39–39, 2000, Dallas, Texas.

9. P. H. Carns, W. B. Ligon III, R. B. Ross, and R. Thakur. PVFS: A parallel file system for Linux clusters. In *Proceedings of the 4th Annual Linux Showcase and Conference—Volume 4*, pages 28–28, Berkeley, CA, USA, 2000. USENIX Association.

10. J. Chen, Y. Kusurkar, and D. Silver. Distributed feature extraction. In *Proceedings of the Conference on Visualization and Data Analysis*, pages 189–195, 2002, San Jose, California.

11. J. Chen, D. Silver, and L. Jiang. The feature tree: Visualizing feature tracking in distributed AMR datasets. In *Proceedings of the IEEE Symposium on Parallel and Large-Data Visualization and Graphics*, pages 103–110, 2003, Seattle, Washington.

12. J. Chen, D. Silver, and M. Parashar. Real time feature extraction and tracking in a computational steering. In *Proceedings of the High Performance Computing Symposium*, pages 155–160, 2003, Orlando, Florida.

13. A. Choudhary, D. Honbo, P. Kumar, B. Ozisikyilmaz, S. Misra, and G. Memik. Accelerating data mining workloads: Current approaches and future challenges in system architecture design. *Wiley Interdisciplinary Reviews: Data Mining and Knowledge Discovery*, 1:41–54, 2011.

14. I. Daubechies. Orthonormal bases of compactly supported wavelets II: Variations on a theme. *SIAM Journal on Mathematical Analysis*, 24:499–519, March 1993.

15. C. Docan, M. Parashar, and S. Klasky. DART: A substrate for high speed asynchronous data IO. In *Proceedings of the 17th International Symposium on High Performance Distributed Computing*, HPDC '08, pages 219–220, New York, NY, USA, 2008. ACM.

16. N. Fabian, K. Moreland, D. Thompson, A. C. Bauer, P. Marion, B. Geveci, M. Rasquin, and K. E. Jansen. The ParaView coprocessing library: A scalable, general purpose in situ visualization library. In *Proceedings of the IEEE Symposium on Large-Scale Data Analysis and Visualization*, October 2011, Providence, Rhode Island.

17. M. Folk, G. Heber, Q. Koziol, E. Pourmal, and D. Robinson. An overview of the HDF5 technology suite and its applications. In *Proceedings of the EDBT/ICDT 2011 Workshop on Array Databases*, AD '11, pages 36–47, New York, NY, USA, 2011. ACM.

18. A. Ganguly, J. Gama, O. Omitaomu, M. Gaber, and R. Vatsavai. *Knowledge Discovery from Sensor Data*. Boca Raton, FL, CRC Press, Taylor & Francis, 2009.

19. B. Goeman, H. Vandierendonck, and K. D. Bosschere. Differential FCM: Increasing value prediction accuracy by improving table usage efficiency. In *Seventh International Symposium on High Performance Computer Architecture*, pages 207–216, 2001, Nuevo Leone, Mexico.

20. Z. Gong, S. Lakshminarasimhan, J. Jenkins, H. Kolla, S. Ethier, J. Chen, R. Ross, S. Klasky, and N. F. Samatova. Multi-level layout optimization for efficient spatio-temporal queries on ISABELA-compressed data. In *Proceedings of the 26th IEEE International Parallel and Distributed Processing Symposium*, IPDPS '12, 2012, Shanghai, China.

21. Z. Gong, T. Rogers, J. Jenkins, H. Kolla, S. Ethier, J. Chen, R. Ross, S. Klasky, and N. F. Samatova. MLOC: Multi-level layout optimization framework for compressed scientific data exploration with heterogeneous access patterns. In *Proceedings of the 41st International Conference on Parallel Processing*, ICPP '12, 2012, Pittsburgh, Philadelphia.

22. R. Gonzalez and R. Woods. *Digital Image Processing (2nd Edition)*. Upper Saddle River, NJ, Prentice Hall, 2002.

23. W. Gropp, E. Lusk, and R. Thakur. *Using MPI-2: Advanced Features of the Message-Passing Interface*. MIT Press, Cambridge, MA, 1999.

24. L. Ibarria, P. Lindstrom, J. Rossignac, and A. Szymczak. Out-of-core compression and decompression of large n-dimensional scalar fields. *Computer Graphics Forum*, 22:343–348, 2003.

25. J. Jenkins, I. Arkatkar, S. Lakshminarasimhan, N. Shah, E. R. Schendel, S. Ethier, C. Chang, J. Chen, H. Kolla, S. Klasky, R. Ross, and N. F. Samatova. Analytics-driven lossless data compression for rapid in-situ indexing, storing, and querying. *Proceedings of the 23rd International Conference on Database and Expert Systems Applications (DEXA)*, 2012, Vienna, Austria.

26. M. Kandemir, T. Yemliha, S. Muralidhara, S. Srikantaiah, M. Irwin, and Y. Zhang. Cache topology aware computation mapping for multicores. *Programming Language Design and Implementation*, 74–85, 2010.

27. S. Klasky, H. Abbasi, J. Logan, et al. In situ data processing for extreme scale computing. In *Proceedings of the Scientific Discovery through Advanced Computing Program (SciDAC)*, Denver, CO, USA, 2011.

28. S. Lakshminarasimhan, N. Shah, S. Ethier, S. Klasky, R. Latham, R. Ross, and N. F. Samatova. Compressing the incompressible with ISABELA: In-situ

reduction of spatio-temporal data. In *Euro-Par*, pages 366–379, 2011, Parallel Processing, Bordeaux, France.

29. S. Lakshminarasimhan, N. Shah, S. Ethier, S.-H. Ku, C. Chang, S. Klasky, R. Latham, R. Ross, and N. F. Samatova. ISABELA for effective in-situ compression of scientific data. *Concurrency and Computation: Practice and Experience*, 2012 (In Press).

30. J. Li, W.-K. Liao, A. Choudhary, R. Ross, R. Thakur, W. Gropp, R. Latham, A. Siegel, B. Gallagher, and M. Zingale. Parallel netCDF: A high-performance scientific I/O interface. In *Proceedings of the 2003 ACM/IEEE Conference on Supercomputing*, SC '03, pages 39–49, New York, NY, USA, 2003. ACM.

31. J. Li, Y. Liu, W. K. Liao, and A. Choudhary. Parallel data mining algorithms for association rules and clustering. In *Handbook of Parallel Computing: Models, Algorithms, and Applications*. Sanguthevar Rajasekaran, John Reif, eds., Boca Raton, FL, CRC Press, 2007.

32. Y. Li. *Dark Matter Halos: Assembly, Clustering and Sub-halo Accretion*. PhD thesis, University of Massachusetts, Amherst, 2010.

33. D. Lignell, J. Chen, T. Lu, and C. Law. Direct numerical simulation of extinction and reignition in a nonpremixed turbulent ethylene jet flame. *Western States Section Meeting of the Combustion Institute*, 2007.

34. P. Lindstrom and M. Isenburg. Fast and efficient compression of floating-point data. *IEEE Transactions on Visualization and Computer Graphics*, 12:1245–1250, 2006.

35. J. Liu, Y. Zhang, W. Ding, and M. Kandemir. On-chip cache hierarchy-aware tile scheduling for multicore machines. In *Code Generation and Optimization (CGO), 2011 9th Annual IEEE/ACM International Symposium on*, pages 161–170, 2011, Chamonix, France.

36. Y. Liu, W. K. Liao, and A. Choudhary. Design and evaluation of a parallel HOP clustering algorithm for cosmological simulation. In *Proceedings of the IEEE International Parallel & Distributed Processing Symposium*, pages 82–89, 2003, Boston, Massachusetts.

37. J. F. Lofstead, S. Klasky, K. Schwan, N. Podhorszki, and C. Jin. Flexible IO and integration for scientific codes through the adaptable IO system (ADIOS). In *Proceedings of the 6th International Workshop on Challenges of Large Applications in Distributed Environments*, CLADE '08, pages 15–24, New York, NY, USA, 2008. ACM.

38. K. L. Ma. In situ visualization at extreme scale: Challenges and opportunities. *IEEE Computer Graphics and Application*, 29:14–19, 2009.

39. K. L. Ma, C. Wang, H. Yu, and A. Tikhonova. In-situ processing and visualization for ultrascale simulations. *Journal of Physics: Conference Series*, 78(1):012043, 2007.

40. A. Mascarenhas, R. W. Grout, P. T. Bremer, E. R. Hawkes, V. Pascucci, and J. H. Chen. Topological feature extraction for comparison of terascale combustion simulation data. In *Topological Methods in Data Analysis and Visualization*. Robert Peikert, Hamish Carr, Helwig Hauser, Raphael Fuchs, eds., Springer, Berlin Heidelberg, 2011.

41. E. L. Miller and R. H. Katz. Input/output behavior of supercomputing applications. In *Proceedings of the 1991 ACM/IEEE Conference on Supercomputing*, Supercomputing '91, pages 567–576, New York, NY, USA, 1991. ACM.

42. B. Moon, H. V. Jagadish, C. Faloutsos, and J. H. Saltz. Analysis of the clustering properties of the Hilbert space-filling curve. *IEEE Transactions on Knowledge and Data Engineering*, 13:124–141, 2001.

43. C. Muelder and K. L. Ma. Rapid feature extraction and tracking through region morphing. Technical Report CSE-2007-25, University of California at Davis, 2007.

44. H. Nagesh, S. Goil, and A. Choudhary. A scalable parallel subspace clustering algorithm for massive datasets. In *Proceedings of the International Conference on Parallel Processing*, pages 477–483, 2000, Toronto, Canada.

45. H. Nagesh, S. Goil, and A. Choudhary. Parallel algorithms for clustering high-dimensional large-scale datasets. In *Data Mining for Scientific and Engineering Applications*, Robert L. Grossman, Chandrika Kamath, Philip Kegelmeyer, Vipin Kumar, Raju R. Namburu, eds., Dordrecht, The Netherlands, Kluwer Academic Publishers, 2001.

46. R. Narayanan, B. Ozisikyilmaz, J. Zambreno, G. Memik, and A. Choudhary. MineBench: A benchmark suite for data mining workloads. In *Proceedings of the IEEE International Symposium on Workload Characterization*, pages 182–188, 2006, San Jose, California.

47. A. Nisar, W.-K. Liao, and A. Choudhary. Scaling parallel I/O performance through I/O delegate and caching system. In *Proceedings of the 2008 ACM/IEEE Conference on Supercomputing*, SC '08, pages 9:1–9:12, Piscataway, NJ, USA, 2008. IEEE Press.

48. B. Ozisikyilmaz, R. Narayanan, J. Zambreno, G. Memik, and A. Choudhary. An architectural characterization study of data mining and bioinformatics workloads. In *Proceedings of the IEEE International Symposium on Workload Characterization*, pages 61–70, 2006, San Jose, California.

49. V. Pascucci and R. J. Frank. Global static indexing for real-time exploration of very large regular grids. In *Proceedings of the 2001 ACM/IEEE Conference on Supercomputing (CDROM)*, Supercomputing '01, pages 2–2, New York, NY, USA, 2001. ACM.

50. C. M. Patrick, S. Son, and M. Kandemir. Comparative evaluation of overlap strategies with study of I/O overlap in MPI-IO. *SIGOPS Operating Systems Review*, 42:43–49, October 2008.

51. P. Pebay. Formulas for robust, one-pass parallel computation of covariances and arbitrary-order statistical moments. Technical Report SAND2008-6212, Sandia National Laboratories, 2008.

52. S. Lakshminarasimhan, J. Jenkins, I. Arkatkar, Z. Gong, H. Kolla, S.-H. Ku, S. Ethier, J. Chen, C. S. Chang, S. Klasky, R. Latham, R. Ross, and N. F. Samatova. ISABELA-QA: Query-driven data analytics over ISABELA-compressed scientific data. In *International Conference for High Performance Computing, Networking, Storage and Analysis (SC 2011)*, Seattle, Washington, 2011.

53. H. Sagan. *Space-Filling Curves*. Springer-Verlag, New York, NY, 1994.

54. R. Samtaney, D. Silver, N. Zabusky, and J. Cao. Visualizing features and tracking their evolution. *Computer*, 27:20–27, 1994.

55. E. Schendel, S. Pendse, J. Jenkins, D. Boyuka II, Z. Gong, S. Lakshminarasimhan, Q. Liu, H. Kolla, J. Chen, S. Klasky, R. Ross, and N. Samatova. *ISOBAR Hybrid Compression-I/O Interleaving for Large-Scale Parallel I/O Optimization.* Delft, The Netherlands, HPDC, pages 61–72, 2012.

56. E. R. Schendel, Y. Jin, N. Shah, J. Chen, C. S. Chang, S. H. Ku, S. Ethier, S. Klasky, R. Latham, R. Ross, and N. F. Samatova. ISOBAR preconditioner for effective and high-throughput lossless data compression. *International Conference on Data Engineering (ICDE)*, 2012, Washington, DC.

57. F. Schmuck and R. Haskin. GPFS: A shared-disk file system for large computing clusters. In *First USENIX Conference on File and Storage Technologies (FAST'02)*, pages 231–244, Monterey, CA, 2002.

58. P. Schwan. Lustre: Building a file system for 1,000-node clusters. In *Proceedings of the Linux Symposium*, 2003, Ottawa, Canada.

59. N. Shah, Y. Shpanskaya, C. S. Chang, S. H. Ku, A. V. Melechko, and N. F. Samatova. Automatic and statistically robust spatio-temporal detection and tracking of fusion plasma turbulent fronts. In *Proceedings of the Scientific Discovery through Advanced Computing Conference*, 2010, Chattanooga, Tennessee.

60. S. Shekhar, V. Gandhi, P. Zhang, and R. R. Vatsavai. *Availability of Spatial Data Mining Techniques.* In *The SAGE Handbook of Spatial Analysis.* A. S. Fotheringham, P. A. Rogerson, eds. pages 63–85. SAGE Publications, Thousand Oaks, California, 2009.

61. S. Shekhar, R. R. Vatsavai, and M. Celik. Spatial and spatiotemporal data mining: Recent advances. In *Next Generation of Data Mining,* Hillol Kargupta, Jiawei Han, Philip S. Yu, Rajeev Motwani, Vipin Kumar, eds. Boca Raton, FL, CRC Press, 2008.

62. S. Shekhar, P. Zhang, Y. Huang, R. Vatsavai. Trends in Spatial Data Mining. *In Data Mining: Next Generation Challenges and Future Directions*, H. Kargupta, A. Joshi, K. Sivakumar, and Y. Yesha, eds. MIT Press, Cambridge, MA, 2003.

63. D. Silver and X. Wang. Tracking and visualizing turbulent 3D features. *IEEE Transactions on Visualization and Computer Graphics*, 3:129–141, 1997.

64. D. Silver and X. Wang. Tracking scalar features in unstructured datasets. In *Proceedings of the IEEE Conference on Visualization*, pages 79–86, Research Triangle Park, North Carolina, 1998.

65. E. Soroush, M. Balazinska, and D. Wang. ArrayStore: A storage manager for complex parallel array processing. In *Proceedings of the 2011 International Conference on Management of Data*, SIGMOD '11, New York, NY, USA, 2010. ACM.

66. N. T. B. Stone, D. Balog, B. Gill, B. Johanson, J. Marsteller, P. Nowoczynski, D. Porter, R. Reddy, J. R. Scott, D. Simmel, J. Sommerfield, K. Vargo, and C. Vizino. PDIO: High-performance remote file I/O for Portals enabled compute nodes. In *Proceedings of the 2006 Conference on Parallel and Distributed Processing Techniques and Applications,* Las Vegas, NV, 2006.

67. R. Thakur, W. Gropp, and E. Lusk. Data sieving and collective I/O in ROMIO. In *Proceedings of the The 7th Symposium on the Frontiers of Massively Parallel Computation*, FRONTIERS '99, pages 182–189, Washington, DC, USA, 1999. IEEE Computer Society.

68. R. Thakur, W. Gropp, and E. Lusk. On implementing MPI-IO portably and with high performance. In *Proceedings of the 6th Workshop on I/O in Parallel and Distributed Systems*, pages 23–32, May 1999. Atlanta, GA, ACM Press.

69. D. Thompson, R. Grout, N. Fabian, and J. Bennett. Detecting combustion and flow features in situ using principal component analysis. Technical Report SAND2009-2017, Sandia National Laboratories, 2009.

70. W. X. Wang, Z. Lin, W. M. Tang, W. W. Lee, S. Ethier, J. L. V. Lewandowski, G. Rewoldt, T. S. Hahm, and J. Manickam. Gyro-kinetic simulation of global turbulent transport properties in Tokamak experiments. *Physics of Plasmas*, 13(9):092505, 2006.

71. C. Wang, H. Yu, and K. L. Ma. Importance-driven time-varying data visualization. *IEEE Transactions on Visualization and Computer Graphics*, 14:1547–1554, 2008.

72. J. Wei, H. Yu, R. Grout, J. Chen, and K. L. Ma. Dual space analysis of turbulent combustion particle data. In *Proceedings of the IEEE Pacific Visualization Symposium*, pages 91–98, 2011, Hong Kong, China.

73. J. Wei, H. Yu, and K. L. Ma. Parallel clustering for visualizing large scientific line data. In *Proceedings of the IEEE Symposium on Large Data Analysis and Visualization*, pages 47–56, 2011, Providence, Rhode Island.

74. J. Woodring, K. Heitmann, J. Ahrens, P. Fasel, C. H. Hsu, S. Habib, and A. Pope. Analyzing and visualizing cosmological simulations with ParaView. *The Astrophysical Journal Supplement Series*, 195(1):11, 2011.

75. K. Wu. FastBit: An efficient indexing technology for accelerating data-intensive science. *Journal of Physics: Conference Series*, 16(1):556, 2005.

76. K. Wu, W. Koegler, J. Chen, and A. Shoshani. Using bitmap index for interactive exploration of large datasets. In *Proceedings of the International Conference on Scientific and Statistical Database Management*, pages 65–74, 2003, Cambridge, Massachusetts.

77. K. Wu, E. J. Otoo, and A. Shoshani. Compressing bitmap indexes for faster search operations. In *Proceedings of the 14th International Conference on Scientific and Statistical Database Management*, SSDBM '02, pages 99–108, Washington, DC, USA, 2002. IEEE Computer Society.

78. S. Yiannakis and J. E. Smith. The predictability of data values. In *Proceedings of the 30th Annual ACM/IEEE International Symposium on Microarchitecture*, MICRO 30, pages 248–258, Washington, DC, USA, 1997. IEEE Computer Society.

79. H. Yu, C. Wang, R. Grout, J. Chen, and K. L. Ma. In situ visualization for large-scale combustion simulations. *IEEE Computer Graphics and Application*, 30:45–57, 2010.

80. F. Zheng, H. Abbasi, C. Docan, J. Lofstead, Q. Liu, S. Klasky, M. Parashar, N. Podhorszki, K. Schwan, and M. Wolf. PreDatA: Preparatory data analytics on peta-scale machines. In *Parallel Distributed Processing (IPDPS), 2010 IEEE International Symposium on*, pages 1–12, April 2010, Atlanta, Georgia.

CHAPTER 14

Interactive Data Exploration

Brian Summa, Attilay Gyulassy, Peer-Timo
Bremer, and Valerio Pascucci

CONTENTS

14.1 INTRODUCTION

Driven by the continued increase in available computing resources, scientists are simulating ever more complex phenomena, at higher spatial and temporal resolutions. The resulting massive datasets are increasingly difficult to handle, and consequently, advanced data management, analysis, and visualization techniques are becoming mandatory to gain new scientific insights. This indicates a shift in overall priorities that has the potential to reshape the landscape of scientific computing.

In the past, the focus of scientific computing as a whole had been on data creation to produce more accurate simulations of natural phenomena. The subsequent analysis could often be restricted to global or local averages and profiles or comparatively straightforward creation of conditional distributions. Similarly, the visualization problem could often be addressed through sufficiently powerful secondary computing resources dedicated to extremely parallel visualization packages [1]. However, as simulations are becoming more detailed, scientists are increasingly interested in answering questions about more complex, and often intermittent, phenomena. For example, instead of a global flame profile, one may be interested in individual ignition or extinction kernels, or instead of an average entropy or density in a mixing simulation, one may like to study the nature of individual pockets of material. These questions require a more detailed analysis of the data with respect to some features of interest, which raises new challenges, such as—how to define the features in a scientifically meaningful and mathematically correct manner; how to develop robust and efficient algorithms to extract and subsequently study the features; how to determine the characteristic parameters, for example, iso-values or threshold? Simultaneously, the data have grown to the point where permanently storing it is becoming a challenge and the cost of file I/O is often preventing scientists from saving all the desired data. Even if a copy of the data exists, the act of reading through a dataset can take days or even weeks without extremely parallel resources, which makes a repeated analysis, for example, for parameter studies, difficult.

As a result, new paradigms for data analysis and visualization are necessary that are more tightly integrated with simulations and/or provide more efficient and flexible means to explore and analyze the resulting data. The former is often referred to as in situ processing [2], in which the data are handled on the fly, circumventing the need for any file I/O while allowing for a much more detailed (especially, in time) analysis. The latter typically relies on multiresolution techniques and/or intelligent preprocessing that allow the user to quickly extract an overview of the data, and subsequently zoom into regions of interest. In this chapter, we discuss two classes of techniques aimed at addressing the challenges discussed earlier using some of the biggest simulations performed to date as case studies. For example, the hydrodynamic instability simulations performed at Lawrence Livermore National Laboratory a decade ago [3,4] already produced several tens of terabytes of data (see Figure 14.1a). Data generation at this scale or higher is now becoming mainstream with high-performance computing codes, such as the adaptive mesh refinement (AMR) combustion simulations of low-swirl hydrogen flames that burn at low temperature with high fuel efficiency and low emission [5,6], the S3D simulations of non-premixed hydrogen flames with accurate representation of extinction and reignition phenomena [7], or the use of molecular dynamics simulations of the impact of a dense grain on low-density foam [8] (see Figure 14.1b–d).

These data must be visualized and analyzed to verify and validate the underlying model, to understand the phenomenon in detail, and to develop new insights into fundamental physics. Both data visualization and data analysis are vibrant research areas and much effort is being put to developing advanced, new techniques to process the massive amounts of data produced by scientists.

In this chapter, we focus on two key components that aid the visualization and analysis process: (1) multiresolution data layouts that provide the ability for quick access of coarse-to-fine resolutions of massive amounts of data, and (2) robust topological analysis techniques that provide the ability to detect and quantify features in data at multiple scales.

To provide context, we highlight these two components in a typical visualization and analysis pipeline in Figure 14.2. We assume that raw

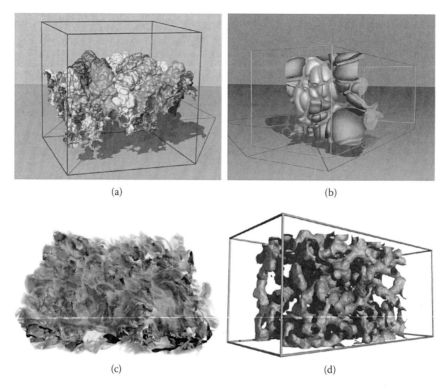

(a)   (b)

(c)   (d)

FIGURE 14.1 **(See color insert.)** Examples of simulations producing routinely massive amounts of data. (a) Miranda hydrodynamic instabilities. (b) AMR combustion simulation of low-swirl hydrogen flames. (c) S3D simulation of non-premixed hydrogen flames with extinction and reignition phenomena. (d) Molecular dynamics simulations of the impact of a dense grain into low-density foam.

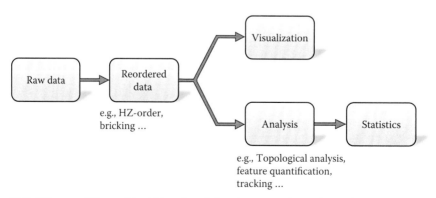

FIGURE 14.2 Hierarchical Z-ordered data layout and topological analysis components in the context of an advanced interactive visualization and analysis pipeline.

data from simulations are available as real-valued, regular samples of space–time. Due to the large size of datasets, we emphasize that the data samples cannot all be loaded in main memory. As a result, it is not feasible to use standard implementations of visualization and analysis algorithms on commodity hardware.

Reordering this raw data into a suitable multiresolution data layout can improve the efficiency of both visualization and analysis. Multiresolution layouts enable interactive visualization by allowing the user to first load the data at a coarse level, then progressively refine it by adding more samples to obtain a more detailed view. Classical schemes, for example, those based on bricking or chunking, do not readily support the type of data access required for progressive or multiresolution techniques.

In the following sections, we first describe a hierarchical Z-order data layout scheme. It builds on the coherent layout provided by the Z-order space-filling curve by incorporating a coarse-to-fine hierarchy on the ordering. The resulting system forms a versatile core technology that is simple to implement and has been applied to a variety of visualization algorithms, such as slicing, isosurfacing, and volume rendering on massive amounts of scientific simulation data.

Subsequently, we describe topology-based techniques that rather than providing access to the entire raw data allow an analysis and exploration of specific features, such as bubbles in turbulent mixing [9], burning cells in combustion [10], or filament structures in material science [11]. In our discussion, we show the practical use of these techniques in different scientific simulations. We underscore both their relevance in enabling specific scientific insight while retaining the generic character needed for their application on a variety of domains.

The contents of this section are a summary of several publications and prior research. For a more detailed description on the data layout scheme, see Refs. [12–14]; for topology-based analysis, see Refs. [15–20]; and for application of these techniques to the analysis of simulation data, see Refs. [9–11,21–23].

14.2 HIERARCHICAL INDEXING FOR OUT-OF-CORE DATA ACCESS

Out-of-core computing [24] specifically addresses the issues of algorithm redesign and data layout restructuring that are necessary to enable data access patterns having minimal out-of-core processing performance degradation. Research in this area is also valuable in parallel and distributed computing, where one has to deal with a similar issue of balancing the processing time with the time required for data access and movement among elements of a distributed or parallel application.

The solution to the out-of-core processing problem is typically divided into two parts: (1) algorithm analysis, to understand data access patterns and, when possible, redesign to maximize data locality; (2) storage of data in secondary memory using a layout consistent with the access patterns of the algorithm, amortizing the cost of individual I/O operations over several memory access operations.

In the case of hierarchical visualization algorithms for volumetric data, the 3D input hierarchy is traversed from a coarse grid to the fine grid levels to build derived geometric models having adaptive levels of detail. The shape of the output models are then modified dynamically with incremental updates of their level of detail. The parameters that govern this continuous modification of the output geometry are dependent on runtime user interaction, making it impossible to determine, a priori, what levels of detail will be constructed. For example, the parameters can be external, such as the viewpoint of the current display window, or internal, such as the isovalue of a contour or the position of a slice plane. The general structure of the access pattern can be summarized in two main points: (1) the input hierarchy is traversed from coarse-to-fine and level by level so that data in the same level of resolution are accessed at the same time, and (2) within each level of resolution, the regions that are in close geometric proximity are stored as much as possible in close memory locations and also traversed at the same time.

In this section, we describe a static indexing scheme that induces a data layout satisfying both requirements (1) and (2) for the hierarchical traversal of n-dimensional regular grids. The scheme has three key features that make it particularly attractive. First, the order of the data is independent of the out-of-core block structure so that its use in different settings (e.g., local disk access or transmission over a network) does not require any large data reorganization. Second, conversion from the Z-order indexing [25] used in classical database approaches to this indexing scheme can be implemented with a simple sequence of bit-string manipulations, making it appealing for a possible hardware implementation. Third, since there is no data replication, it avoids the performance penalties associated with dynamic updates as well as increased storage requirements typically associated with most hierarchical and out-of-core schemes.

Beyond the theoretical interest in developing hierarchical indexing schemes for n-dimensional space-filling curves, this approach targets practical applications in out-of-core visualization algorithms. For details on related work, algorithmic analysis, and experimental results, see Ref. [13].

14.2.1 Hierarchical Subsampling Framework

This section discusses the general framework for an efficient definition of a hierarchy over the samples of a dataset.

Consider a set S of n elements decomposed into a hierarchy $\mathcal{H}$ of k levels of resolution $\mathcal{H} = \{S_0, S_1, \ldots, S_{k-1}\}$ such that $S_0 \subset S_1 \subset \cdots \subset S_{k-1} = S$, where S_i is said to be coarser than S_j if $i < j$. The order of the elements in S is defined by a cardinality function $I: S \to \{0 \ldots n - 1\}$. This means that the following identity always holds $S[I(s)] \equiv s$, where square brackets are used to index an element in a set.

One can define a derived sequence $\mathcal{H}'$ of sets S_i' as follows: $S_i' = S_i \backslash S_{i-1}$ $i = 0, \ldots, k-1$, where formally, $S_{-1} = ;$. The sequence $\mathcal{H}' = \{S_0', S_1', \ldots, S_{k-1}'\}$ is a partitioning of S. A derived cardinality function $I': S \to \{0 \ldots n - 1\}$ can be defined on the basis of the following two properties: (1) $\forall s, t \in S_i': I'(s) < I'(t) \Leftrightarrow I(s) < I(t)$ and (2) $\forall s \in S_i', \forall t \in S_j': i < j \Rightarrow I'(s) < I'(t)$.

If the original function I has strong locality properties when restricted to any level of resolution S_i, then the cardinality function I' generates the desired global index for hierarchical and out-of-core traversals. The scheme has strong locality if elements with close indices are also close in geometric position. These locality properties are studied in detail in Ref. [26].

The function I' can be constructed as follows: (1) determine the number of elements in each derived set S_i', and (2) determine a cardinality function $I_i'' = I'|_{S_i'}$, restriction of I' to each set S_i'. In particular, if c_i is the number of elements of S_i', one can predetermine the starting index of the elements in a given level of resolution by building the sequence of constants $C_0, \ldots, C_{k-1}$ with

$$C_i = \sum_{j=0}^{i-1} c_j. \tag{14.1}$$

Next, one must determine a set of local cardinality functions $I_i'': S_i' \to \{0 \ldots c_i - 1\}$ so that

$$\forall s \in S_i': I'(s) = C_i + I_i''(s). \tag{14.2}$$

The computation of the constants C_i can be performed in a prepro-cessing stage so that the computation of I' is reduced to the following two steps:

1. Given s, determine its level of resolution i (i.e., the i such that $s \in S'_i$).

2. Compute $I''_i(s)$ and add it to C_i.

These two steps must be performed very efficiently as they will be exe-cuted repeatedly at run-time. The following section reports a practical realization of this scheme for rectilinear cube grids in any dimension.

14.2.2 Binary Trees and the Lebesgue Space-Filling Curve

This section reports the details on how to derive the local cardinality func-tions I''_i for a binary tree hierarchy in any dimension from the Z-order space-filling curve and their remapping to the new index I'.

14.2.2.1 Indexing the Lebesgue Space-Filling Curve

The Lebesgue space-filling curve, also called Z-order space-filling curve for its shape in the 2D case, is depicted in Figure 14.3a–e. The Z-order space-filling curve can be defined inductively by a Z-shaped base of size 1 (Figure 14.3a), that is, by the vertices of a square of side 1 that are connected along a Z pattern. Such vertices can then be replaced each by a Z shape of size $\frac{1}{2}$, as in Figure 14.3b. The vertices obtained in this way are then replaced

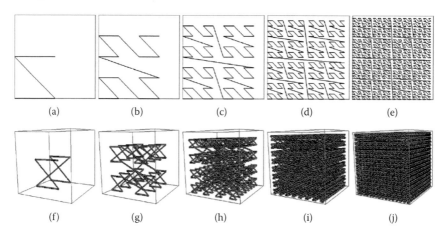

(a) (b) (c) (d) (e)

(f) (g) (h) (i) (j)

FIGURE 14.3   (a–e) The first five levels of resolution of the 2D Lebesgue's space-filling curve. (f–j) The first five levels of resolution of the 3D Lebesgue's space-filling curve.

by Z shapes of size $\frac{1}{4}$, as in Figure 14.3c, and so on. In general, the ith level of resolution is defined as the curve obtained by replacing the vertices of the $(i-1)^{th}$ level of resolution with Z shapes of size $\frac{1}{2^i}$. The 3D version of this space-filling curve has the same hierarchical structure, with the only difference being that the basic Z shape is replaced by a connected pair of Z shapes lying on the opposite faces of a cube, as shown in Figure 14.3f. Figure 14.3f–j show five successive refinements of the 3D Lebesgue space-filling curve. The d-dimensional version of the space-filling curve also has the same hierarchical structure, where the basic shape (the Z of the 2D case) is defined as a connected pair of $(d-1)$-dimensional basic shapes lying on the opposite faces of a d-dimensional cube.

The property that makes the Lebesgue's space-filling curve particularly attractive is the easy conversion from the d indices of a d-dimensional matrix to the 1D index along the curve. If one element e has d-dimensional reference $(i_1, \ldots, i_d)$, its 1D reference is built by interleaving the bits of the binary representations of the indices $i_1, \ldots, i_d$. In particular, if i_j is represented by the string of h bits $b_j^1 b_j^2 \cdots b_j^h$ (with $j = 1, \ldots, d$), then the 1D reference I of e is represented by the string of hd bits $I = "b_1^1 b_1^1 \cdots b_d^1 b_1^2 b_2^2 \cdots b_d^2 \cdots b_1^h b_2^h \cdots b_d^h"$.

The 1D order can be structured in a binary tree by considering the elements of level i, those that have the last i bits all equal to 0. This yields a hierarchy where each level of resolution has twice as many points as the previous level. From a geometric point of view, this means that the density of the points in the d-dimensional grid is doubled, alternating along each coordinate axis. Figure 14.4 shows the binary hierarchy in the 2D case

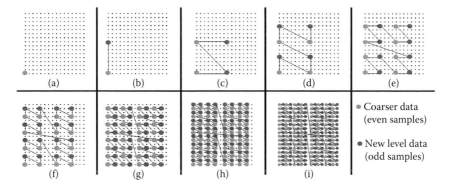

• Coarser data (even samples)

• New level data (odd samples)

FIGURE 14.4 The nine levels of resolution of the binary tree hierarchy defined by the 2D space-filling curve applied on 16×16 rectilinear grid. The coarsest level of resolution (a) is a single point. The number of points that belong to the curve at any level of resolution (b) to (i) is double the number of points of the previous level.

where the resolution of the space-filling curve is doubled alternately along the x and y axis. The coarsest level (a) is a single point, the second level (b) has two points, the third level (c) has four points (forming the Z shape), and so on.

14.2.2.2 Index Remapping

The cardinality function discussed in Section 14.2.1 for the binary tree case has the structure shown in Table 14.1. Note that this is a general structure suitable for out-of-core storage of static binary trees. It is independent of the dimension d of the grid of points or of the Z-order space-filling curve.

The structure of the binary tree defined on the Z-order space-filling curve allows one to easily determine the three elements necessary for the computation of the cardinality. They are: (1) the level i of an element, (2) the constants C_i of Equation 14.1, and (3) the local indices I_i''. Here,

i—if the binary tree hierarchy has k levels, then the element of Z-order index j in the Z-order belongs to the level $k - h$, where h is the number of trailing zeros in the binary representation of j;

C_i—the total number of elements in the levels coarser than i, with $i > 0$, is $C_i = 2^{i-1}$ with $C_0 = 0$;

I_i''—if an element has index j and belongs to the set S_i', then $\frac{j}{2^{k-i}}$ must be an odd number, by definition of i. Its local index is then:

$$I_i''(j)=\left\lfloor \frac{j}{2^{k-i+1}} \right\rfloor.$$

TABLE 14.1 Structure of the Hierarchical Indexing Scheme for Binary Tree Combined with the Order Defined by the Lebesgue Space-Filling Curve

Level	0	1	2	3		4								
Z-order index (2 levels)	0	1												
Z-order index (3 levels)	0	2	1	3										
Z-order index (4 levels)	0	4	2	6	1	3	5	7						
Z-order index (5 levels)	0	8	4	12	2	6	10	14	1	3	5	7	9	11	13	15
Hierarchical index	0	1	2	3	4	5	6	7	8	9	10	11	12	13	14	15

The computation of the local index I_i'' can be explained easily by looking at the bottom right part of Table 14.1, where the sequence of indices (1, 3, 5, 7, 9, 11, 13, 15) needs to be remapped to the local index (0, 1, 2, 3, 4, 5, 6, 7). The original sequence is made of a consecutive series of odd numbers. A right shift of one bit (or rounded division by two) turns them to the desired index.

These three elements can be put together to build an efficient algorithm that computes the hierarchical index $I'(s) = C_i + I_i''(s)$ in the two steps shown in Figure 14.5:

1. Set the bit in position $k + 1$ to 1.

2. Shift to the right until a 1 comes out of the bit-string.

This algorithm could have a very simple and efficient hardware implementation. The software C++ version can be implemented as follows:

```
inline adhocindex remap(register adhocindex i){
i |  = last_bit_mask;//set leftmost one
i/= i&-i;                //remove trailing zeros
return (i>>1);           //remove rightmost one
}
```

This code would work only on machines with two's complement representation of numbers where a negative numbers is represented by the two's complement of its absolute value. In a more portable version, one needs to replace i/= i&-i with i/= i&((~i)+1).

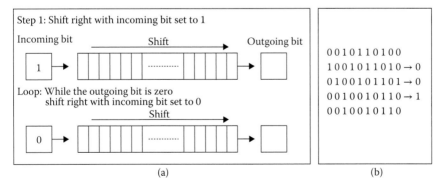

(a)   (b)

FIGURE 14.5 (a) Diagram of the algorithm for index remapping from Z-order to the hierarchical out-of-core binary tree order. (b) Example of the sequence of shift operations necessary to remap an index. The top element is the original index and the bottom is the remapped, output index.

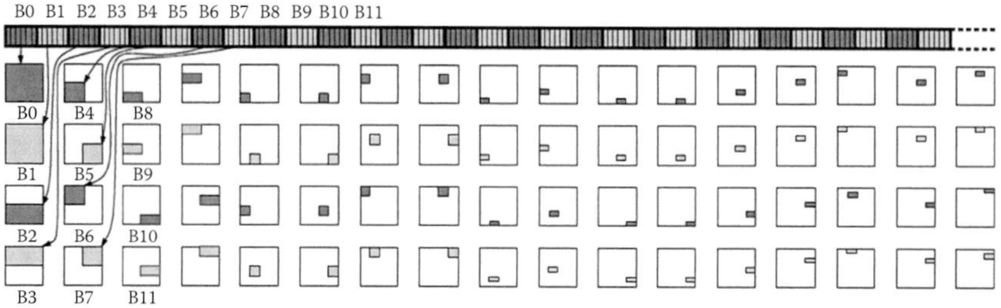

FIGURE 14.6 Data layout obtained for a 2D matrix reorganized using the index I' (1D array at the top). The 2D image of each block in the decomposition of the 1D array is shown below. Each gray region (odd blocks dark gray, even blocks light gray) shows where the block of data is distributed in the 2D array. In particular, the first block is the set of coarsest levels of the data distributed uniformly on the 2D array. The next block is the next level of resolution still covering the entire matrix. The next two levels are finer data covering each half of the array. The subsequent blocks represent finer resolution data distributed with increasing locality in the 2D array.

Figure 14.6 shows the data layout obtained for a 2D matrix when its elements are reordered following the index I'. The data are stored in this order and divided into blocks of constant size. The 2D image of such decomposition has the first block corresponding to the coarsest level of data resolution. The subsequent blocks correspond to finer and finer data resolution, which is distributed more and more locally.

14.2.3 Performance

In this section, we describe the experimental results for a simple, fundamental visualization technique—the orthogonal slicing of a 3D rectilinear grid. Slices can be at different resolutions to allow interactivity—as the user manipulates the slice parameters, we compute and display a coarse resolution slice, then refine it progressively. We compare this layout with two common array layouts—row major, and $h \times h \times h$ brick decomposition.

14.2.3.1 Data I/O Requirements

As we shall see, the amount of data required to be read from the disk varies substantially from one array layout to another. By way of example, consider the case of an 8GB dataset (a $2K^3$ mesh of unsigned char data values). An orthogonal slice of this mesh consists of 4M points/bytes. In this example, disk pages are 32KB in size (see Figure 14.7a). For the brick

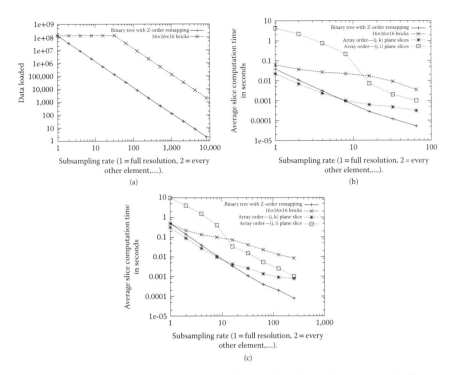

FIGURE 14.7 Comparison of static analysis and real performance in different conditions. For an 8GB dataset, (a) compares the amount of data loaded from disk (vertical axis) per slice while varying the level of subsampling and using two different access patterns/storage layouts—Z-order remapping and brick decomposition. The values on the vertical axis are reported using a logarithmic scale to highlight the performance difference—orders of magnitude—at any level of resolution. (b–c) Two comparisons of slice computations times (log scale) of four different data layout schemes with slices parallel to the (j,k) plane (orthogonal to the x axis). The horizontal axis is the level of subsampling of the slicing scheme, where values on the left are of finer resolution. Note how the practical performance of the Z-order versus brick layout is predicted very well by the static analysis. The only layout that can compete with the hierarchical Z-order is the row-major array layout optimized for (j,k) slices (orthogonal to the x axis). Of course, the row-major layout performs poorly for (j,i) slices (orthogonal to the z axis), while the hierarchical Z-order layout would maintain the same performance for access at any orientation.

decomposition case, one would use 32 × 32 × 32 blocks of 32KB for the entire dataset. The data loaded from the disk for a slice is 32 times larger than the output, or 128 MB bytes. As the subsampling increases up to a value of 32 (one sample out of 32), the amount of data loaded does not decrease because each 32 × 32 × 32 brick needs to be completely loaded.

At lower subsampling rates, the data overhead remains the same—the data loaded are 32,768 times larger than the data needed. In the binary tree with Z-order remapping, the data layout is equivalent to a *KD*-tree, constructing the same subdivision as an octree. For a 2D slice, the *KD*-tree mapping is equivalent to a quadtree layout. The data loaded are grouped into blocks along the hierarchy that gives an overhead factor in number of blocks of $1 + \frac{1}{2} + \frac{1}{4} + \frac{1}{16} + \cdots < 2$ (as for one added to a geometric series), while each block is 32 KB.

14.2.3.2 Tests with Memory Mapped Files

A series of basic tests was performed to verify the performance of the approach using a general purpose paging system. The out-of-core component of the scheme was implemented simply by mapping a 1D array of data to a file on disk using the mmap function. In this way, the I/O layer is implemented by the operating system virtual memory subsystem, paging in and out a portion of the data array, as needed. No multithreaded component is used to avoid blocking the application while retrieving the data. The blocks of data defined by the system are typically 4KB. Figure 14.7b shows the performance tests executed on a Pentium III laptop. The proposed scheme shows the best scalability in performance. The brick decomposition scheme with 16^3 chunks of regular grids shows the next best performance. The (i,j,k) row-major storage scheme has the worst performance because of its dependency on the slicing direction—best for (j,k) plane slices and worst for (j,i) plane slices. Figure 14.7c shows the performance results for a test on a larger, 8 GB dataset, run on an SGI Octane. The results are similar.

14.2.3.3 Parallel Write

The multiresolution data layout outlined earlier is a progressive, linear format, and therefore has a write routine that is inherently serial. As large simulations run, it would be ideal for in situ visualization for each node in the large supercomputer to be able to write out its piece of the simulation data directly in this layout. Therefore, a parallel write strategy must be employed. Figure 14.8 illustrates different possible parallel strategies. As shown in Figure 14.8a, each process can naively write its own data directly to the underlying binary file. This is inefficient due to the large number of small file accesses. As data get large, they become disadvantageous to store the entire dataset as a single, large file and, typically,

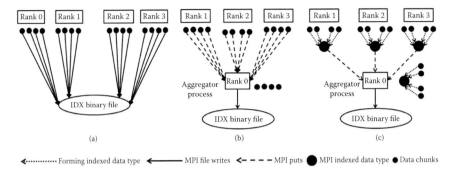

FIGURE 14.8 (a) Naive parallel strategy where each process writes its piece of the overall dataset into the underlying file, (b) Each process transmits each contiguous data segment to an intermediate aggregator. Once the aggregator's buffer is complete, the data are written to disk, (c) Several noncontiguous memory accesses are bundled into a single message to decrease communication overhead.

the entire dataset is partitioned into a series of smaller, more manageable files. This disjointedness can be used by a parallel write routine. As each simulation process produces simulation data, it can store its piece of the overall dataset locally and pass the data on to an aggregator process. These aggregator processes can be used to gather the individual pieces and composite the entire dataset. In Figure 14.8b, each process transmits each contiguous data segment to an intermediate aggregator. Once the aggregators buffer is complete, the data are written to disk using a single large I/O operation. Figure 14.8c illustrates a strategy where several noncontiguous memory accesses from each process are bundled into a single message. This approach reduces the number of small network messages needed to transfer data to the aggregators. This last strategy has been shown to exhibit good throughput performance and weak scaling for S3D combustion simulation applications when compared to a standard Fortran I/O benchmark [12,27].

14.2.3.4 On-Demand, Stream Processing

Even simple manipulations can be overly expensive if they are to be applied to each variable in a large dataset. For data of this size, it would be ideal to process the data as requested and pushed to a user and, most ideally, only operate on the portion of the data needed for display. The multiresolution data layout outlined earlier enables the efficient access of a subset of a large dataset. Moreover, these data will be accessed in a progressive hierarchy. Operations such as binning, clustering, or rescaling are trivial to implement on this hierarchy, given some known statistics on the data, such as the function value range, and so on. These operators can be applied to the data stream as-is while the data are moving to the user, progressively refining the operation as more data arrives. Much more complex operations can be reformulated to work well using the hierarchy. For instance, using the layout for a two-dimensional image, the data produce a hierarchy that is identical to a subsampled image pyramid on the data. Moreover, as data are requested, the progressivity of the transfer will traverse this pyramid in a coarse-to-fine manner. Techniques such as gradient–domain image editing can be reformulated to use this progressive stream and produce visually acceptable solutions [14,28]. These adaptive, progressive solutions allow the user to explore a full resolution solution as if it was fully available, without its expensive computation.

Therefore, the hierarchical Z-order is the layout of choice for efficient data management for visualizing and analyzing large-scale scientific simulation data. In the next section, we describe some of the fundamental mathematical techniques that we use for interactive exploration of such datasets.

14.3 FEATURE-BASED EXPLORATION AND ANALYSIS OF LARGE SCALE DATA

As the resolution of simulations increases, scientists are becoming more interested in relatively smaller scale phenomena. Rather than global statistical analysis or visualization, the focus is often on highly localized features corresponding to some region of interest, for example, material core lines [11], extinction regions [23], or burning cells [10]. Traditionally, such features are analyzed by extracting them in an (often costly) postprocessing step, and subsequently characterizing them through various statistics, for example, number of features, size, and/or conditional statistics. However, typically, the feature definition depends on multiple parameters and thresholds which, depending on the application area, are picked more or less ad hoc. This introduces problems in the analysis as the outcomes might depend heavily on the choice of input parameters; yet, few or only one input parameter is ever explored due to computational costs. Instead, we present alternative techniques to represent entire feature families for a wide range of parameters based on topological feature definitions. These techniques encode a wide range of threshold- and gradient-based features in a parameter-independent manner. Furthermore, various additional attributes can be computed for all features, such as characteristic length scales [21] and descriptive statistics [10]. In the following section, we will first describe some theoretical concepts of Morse theory that are used to define and represent features, and subsequently discuss various applications of the theory.

14.3.1 Morse Theory for Robust, Multi-Scale Feature Analysis

To begin, we present some background from Morse theory [29,30] and from combinatorial and algebraic topology [31,32].

14.3.1.1 Smooth Maps on Manifolds

Let $f:M{\rightarrow}\mathbb{R}$ be a smooth map. A point $x \in M$ is a critical point of f if the gradient of f vanishes at x, and the value $f(x)$ is a critical value. Noncritical points and noncritical values are called regular points and regular values, respectively. A critical point x is nondegenerate if the Hessian (matrix of second-order partial derivatives) at x is nonsingular. The index of a critical point x is the number of negative eigen values of the Hessian. For $d = 3$, there are four types of nondegenerate critical points—the minima (index 0), the *1-saddles* (index 1), the *2-saddles* (index 2), and the maxima (index 3). A function f is Morse if all critical points are nondegenerate with distinct values.

14.3.1.2 Morse-Smale Complex

An integral line is a path on M whose tangent vectors are parallel to the gradient of f. The stable manifold of a critical point x is the union of x and all integral lines that end at x. The unstable manifold of x is defined symmetrically as the union of a critical point x and all integral lines that start at x. One can superimpose the stable and unstable manifolds of all critical points to create the Morse–Smale complex (or MS-complex) of f [16,18]; see Figure 14.9a–d. The nodes of this complex are the critical points of f,

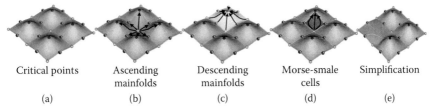

Critical points	Ascending mainfolds	Descending mainfolds	Morse-smale cells	Simplification
(a)	(b)	(c)	(d)	(e)

FIGURE 14.9   MS-complex construction, simplification, and topologically valid approximation: (a) Morse function with critical points shown; (b) Stable manifolds; (c) Unstable manifolds; (d) MS-complex; (e) MS-complex and manifold after simplification. Maxima are solid black, minima are solid white and saddles are black and white.

its arcs are integral lines starting or ending at the saddles and its regions are the nonempty intersections of stable and unstable 2-manifolds. More details on the definition of the MS-complex on the 2-manifold triangle meshes and algorithms to compute it are given by Bremer et al. [16].

14.3.1.3 Simplification

It is often useful to simplify an MS-complex to remove noise and to perform multi-scale function analysis. Following [16], cancellations of arc-connected maximum-saddle and minimum-saddle critical point pairs are performed to simplify an MS-complex. Cancellations are ranked by their persistence, the absolute difference in function value between the canceled critical point pair. Figures 14.9e and f show an example of a topological simplification and a corresponding approximation of f.

14.3.1.4 Computation

Conceptually, computing the MS-complex is simple. The boundaries of the regions are defined by integral lines/surfaces that start or end at saddles, and thus, the complex can be constructed by tracing these lines/surfaces. In practice, however, directly tracing these boundaries is costly and numerically unstable. As a result straightforward implementations typically result in invalid complexes that violate the fundamental invariants that make, for example, simplification possible. Instead, both piece-wise linear and discrete equivalents of smooth Morse theory have been developed [16,18,33] that do not rely on any numerical calculations and guarantee valid results. In the piece-wise linear case, these are based on tracing the steepest lines while carefully maintaining consistency through local validity checks [16,18,34]. The discrete approach uses a region-growing paradigm instead, which has the advantage of being extensible to any dimensions [33]. In both cases, the ability to formally guarantee the consistency of the results enables the application of these algorithms to massive and highly complex datasets since no parameter tuning or other application-specific adaptations are necessary.

14.3.2 Analysis of Turbulent Mixing in Hydrodynamic Instabilities

Understanding the turbulent mixing of fluids is one of the fundamental research problems in the area of fluid dynamics. Turbulent mixing occurs in a broad spectrum of phenomena, ranging from boiling water to astrophysics and nuclear fusion. Rayleigh–Taylor instability (RTI) occurs when

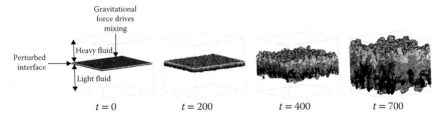

Gravitational force drives mixing

Perturbed interface

Heavy fluid

Light fluid

$t = 0$ $t = 200$ $t = 400$ $t = 700$

FIGURE 14.10 **(See color insert.)** An overview of the 11523 simulation (periodic in x and y) of the Rayleigh–Taylor instability at start ($t = 0$), early ($t = 200$), middle ($t = 400$), and late ($t = 700$) time. The light fluid has a density of 1.0, the heavy fluid has a density of 3.0. Two envelope surfaces (at densities 1.02 and 2.98) capture the mixing region. The boundaries of the box show the density field in pseudo-color. The upper envelope is analyzed to study bubble structures and the midplanes to study mixing trends.

two fluids of different density are accelerated opposite the mean density gradient. That is, a heavier fluid is accelerated against a lighter fluid. Figure 14.10 shows the mixing layers and the progression of the mixing process for the canonical case of a heavy fluid on top of a light fluid being accelerated by gravity. The heavy fluid accelerates downward, forming spikes, while the light fluid moves upward forming bubbles. The bubbles and spikes are thought to be one way to characterize the large-scale behavior of the mixing process. Scientists analyzing these simulations are particularly interested in the number of bubbles (and spikes) and their respective evolution. Large-scale models have been proposed based on bubble dynamics in which bubble growth, movement, and interaction are modeled [35]. Topology-based analysis is performed on the envelope surfaces, describing the boundary between undisturbed and mixed fluids. The techniques described in Section 14.2 are used for efficient data management and those described in Section 14.3.1 are used for the analysis. The Rayleigh–Taylor simulations were developed at Lawrence Livermore National Laboratory [3,4], and a detailed analysis is described by Laney et al. [9].

14.3.2.1 Segmentation of Bubbles

One of the challenges in analyzing mixing behavior is that there exists no prevalent mathematical definition for what constitutes a bubble/spike. In general, a bubble can be understood as a three-dimensional feature composed of lighter density fluid moving upwards (in the Z-direction) into a heavier density fluid. One can use the topological concepts introduced in Section 14.3.1 to define bubbles, spikes, and other features of interest.

Consider the images of the segmented mixing envelope surface at different times shown to the left, bottom, and right of the plot in Figure 14.11. During early time steps (Figure 14.11 upper left and middle left), it is natural to consider the mixing envelope as a time-varying functional surface defined over the XY-plane (i.e., a function with a single value per XY coordinate, such as in a terrain) and to associate local maxima with bubbles. This analogy fails at later time steps because the surface becomes nonfunctional.

However, it is possible to generalize this approach by treating the envelope surface as the domain of a function whose value at a point x is the Z-coordinate of x. It is natural to connect the maxima of this function to bubbles and compute the stable manifold of each maximum as a segmentation of the surface into bubbles. As can be seen in Figure 14.11, this segmentation corresponds very well to the notion of a bubble. Symmetrically, one can define spikes using the unstable manifolds of minima. Potentially, other functions could be defined on the envelope surface that would result in a robust segmentation. For example, the Z-velocity at all of the points on the envelope surface could be incorporated to capture the fact that the bubbles should be moving upwards into the heavy fluid.

In general, topology-based segmentations are often linked to important features—maximal and minimal Z-velocities on the midplanes correspond to the cores of rising and falling sections of fluids; density extrema correspond to pockets of unmixed fluids. Topological methods are flexible

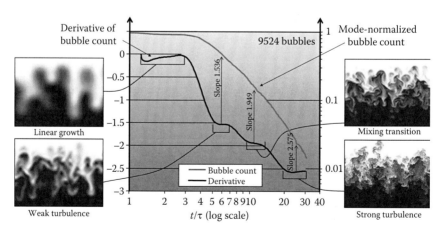

FIGURE 14.11 The plot depicts bubble counts of the envelope surface for three persistence values. To the right of the plot, the MS-segmentation at three persistence values for time 700. To the left and below the plot, the bubble segmentation along the curve of medium persistence at various times. Each maximum along with the Morse cells of its child-maxima are colored the same.

and enable the analysis of these phenomena using a uniform methodology. Furthermore, the MS-complex can be computed combinatorially [16,18], which translates into provably correct and stable algorithms that are crucial when dealing with large and complex data.

14.3.2.2 Multiscale Analysis and Persistence Selection

The MS-complex, just as any other segmentation, captures noise as well as features. Topological simplification removes noise and produces a series of approximations at decreasing resolution. Unlike many other techniques, topological segmentations allow a simplification scheme that is optimal in the L_∞-norm. One can formulate the problem of coarsening a segmentation in the following manner—given a function f and a segmentation S of the domain of f, what is the minimal change on f such that S is coarsened? If the segmentation one considers is the MS-complex of f, then it can be shown [16] that canceling a critical point pair with persistence p in f requires an approximation $\hat{f}$ with $\| f - \hat{f} \|_\infty \geq p/2$. Therefore, canceling critical points in order of increasing persistence corresponds to an L_∞-optimal simplification.

Cancellations of critical points can be used both to remove noise and, more in general, to analyze the trends in the data at multiple scales. This type of multiscale analysis is not readily available in classical image processing techniques and constitutes a fundamental advantage of a topology-based approach. For each MS-complex, a sequence of cancellations is computed that optimally simplifies the complex down to its minimal configuration. Thus, a family of segmentations of the envelope surface is defined, ranging from persistence $p = 0.0$, where both signal and noise features are segmented, to persistence $p = 1.0$ (full function range), where the entire surface is collapsed into a single topological feature. It is then possible to create statistics showing the number of bubbles over time using a range of persistence thresholds. As shown in Section 14.3.2, the mixing behavior can differ significantly across scales. Using the simplification sequences, one can capture the behavior on all scales without recomputing the MS-complex. Domain scientists interact with a visualization of the segmented surface and select an appropriate persistence value, based on their physical intuition of a correct segmentation of bubbles.

14.3.2.3 Results: Bubble Counts and Quantification of Mixing Phases

The input data consist of 758 time steps, each time step containing about 5.8 GB of density data defined on a 1152^3 grid. This analysis was performed on 68 dual-processor nodes of a cluster running Linux. For each time step, an isosurface at density value 2.98 is extracted and the MS-complex is computed with function f set to the Z-coordinate of each point. The isosurface extraction takes about 15 seconds at early time steps and about 30 seconds at the late time steps, where the surface is more complex. The MS-complex computation takes about 2 seconds at early time steps and about 25 seconds at late time steps. A digest of all required data is computed from the MS-complex, which enables efficient calculation of bubble counts for any choice of persistence value.

Figure 14.11 describes the count for three choices of persistence values. The inset figures show the segmentation of the isosurface into bubbles. Initially, there existed no clear choice for the persistence. However, the analysis pipeline is flexible enough to compute the MS-complex and bubble count at any persistence from the digest. Thus, this approach can provide scientists with a variety of visualizations and statistics for different choices. Ultimately, they can study the results and choose the parameter most closely aligned with their understanding of the phenomenon. It is important to note that the points of inflection, where the slope of each curve changes, occur at the same time step for all three scales. This fact indicates that the key results of the analysis are fundamentally insensitive to the choice of persistence value—the measurements of interest for this data are primarily focused on the trends of the derivative rather than the absolute value of the bubble count.

After an extensive evaluation of visual and numerical information by expert scientists, a bubble count was settled upon, corresponding to a persistence value of 2.39% of the function range at the end of the simulation. Figure 14.12 shows the result of this detailed analysis summarized by the final bubble count curve and its derivative. The plot reveals four phases of the mixing process providing, for the first time, a quantification of the mixing rates and the transition times between the different stages.

14.3.3 Topological Analysis of Filament Structures in Porous Media

Having a concise representation can enable interactive exploration of the feature space of a dataset. We illustrate this with an example from the physical sciences [11]. Physicists and material scientists conducted an experiment where a simulated porous solid is struck by an object travelling at high speed. The impact forms a crater over time, changing the properties of the material. The porous solid is made of copper filaments with 25% density.

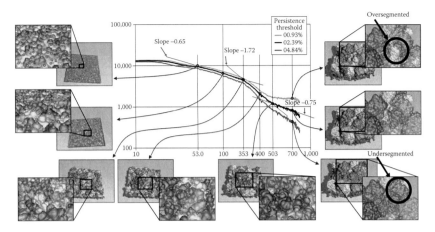

FIGURE 14.12   (**See color insert.**) Bubble count and its derivative. The derivative curve shows four regions of constant slope, each corresponding to distinct stages of the turbulent mixing process.

The data are represented as a signed distance field from an interface surface that is the outer boundary of a solid material. Positive values indicate locations inside and negative values outside the solid material of the copper foam. The analysis of these data must answer the following questions—How can one quantify the loss of porosity of the material? How does the filament density profile of the material change? What is the portion of the material that is affected by the impact crater? How does the structure around the impact crater change? Furthermore, the answers to these questions also should account for uncertainty in the feature definition, that is, how a filament is defined. We show a topology-based technique to answer these questions that computes a curved skeleton representation of the data using the MS-complex. Figure 14.13 illustrates the porous material at different times in the simulation as well as the final results of the filament extraction.

Several factors make the analysis of the filament structure difficult. First, the scale of the structure is not known. While scientists may have a general idea of the size and distribution of the structure, any analysis based on such guesses would be skewed. For example, the scale of porosity is locally affected by the impact crater, and an analysis technique must capture both large- and small-scale features. Second, the initial interface surface, from which the distance field is generated, is constructed with uncertainty in its initial position. The results of the analysis should be stable to small perturbations in this position, and additionally, quantify the level of stability of the results.

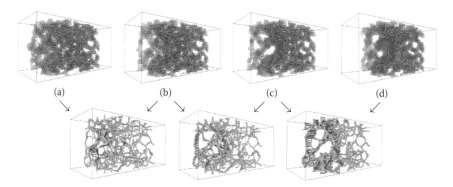

FIGURE 14.13   Top: A volume rendering of the positive values of the distance field for four time steps. The ball, shown in the starting configuration (a), strikes the porous solid and forms an impact crater (b–d). Bottom: A visual comparison of the filament structures computed using the MS-complex for subsequent time steps. The segments indicate where the filament structure has moved.

This topology-based approach constructs the MS-complex for the distance field, enabling interactive analysis of its 1-skeleton to find stable thresholds from which to construct a filament representation of the porous solid. The filament representation is used to answer the underlying science questions.

14.3.3.1 Finding a Stable Filament Structure

First, the three-dimensional MS-complex is computed for the full resolution data and its one-skeleton, critical points, and arcs are stored. These represent all possible filaments at the finest resolution. The signed distance field has a string of maxima located along potential filament locations, and these maxima are connected to each other by 2-saddle-maximum arcs of the complex. By representing the function concisely in a graph structure, subsequent queries become fast and combinatorial in nature, enabling interactive exploration of feature definitions.

It is through simplification and filtering that filaments of the porous material are extracted from the feature space encoded by the MS-complex. Similar to the over-segmentation of bubbles in Section 14.3.2, the finest-scale MS-complex encodes unimportant features as well as the important ones. Persistence-based simplification is used to remove noise; however, further analysis is first performed to determine what can be removed safely as noise, and what must to be kept as a feature. To do this, the distribution of 2-saddle-maximum arcs is plotted, as illustrated in Figure 14.14.

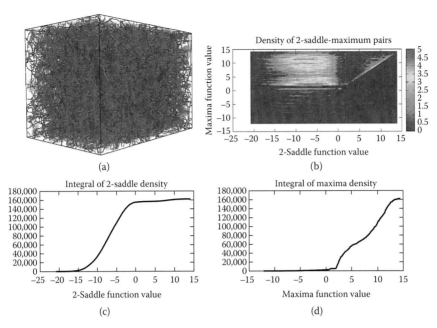

FIGURE 14.14 The fine-scale 1-skeleton of the initially computed Morse–Smale complex for the full dataset (a). Each 2-saddle-maximum arc is plotted in a density function by the values of the 2-saddle and maximum (b). Integrals along each axis give cumulative density functions (c,d). Flat regions in these indicate stable thresholds—moving a threshold would not change the number of 2-saddles or maxima, and hence the topology analysis would be the same.

A high-valued maximum should only be cancelled with another high-valued maximum to ensure that a cancellation does not artificially connect two filaments that are sufficiently separated in value in the distance function. By inspecting the plot and its cumulative density functions (along x and y directions), one can determine that there is sufficient separation to designate a stable threshold for simplification. In particular, it is possible to cancel all arcs that do not entirely cross the range $[-0.8, 1.5]$. Having flexible ranges helps overcome any sensitivity to the initial location of the interface surface. In this case, the extracted filament structure is stable with the input surface moving within that range of values. Note that the analysis is entirely performed using the 1-skeleton of the complex, a concise representation with enough semantic information for a detailed analysis.

The filaments are finally extracted as 2-saddle-maximum arcs that survive the simplification and are entirely contained above the isosurface for isovalue zero. Selection and display of these arcs with different threshold values is performed interactively, allowing a scientist not only to see the quantitative results change with threshold values, but also understand how the thresholds affect the extracted structure.

14.3.3.2 Comparison of Filament Structures

Comparison of two filament structures is performed using a number of heuristics: (1) Hausdorff distance, (2) average distance between closest pairs on the two graphs, (3) number of simple cycles in each graph, to estimate connectivity, and (4) total length of edges in each graph.

The filament structure is computed using the MS-complex for each time step in the simulation. Using these distance measures, one can compare qualitatively how the structure of the material changes between time steps. Figure 14.13 is a visual depiction of the displacement of filament segments at different times after impact. Note the large displacements near the crater, and the nearly zero displacement well below the crater.

It can be seen from a visual inspection that the density of the material has limited changes in the bottom one-third of the sample. This is partly due to the fact that the foam is extremely efficient at absorbing the impact shock wave. The key statistics of the filament structure for each time step are summarized in Table 14.2.

The ratio of cycle counts before and after the impact supports this observation, as approximately two-thirds of the cycles are destroyed. The ratio of the total length of the filaments before and after the particle impact implies that the volume of material displaced by the crater is approximately one-half the volume of the rest of the material. Since this ratio is fairly close to the ratio of the cycle counts, one can say that the majority of the filaments that were broken happened to be in the interior of the crater. The sum of the Hausdorff distances between the time steps is 98.6, giving the maximum distance that any element of the material traveled during the impact. This number is surprisingly high, corresponding to the entire depth of the crater; it indicates that the material of the filaments first hit by the particle was displaced along the trajectory of the particle. The average distance between the closest pairs in the graphs of the

TABLE 14.2 Statistics for Individual Time Steps

Metric	$t = 500$	$t = 12,750$	$t = 25,500$	$t = 51,000$
No. of cycles	762	340	372	256
Total length	34,756	24,316	23,798	18,912

consecutive time steps was less than 5.0, indicating that the displacement did not propagate into the material, outside the direct path of the particle.

This approach shows that the concise representation provided by the 1-skeleton of the MS-complex was powerful enough for sophisticated analysis of the dataset. Furthermore, features may arise in locations not initially predicted. The MS-complex is a useful tool in identifying such features since it provides a full characterization of the gradient flow behavior. Therefore, analysis of the critical point pairs and arcs of the complex can lead to a better understanding of the actual locations of the features, and indicate where to apply topological simplification.

14.4 DATA EXPLORATION IN THE FUTURE

Given the ever widening gap between the data produced by the fastest machines and file I/O rates, techniques like the ones described earlier will be vital to ensure that scientists can explore and analyze their data in the future. In particular, going forward, an ever increasing portion of analysis and visualization related tasks will have to be incorporated into the simulations themselves. We contend that this will be the only practical means to access sufficiently large portions of the data at sufficiently high temporal frequencies to allow for a reliable analysis. As indicated earlier, the processing will likely take one of two forms: (1) create efficient data layouts and/or summaries for later exploration; and (2) perform a large set of meta-analysis steps, such as extracting the topological structure, to enable more detailed postprocessing with less overall data.

For the former, we believe the next generation of techniques will concentrate on storing only "interesting" subsets identified on the fly by appropriate analysis techniques while allowing direct access to a running simulation for computational steering, on-the-fly visualization, and related approaches. For the latter class of meta-analysis, the focus will be on the development of abstract data representations, such as the MS-complex, Jacobi sets [36], or Ridge-Valley graphs [37], which can compactly represent a broad class of features for subsequent analysis. Furthermore, new large-scale, parallel algorithms will be required, which can extract such structures from a running simulation with minimal impact to the performance of the overall code. Some initial results exist, for example, to compute MS-complexes in parallel [38] or to separate algorithms into a fast in situ and an asynchronous in-transit part [2]. However, for other structures, such as Jacobi sets, even serial implementations remain challenging today.

14.5 CONCLUSION

Visualization, which is the transformation of abstract data into images, plays a central role in virtually all fields of scientific endeavor. In particular, it is key for interactive data exploration, which is an indispensable part of hypothesis testing and knowledge discovery. Like most fields discussed in this book, visualization faces substantial scientific data management challenges that are the result of growth in size and complexity of the data being produced by simulations and collected from experiments. In this chapter, we have indicated some effective directions in scientific data management that play a unique role in interactive data exploration.

Visualization applications that run on large-scale parallel machines can derive great benefit from the tight integration of data management, analysis, and visualization techniques. Staple operations, like slicing and isosurfacing, can be vastly accelerated by taking advantage of proper data layouts and associated metadata. In some cases, the simulation or experiment produces such metadata as part of the data production process. In other cases, we must generate that data ourselves. A significant fraction of the code in these production visualization applications is dedicated to scientific data management—they must support a plethora of input data formats, and therefore, they contain a number of data loader modules; they must create an internal data structure that is suitable for use by a potentially large collection of visualization, analysis, and rendering modules—all of which may potentially run in parallel on shared and/or distributed memory machines. An open problem is one of data models and semantics, where meaning is assigned to arrays of data stored in data files.

With data of massive scale, it is often useful to perform a multiresolution analysis, working first with a smaller, coarser version of the data, then progressively refine the analysis as interesting features are revealed. We saw that a space-filling curve model has proven to be highly efficient for the interactive analysis of massive data. However, direct output of such a data model and layout are still in an experimental stage. This is a good example of a data model and layout that works very well for multiresolution analysis, but that will require more research and development before it will be used regularly by simulations for output. Topological methods for multiscale, quantitative feature detection, and analysis have also been demonstrated as highly effective in scientific knowledge discovery. They provide higher-level abstractions that scientists can explore interactively even if the raw data are prohibitively large in size. They can, therefore, play a central role in the future development of interactive data exploration techniques.

The major barrier that data management, analysis, and visualization techniques will face in the future is going beyond the current design as postprocessing computations. Only a new generation of tools that can be systematically deployed in situ can address the needs of future simulation environments. In fact, the massive sizes of the models generated will make it impossible to process the data effectively after the simulations are completed. Moreover, the ratio between data stored and computed will continue to get smaller and the analysis techniques would be more likely to miss important phenomena since they may not be captured by the data stored. An effective parallelization of all these techniques will be essential for their practical use in knowledge discovery in a paradigm-changing approach that requires tight integration with the simulation environments.

REFERENCES

1. Lawrence Livermore National Laboratory. *VisIt Visualization Software*, 2008. http://www.llnl.gov/visit/.
2. J. Bennett, H. Abbasi, P.-T. Bremer, R. Grout, A. Gyulassy, T. Jin, S. Klasky, H. Kolla, M. Parashar, V. Pascucci, P. Pebay, D. Thompson, H. Yu, F. Zhang, and J. Chen. Combining in-situ and in-transit processing to enable extreme-scale scientific analysis. In *Proceedings of ACM/IEEE Conference on Supercomputing (SC12)*, IEEE Computer Society Press, Los Alamitos, CA, 2012.
3. A. W. Cook, W. Cabot, and P. L. Miller. The mixing transition in rayleigh-taylor instability. *J. Fluid Mech.*, 511:333–362, 2004.
4. A. W. Cook and P. E. Dimotakis. Transition stages of rayleigh-taylor instability between miscible fluids. *J. Fluid Mech.*, 443:66–99, 2001.
5. J. B. Bell, R. K. Cheng, M. S. Day, and I. G. Shepherd. Numerical simulation of Lewis number effects on lean premixed turbulent flames. *Proc. Combust. Inst.*, 31:1309–1317, 2007.
6. M. S. Day and J. B. Bell. Numerical simulation of laminar reacting flows with complex chemistry. *Combust. Theory Modelling*, 4(4):535–556, 2000.
7. E. R. Hawkes, R. Sankaran, J. C. Sutherland, and J. H. Chen. Scalar mixing in direct numerical simulations of temporally evolving plane jet flames with skeletal co/h2 kinetics. *Proc. Combust. Inst.*, 31(1):1633–1640, 2007.
8. E. M. Bringa, J. U. Cazamias, P. Erhart, J. Stölken, N. Tanushev, B. D. Wirth, R. E. Rudd, and M. J. Caturla. Atomistic shock Hugoniot simulation of single-crystal copper. *J. Appl. Phys.*, 96:3793–3799, October 2004.
9. D. Laney, P. T. Bremer, A. Mascarenhas, P. Miller, and V. Pascucci. Understanding the structure of the turbulent mixing layer in hydrodynamic instabilities. *IEEE Trans. Vis. Comput. Graph.*, 12(5):1053–1060, 2006.

10. P.-T. Bremer, G. Weber, J. Tierny, V. Pascucci, M. Day, and J. B. Bell. Interactive exploration and analysis of large scale simulations using topology-based data segmentation. *IEEE Trans. Vis. Comput. Graph.*, 17(99):1307–1324, 2010.

11. A. Gyulassy, M. Duchaineau, V. Natarajan, V. Pascucci, E. Bringa, A. Higginbotham, and B. Hamann. Topologically clean distance fields. *IEEE Trans. Vis. Comput. Graph.*, 13(6):1432–1439, November/December 2007.

12. S. Kumar, V. Vishwanath, P. Carns, B. Summa, G. Scorzelli, V. Pascucci, R. Ross, J. Chen, H. Kolla, and R. Grout. Pidx: Efficient parallel i/o for multi-resolution multi-dimensional scientific datasets. In *Proceedings of IEEE Cluster 2011*, IEEE Computer Society Press, Los Alamitos, CA, September 2011.

13. V. Pascucci and R. J. Frank. Global static indexing for real-time exploration of very large regular grids. In *Supercomputing '01: Proceedings of the 2001 ACM/IEEE Conference on Supercomputing (CDROM)*, pages 2–2. ACM, New York, NY, 2001.

14. B. Summa, G. Scorzelli, M. Jiang, P.-T. Bremer, and V. Pascucci. Interactive editing of massive imagery made simple: Turning atlanta into atlantis. *ACM Trans. Graph.*, 30:7:1–7:13, April 2011.

15. P.-T. Bremer, H. Edelsbrunner, B. Hamann, and V. Pascucci. A multi-resolution data structure for two-dimensional Morse-Smale functions. In G. Turk, J. J. van Wijk, and R. Moorhead, editors, *Proc. IEEE Visualization '03*, pages 139–146. IEEE, IEEE Computer Society Press, Los Alamitos, CA, 2003.

16. P.-T. Bremer, H. Edelsbrunner, B. Hamann, and V. Pascucci. A topological hierarchy for functions on triangulated surfaces. *IEEE Trans. Vis. Comput. Graph.*, 10(4):385–396, 2004.

17. K. Cole-McLaughlin, H. Edelsbrunner, J. Harer, V. Natarajan, and V. Pascucci. Loops in reeb graphs of 2-manifolds. In *Proceedings of the 19th Annual Symposium on Computational Geometry*, pages 344–350. ACM Press, 2003.

18. H. Edelsbrunner, J. Harer, and A. Zomorodian. Hierarchical Morse-Smale complexes for piecewise linear 2-manifolds. *Discrete Comput. Geom.*, 30: 87–107, 2003.

19. H. Edelsbrunner, D. Letscher, and A. Zomorodian. Topological persistence and simplification. In *FOCS '00: Proceedings of the 41st Annual Symposium on Foundations of Computer Science*, page 454. IEEE Computer Society, Washington, DC, 2000.

20. V. Natarajan, Y. Wang, P.-T. Bremer, V. Pascucci, and B. Hamann. Segmenting molecular surfaces. *Comput. Aided Geom. Des.*, 23(6):495–509, 2006.

21. J. Bennett, V. Krishnamurthy, S. Liu, V. Pascucci, R. Grout, J. Chen, and P.-T. Bremer. Feature-based statistical analysis of combustion simulation data. *IEEE Trans. Vis. Comp. Graph.*, 17(12):1822–1831, 2011.

22. P.-T. Bremer, G. Weber, V. Pascucci, M. Day, and J. Bell. Analyzing and tracking burning structures in lean premixed hydrogen flames. *IEEE Trans. Vis. Comput. Graph.*, 16(2):248–260, 2010.

23. A. Mascarenhas, R. W. Grout, P.-T. Bremer, E. R. Hawkes, V. Pascucci, and J. H. Chen. Topological feature extraction for comparison of terascale combustion simulation data. In V. Pascucci, X. Tricoche, H. Hagen, and J. Tierny, editors, *Mathematics and Visualization*, pages 229–240. Springer, 2011.

24. J. S. Vitter. External memory algorithms and data structures: Dealing with massive data. *ACM Computing Surveys*, 33:2001, 2000.

25. J. K. Lawder and P. J. H. King. Using space-filling curves for multi-dimensional indexing. In *Lecture Notes in Computer Science*, pages 20–35, Springer Verlag, Berlin, Germany, 2000.

26. B. Moon, H. Jagadish, C. Faloutsos, and J. Saltz. Analysis of the clustering properties of hilbert spacefilling curve. *IEEE Trans. Knowl. Data Eng.*, 13(1):124–141, 2001.

27. S. Kumar, V. Pascucci, V. Vishwanath, P. Carns, R. Latham, T. Peterka, M. Papka, and R. Ross. Towards parallel access of multi-dimensional, multiresolution scientific data. In *Proceedings of 2010 Petascale Data Storage Workshop*, New Orleans, LA, November 2010.

28. S. Philip, B. Summa, P.-T. Bremer, and V. Pascucci. Parallel Gradient Domain Processing of Massive Images. In T. Kuhlen, R. Pajarola, and K. Zhou, editors, *Eurographics Symposium on Parallel Graphics and Visualization*, pages 11–19. Eurographics Association, Llandudno, Wales, UK, 2011.

29. Y. Matsumoto. *An Introduction to Morse Theory. Translated from Japanese by K. Hudson and M. Saito*. American Mathematical Society, Providence, RI, 2002.

30. J. W. Milnor. *Morse Theory*. Princeton Univ. Press, New Jersey, 1963.

31. P. S. Alexandrov. *Combinatorial Topology*. Dover, Mineola, NY, 1998.

32. J. R. Munkres. *Elements of Algebraic Topology*. Addison-Wesley, Redwood City, CA, 1984.

33. A. Gyulassy, V. Natarajan, V. Pascucci, and B. Hamann. Efficient computation of Morse-Smale complexes for three-dimensional scalar functions. *IEEE Trans. Vis. Comput. Graph. (TVCG)*, 13(6):1440–1447, 2007.

34. H. Edelsbrunner, J. Harer, V. Natarajan, and V. Pascucci. Morse-Smale complexes for piecewise linear 3-manifolds. In *Proceedings of 19th Symposium on Computational Geometry*, pages 361–370. ACM Press, San Diego, 2003.

35. U. Alon and D. Shvarts. Two-phase flow model for rayleigh-taylor and richtmeyer-meshkov mixing. In *Proceedings of Fifth International Workshop on Compressible Turbulent Mixing*. World Scientific, Singapore, 1996.

36. H. Edelsbrunner and J. Harer. Jacobi sets of multiple Morse functions. In F. Cucker, R. DeVore, P. Olver, and E. Sueli, editors, *Foundations of Computational Mathematics, Minneapolis 2002*, pages 37–57. Cambridge Univ. Press, England, 2002.

37. G. Norgard and P.-T. Bremer. Ridge-valley graphs: Combinatorial ridge detection using jacobi sets. *Comput. Aided Geom. Des.*, 2012, http://dx.doi.org/10.1016/j.cagd.2012.03.015.

38. A. Gyulassy, T. Peterka, V. Pascucci, and R. Ross. Characterizing the parallel computation of morse-smale complexes. In *Proceedings of IPDPS '12*, Shanghai, China, 2012.

CHAPTER 15

Linked Science

Interconnecting Scientific Assets

Tomi Kauppinen, Alkyoni Baglatzi,
and Carsten Keßler

CONTENTS

15.1 INTRODUCTION

The way scientific results are published needs to be improved. First of all, scientific research settings have changed dramatically toward a digital environment [1], which calls for changes to traditional, paper-based publishing. The idea of "nanopublications" [2], for example, suggests a transition from the one-sided text publishing of scientific results to a more structured, machine-understandable, and data-oriented perspective of research and findings.

There is a need for change because the state of the art is insufficient in too many ways: methods, datasets, results, and even publications are not described in a machine-understandable way [3,4,5]. Moreover, they are not openly accessible, and as a result, scientific settings can hardly be reproduced. It is, therefore, difficult and time consuming to validate results of any particular scientific effort. The problem is that either or both the implementation of *methods* and the *data* behind a scientific article are not openly available to assist a reviewer—or anyone else trying to reproduce a scientific setting—in his/her task. As a result, it takes too much time to get scientific results into practice [1] or to produce new knowledge on top of existing knowledge. Communities need better, faster, and more open science to cope with, for example, the huge challenges related to our environment (climate change, natural disasters) and society (health, food, poverty).

Linked Science (see http://linkedscience.org)—or Linked Open Science to emphasize the need for transparency—is an approach to interconnect all scientific assets. By doing this, Linked Science seeks to revolutionize the way thousands of research organizations and millions of scientists in them work and produce new knowledge. Linked Science is defined as a combination of Linked Data (see http://linkeddata.org) [6,7], Semantic Web [8] and web standards, open source and web-based online environments, cloud computing (http://en.wikipedia.org/wiki/Cloud_computing), and a machine-understandable technical and legal infrastructure.

In this chapter, we first describe the background concerning semantic modeling of scientific resources and introduce the Linked Open Data University of Münster (LODUM) initiative in Section 15.2. In Section 15.3, we present and define the Linked Science approach. Section 15.4 presents the Linked Science Core vocabulary (LSC) for linking research settings and all elements related to them together. We also present a case study concerning geochange and deforestation, and provide a discussion of the related work, as well as a research agenda for the years to come. Section 15.5 concludes the chapter.

15.2 BACKGROUND

This section points to related work in the use of semantic technologies in archiving and interlinking scientific assets. Moreover, we introduce the LODUM project as a specific initiative to foster the future development of the research process.

15.2.1 Semantics for Scientific Linkage

The vision of the Semantic Web introduced by Tim Berners-Lee in 2001 [8] is based on the idea of making content on the web machine understandable. The transition from the level of character sequences to a level of meaning is supposed to provide new opportunities for improving interoperability, search, and intelligent applications.

The basic prerequisite for communication is a basic common and shared understanding. In the vision of the Semantic Web, ontologies play a major role in the chain. According to Gruber, "an ontology is an explicit specification of a conceptualization" [9]. Given this, ontologies aim to assist in formalizing the knowledge of a domain. The goal is to enable efficient sharing and integration of information. To lower the barrier of achieving the Semantic Web outside the scientific research community, Berners-Lee has proposed the Linked Data approach [6,7] that starts from existing datasets (e.g., in relational databases or various flavors of XML) and provides lightweight semantic annotations.

In this way, the contents in millions of datasets—freed from *data silos** that do not allow easy access or reuse—become part of the web of data and will be interlinked and described using vocabularies. In Linked Data, online resources are identified by uniform resource identifiers (URIs), which are in an ideal case dereferenceable, so that users and machines accessing the data can discover more knowledge by following links. Shared vocabularies will assist in linking data in a useful and meaningful way. Thus, the idea is that data published as Linked Data facilitate sharing between heterogeneous domains, such as scientific disciplines.

Vocabularies are a key part of the Linked Data principles as they provide means to overcome semantic interoperability problems [10]. An example

---

* Berners-Lee called for *raw data now* in his 2009 TED talk; see http://on.ted.com/9x4B.

underlying the necessity of vocabularies can be seen in the deforestation domain [11]. In Brazil's National Institute for Space Research (INPE) (see http://www.inpe.br/ingles/index.php), the term "desflorestamento" (in Portuguese) has been introduced to refer to the clearing of vegetation and transition categories, omitting the clearing of cerrado (savanna). However, the state government of Mato Grosso uses the term "deforestation" (desmatamento), referring to all three categories (forest, transition categories, and savanna).

As a result, the deforestation rates of the two organizations are not comparable because of the disagreement on the basic terms. By formalizing such concepts and publishing them as vocabularies, disambiguities could be identified and eventually resolved so that knowledge can be shared more efficiently. Naturally, these semantic interoperability problems expand even more between and across heterogeneous scientific domains. The Linked Data approach aims to provide new perspectives and solutions for knowledge interconnection to overcome such problems.

15.2.2 LODUM: Linked Open Data University of Münster

The University of Münster, Germany, is one of the first universities to commit to an institution-wide Linked Open Data program. The goal of the LODUM initiative (see http://lodum.de and http://data.uni-münster.de) is to increase transparency and comprehensibility both for research and for administrative matters. Beyond the technical infrastructure required to put this undertaking into practice—server facilities to run a triple store, SPARQL endpoint, data dumps, backups, and synchronization jobs—fostering a change in the mindset of the university's research community is key for the success of LODUM. Open Access must be propagated as the preferred way of publishing results, along with the primary data on LODUM (or any other) data platform.

Although the long-term focus is on scientific data and publications, other data, such as class schedules and administrative data, are also already being integrated into the LODUM infrastructure. These additional data complement the LODUM strategy and allow a larger group of users to benefit from the infrastructure by including students and administrative staff. By implementing Open Access and Linked Open Data principles throughout the university, LODUM aims to make the output of the university more visible and foster collaboration, both between the university's faculties and with partners. Figure 15.1 shows an example of how LODUM interconnects different resources—organizations, people, buildings, rooms, courses, research, and publications—coming from heterogeneous sources.

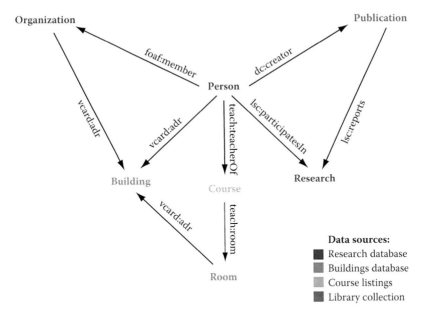

FIGURE 15.1   Example of how LODUM interconnects different resources coming from heterogeneous sources.

Making these raw data accessible to guarantee reproducibility of the results documented in the publication has to become the rule, not the exception. In some fields, this is already a common practice, such as in bioinformatics, where articles can only be submitted to a journal along with an ID for the genome sequence in the focus of the article. This ID is obtained by uploading the sequence to the Gene database hosted at the National Center for Biotechnology Information (see http://www.ncbi.nlm .nih.gov/gene).

15.2.3 Activities by Scientific and Publishing Communities

More and more universities and research-related organizations are likely to increase efforts for opening their data. Projects such as the VIVO Project at Cornell University (see http://vivo.cornell.edu), LUCERO at the Open University (see http://lucero-project.info), the University of Southampton Linked Data (see http://data.southampton.ac.uk), Linked Life Data (see http://linkedlifedata.com), Linked Open Drug Data (see http://www .w3.org/wiki/HCLSIG/LODD), and Bio2RDF (see http://bio2rdf.org/) are all actively producing Linked Data related to science, research, and learning. To increase and enable collaboration and sharing of ideas, the Linked Universities (see http://linkeduniversities.org) network has been founded as an alliance of European universities engaged in exposing their public data.

Publishers and other related communities are also interested in developing new ways of ensuring reproducibility of research and developing new types of citations. These efforts include, for example, workshops such as "Beyond the PDF" at the University of California (see http://sites .google.com/site/beyondthepdf/), "Data Citation Principles" at Harvard University (see http://www.iq.harvard.edu/events/node/2462), and "Apps for Science" (see http://appsforscience.com/) and "Executable Papers" (see http://www.executablepapers.com/) challenges sponsored by Elsevier.

There is a need for efforts in building a community around Linked Science to put this approach into practice. One of these efforts that the authors of this chapter were involved in was the First International Workshop on Linked Science (LISC 2011) (see http://linkedscience.org /events/lisc2011) at the International Semantic Web Conference 2011 in Bonn, Germany. The goal was to "discuss and present results of new ways of publishing, sharing, linking, and analyzing such scientific resources motivated by driving scientific requirements, as well as reasoning over the data to discover interesting new links and scientific insights." Moreover, related to this, the authors also organized a breakout session in Science Online London 2011 (http://www.scienceonlinelondon.org/)—hosted by Nature.com (see http://www.nature.com/) and Digital Science at the British Library in London, UK—to develop vocabularies for scientific data sharing and reuse. These events, along with others to be established, need to continue at a regular pace to build a community around Linked Science. The two meetings mentioned earlier have already shown that there is a vibrant community working on various aspects of Linked Science and that there is a lot of interest to develop this approach further and bring into research practice, curricula, publishers, and funding bodies.

15.3 LINKED SCIENCE

Linked Science is an approach where scientific resources—for example, workflows, processes, models, data, methods, and evaluation metrics— are semantically annotated and interconnected [3]. To achieve linkage, different resources have to be described and connected in an explicit, formalized manner. In practice, this needs shared conceptualizations, well-defined ontologies and vocabularies, and reasoning mechanisms to represent, link, and share scientific knowledge.

The key motivation behind Linked Science is that it is crucial to efficiently communicate information about scientific findings. Opening up scientific data, methods, and results improves transparency, allows for reasoning to find links between researchers (e.g., "John has collaborated with Mary"), supports validation of research results (e.g., "the scientific findings about typical deforestation patterns by John can be verified online by anyone"), and enables new ways of communicating results as media through interactive visualizations. Ethical aspects and impacts on society and environment are highly important when communicating science. To give an example, observing deforestation in rain forests and reasoning about changes can help in understanding and finally reducing deforestation [12].

In Linked Science, data are published on the web and resources are semantically annotated, methods are available as open source for running against the data, copyrights and licenses are clearly explicated, and all bits and pieces are distributed and in the cloud. This means that Linked Science is much more than simply providing scientific data as Linked Data. Namely, as argued earlier by Bechhofer et al. [1], publishing data "has requirements of provenance, quality, credit, attribution, methods in order to provide the reproducibility that allows validation of results."

One challenge that Linked Science targets is how to achieve semantic integration and interlinkage of space and time within the scientific studies. Temporal issues in representing and sharing knowledge are crucial as scientific results often refer to a specific period. By linking knowledge, which has these immanent properties of space and time, new perspectives for the scientific community emerge. Links are useful in many practical settings. On the web, links are used to browse web pages and to jump from one web page to the next. Analysis of links is used by web search engines to rank web pages given a certain query. Links between current and historical places can be used to align historical and current scientific observations together to analyze, for example, climate change. This could be done similarly as a cultural heritage portal can recommend historical content even if a user is querying by a contemporary place name [13]. The idea in the Linked Science approach thus is that through interlinking all of the scientific components together with vocabularies, it is possible to form a huge collection of scientific data. This allows for interesting link and pattern discovery, for example, links between research institutes and research trends.

In Sections 15.3.1 to 15.3.4, we present different aspects of Linked Science.

15.3.1 Distributing, Sharing, and Archiving Data

Linked Science relies heavily on Linked Data technologies. In brief, Linked Data is about "using the web to create typed links between data from different sources" [7]. It allows sharing and use of data, ontologies, and various metadata standards; in fact, a common vision is that it will be the de facto standard for providing metadata, and the data itself on the web.

Linked Data is based on *principles* (http://www.w3.org/DesignIssues /LinkedData) that include using HTTP URIs as names for things so that people can look up those names (and) when someone looks up a URI, providing of useful information, using standards (Resource Description Framework [RDF], SPARQL) and links to other URIs so that they can discover additional information.

In Linked Data, all information is encoded using the RDF (http://www .w3.org/RDF/) as triples of the form `<subject,predicate,object>`. This allows different resources to be linked together using predicates, and also literal values to be defined, such as names, for the resources. All resources in Linked Data are identified using URIs. This allows for requesting more information using the URIs in a machine-processable manner. In practice, this means that if a software agent requests a URI, then, for example, all the triples where this URI is as a subject could be returned, or alternatively even all triples where this URI is either a subject, a predicate, or an object. Linked Science adopts this approach by identifying researchers, research institutes, publications, and research datasets as URIs.

Linked Data thus allows for the efficient distribution of data using web standards. Scientists can make links from their data to existing datasets, thus connecting scientific resources. There are several benefits of this approach. First of all, distribution of data over the web reduces the space needed to store data in a single, local environment. For example, if a scientist is using geographic information—places with their coordinates and polygonal boundaries and so on—he/she just links to existing datasets providing this information rather than downloading everything into his/her local environment.

Thus, with the Linked Data approach, the *size* of the data is a smaller issue because most of the data are already on the web, somewhere, and served automatically by URIs on demand. Linked Data also allows for *executing* and *validating* methods on real data on the (Semantic) web. Because it is based on web standards, it clearly helps to achieve *compatibility*, in both the long and short terms. In Linked Science, *provenance* [14,15] information is published as Linked Data, using, for example, the Open Provenance Model Vocabulary (http://purl.org/net/opmv/ns). For datasets, this means that there is a record of who created the data, or encoded or transformed it, who published it, and also who has used it. Moreover, knowledge about the provenance of links themselves—for example, when a link was created, published, by whom, using which version of which database, or when a link became obsolete [16]—may be used to connect and compare different scientific results.

For each piece, records are also kept of when actions are performed. In addition, the published article serves as documentation for data and methods. Note that the scientific data itself can become a publication, and can be referred and linked to. All this allows us to use methods from the fields of Semantic Computing (see http://en.wikipedia.org/wiki/Semantic_computing) and Machine Learning (see http://en.wikipedia.org/wiki /Machine_learning) to analyze the semantic and statistical similarities of Linked Data sets and other research resources and thus detect *plagiarism* and *copyright* issues.

15.3.2 Open Source for Reproducible Research

The ability to reproduce research reported in publications is one of the key requirements in science. However, practices for documenting are not fully supporting this ideal. There are numerous reasons for this: data are not easily available and reimplementing methods based on abstract descriptions is time consuming, error prone, and sometimes even impossible—the devil is in the details. In Linked Science, the methods and their implementations are provided as an inherent part of the publication.

For example, when a method running a statistical analysis is implemented using the R Project (see http://www.r-project.org/), others can simply run the same method again and thus reproduce results. If they are experts in the field, they can analyze the implementation of the method. In this setting, data are also needed—for this, accessing Linked Data via SPARQL endpoints from within R has recently been made possible by the SPARQL package [17].

15.3.3 Cloud Computing for Virtualization of Research

Cloud computing refers to efforts to execute code on machines around the web, without the user knowing on which machine(s) the execution actually happens—thus, the term cloud computing. It is cost effective because it makes efficient use of machines. Moreover, one pays only for the traffic and computation needed, and not for the whole infrastructure. Cloud computing also enables one to handle large-sized datasets because the user does not have to handle them in their own environment. Furthermore, the user does not have to self-maintain all different kinds of systems: it is likely and also easier to find a needed system environment in the cloud than setting it up from scratch in a local environment. Furthermore, viruses cannot cause that much harm in controlled environments offered by the cloud. The access to data in Linked Science is provided also visually by cloud-based services.

When all the predicates for describing the data are given unique URIs, they enable the creation of more generic browsing and visualization facilities. For example, there are already numerous online applications capable of putting data on a map, if the data use those URIs for latitude and longitude proposed by the World Wide Web Consortium. Similarly, data in Linked Science can be explored on a timeline, and by using other facets such as theme, origin, author, and usage history.

15.3.4 Managing Licenses and Copyrights

In Linked Science, all the research data, copyright and license information, the article itself, actions taken on them, and their provenance information will be publicly available and represented as Linked Data. In addition to checking copyright and license issues of certain datasets, this approach also allows for querying and filtering datasets using the references to copyright schemes and licenses. Hence, this also enables on to find interesting datasets that fulfill the usage permission criteria of a planned research setting.

License issues have widely been recognized as important to ensure transparency. A set of Panton Principles has been published recently to promote a clear explication of licenses and to ensure openness of data. As a summary, the Panton Principles (Quoted from http://pantonprinciples .org) state the following

1. When publishing data make an explicit and robust statement of your wishes.

2. Use a recognized waiver or license that is appropriate for data.

3. If you want your data to be effectively used and added to by others, it should be open as defined by the Open Knowledge/Data Definition—in particular, noncommercial and other restrictive clauses should not be used.

4. Explicit dedication of data underlying published science into the public domain via PDDL or CCZero is strongly recommended and ensures compliance with both the Science Commons Protocol for implementing Open Access Data and the Open Knowledge/Data Definition.

15.4 ENCODING AND LINKING SCIENTIFIC KNOWLEDGE

This section first introduces the Linked Science Core Vocabulary. This is followed by a use case to show how the vocabulary can be used to describe research settings dealing with deforestation and related phenomena within the Brazilian Amazon Rainforest. We conclude by an overview of relevant related work.

15.4.1 Linked Science Core Vocabulary

The LSC (see http://linkedscience.org/lsc/ns/) is designed for describing scientific resources, including elements (see, e.g., http://en.wikipedia.org /wiki/Scientific_method) of research, their context, and interconnecting them. We introduce LSC as an example of building blocks for Linked Science to communicate the linkage between scientific resources in a machine-under-standable way. The "core" in the name refers to the fact that LSC only defines the basic terms for science. We argue that the success of Linked Science—or Linked Data in general—lies in interconnected, yet distributed vocabularies that minimize ontological commitments. More specific terms needed by different scientific communities can, therefore, be introduced as extensions of LSC.

Lightweightiness, simplicity, and intuitiveness have been the main design principles of LSC—it focuses on simple properties that can be used to describe the content of a research article, that is, to relate the research, hypotheses, predictions, experiments, data, and publications together. The main classes include Research, Researcher, Publication, Hypothesis, Prediction, and Conclusion. Figure 15.2 shows the main concepts and

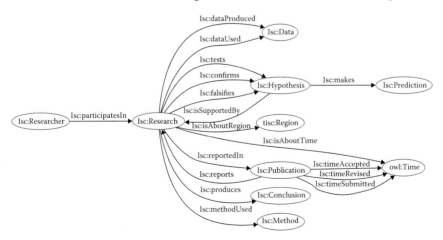

FIGURE 15.2 Graph presenting the concepts and relations of the Linked Science vocabulary.

properties of the vocabulary. The property `<isSupportedBy>` can be used for stating that a hypothesis is supported by a certain research. Moreover, one hypothesis `<makes>` one or more predictions, and a research `<produces>` a conclusion. Further on, `<reportedIn>` and `<reports>` are used to relate a publication, that is, the scientific paper with the research embedded in it. The interconnection between the research and data can be made via the properties `<dataUsed>` and `<dataProduced>`. LSC also provides properties for locating the research in space and time via the properties `<isAboutRegion>` and `<isAboutTime>`. Researcher is related to Research via `<participatesIn>`.

15.4.2 Use Case: Describing and Linking a Research Setting

Both publications and studies reported in them can be located in space and time. In the Linked Science approach, methods, data, and results are opened up and made available in a machine-processable way. A similar rationale applies to the meta-level information of scientific resources, which refer to the metadata of the publication resources and the meta-level information of the research embedded in them.

The spatial location of a publication has two aspects. On the one hand, it may be the place it originates from. This could be, for example, a certain university or research institute. On the other hand, location information is introduced by the spatial extent of the study that is described in it, for example, deforestation in the Brazilian Amazon Rainforest. The same principle applies to the two temporal aspects of a publication—namely, the date of the publication, for example, the year 2006, and the temporal extent of the research it is about, that is, deforestation data from 1987 to 2000.

It is very crucial to be aware of these different aspects when trying to organize and interconnect knowledge not only within one domain but also in heterogeneous ones. Especially, the ability to visualize the spatial and temporal information of the studies sheds light on the spatial and temporal gaps and overlaps, enabling better organization and management of the knowledge.

In the LSC, we propose terms for representing the different spatial and temporal aspects of publications so that scientific data may be published according to the Linked Data principles. For that, the properties `<isAboutTime>` and `<isAboutRegion>` are introduced for relating the research in a publication to its spatial and temporal extend. In combination with the Open Time and Space Core Vocabulary (see http://observedchange.com/tisc/ns/), a pluralism of spatial expressions can be used as objects, describing the region a research is about.

Figure 15.3 shows periods of interest of 13 different research settings. The periods are shown by two granularities, decades (lower time band) and years (upper time band). A lighter area in the lower time band indicates the temporal extent of the upper time band. Visualization of the periods enables one to detect gaps and overlaps between periods. For example, in the figure we can see that data from the year 1999 is of interest in six different studies while data for the year 2002 have been studied only in two research settings. If similar kinds of descriptions were available for all research concerning deforestation in the Brazilian Amazon Rainforest, it would enable researchers and funding agencies to find gaps and overlaps in a similar manner.

Figure 15.4 shows an extract of one of the example research settings modeled, namely, an article entitled "Is deforestation accelerating in the

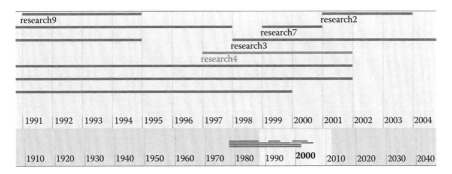

FIGURE 15.3 A timeline presenting the periods of interest of 13 different researches made concerning the Brazilian Amazon Rainforest.

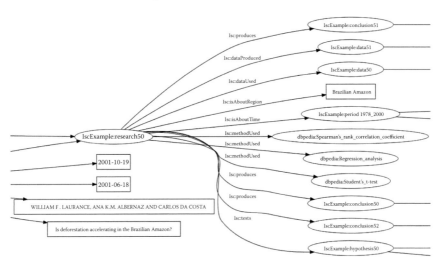

FIGURE 15.4 An example research setting modeled using LSC vocabulary.

Brazilian Amazon?" by Laurance et al. (2001) [18]. Different LSC properties have been in use to model their research, for example, which methods and data they used and which conclusions the research produced.

15.4.3 Discussion and Related Work

There are interesting related efforts for describing science. For example, the Scientific Knowledge Infrastructure Ontology (SKIo) provides means for describing scientific entities, processes, and methods using ontological structures. SKIo is aligned with the foundational ontology DOLCE to provide basic distinctions between objects, processes, abstract entities, and qualities within the science domain. Moreover, the structures from DOLCE to define causes and effects are used. The core idea in SKIo is that it can be used to formally define scientific processes—that is, process semantics—and to link them to publications. In a climate-modeling scenario, these formal models may be used, for example, to examine multiple processes causing the same output. This means that SKIo treats processes as functions for similarity reasoning: if two functions with the same inputs—such as the amount of rainfall—produce the same results, then the processes are essentially the same. Moreover, ontology could be used for querying all the environmental processes described in a given set of publications.

The Semantic Publishing and Referencing Ontologies (SPAR) (see http://purl.org/spar/) is a set of ontologies intended for describing scientific publications. The Citation Typing Ontology (CiTO) [19], for example, can be used to explicate what kind of a reference relation (explicit, implicit, or indirect) is in question between two publications. With CiTO, users can indicate in a machine-processable way whether they agree or disagree with other published research.

DataCite (see http://datacite.org/) is a community-driven effort for enabling citations of datasets using Digital Object Identifiers (DOIs), and to help archiving and accessing research data. The Semantic Web journal (http://www.semantic-web-journal.net/) provides an open review process, meaning that submitted articles are online, enabling anyone to comment on them. Moreover, the reviews are also published online, exposing the reviewer names by default.

Linked Data seems to offer a good basis for building Linked Science. For example, CrossRef (http://www.crossref.org/)—a consortium of roughly 3000 publishers—published millions of pieces of information about publications as Linked Data at the end of April 2011. All these pieces of data may be further linked to research settings they had, methods they used, and scientific knowledge they produced to implement the Linked Science approach.

15.4.4 Research Agenda for the Future

Enabling exact descriptions of all relevant scientific resources is crucial in achieving Linked Science. This means that pieces of well-defined information can and should themselves become publications. A research agenda for the future should, therefore, include finding ways to lower the barrier of publishing small pieces of new knowledge together with verification mechanisms rather than always requiring full 10- to 20-page articles. This will ultimately include a data reviewing process, comparable to the established peer reviewing processes for scientific publications.

In Linked Science, browsing and reproducing results should be as easy as it is today to click pages on the web. The research agenda for the future should thus concentrate on making, sharing, and reusing scientific knowledge as easy as possible.

In this chapter, we proposed and presented the LSC and showed how to use it for describing a research setting. We foresee that the research agenda for years to come should include to further develop and evaluate the LSC in different scientific settings. This includes research about required reasoning mechanisms to interconnect scientific resources.

The listed topics of the First International Workshop on Linked Science (see http://linkedscience.org/events/lisc2011) provide some ideas about where the research on Linked Science should focus. These topics include, for example, formal representations of scientific data, integration of quantitative and qualitative scientific information, ontology-based visualization of scientific data, semantic similarity in scientific applications, semantic integration of crowd sourced scientific data, and connecting scientific publications with underlying research datasets.

To continue this list, provenance, quality, privacy, and trust of scientific information are also crucial. Other goals include enriching scientific data through linking and data integration, and having case studies on linked science, that is, statistics and environmental monitoring. There is also a need for developing Linked Data practices for disseminating and archiving research results, collaboration, and research networks, and for research assessment. The development of application scenarios of Linked Science together with all their legal, ethical, and economic aspects provides interesting research topics.

Finally, the research agenda should also take into account the whole variety of information in academic settings. For example, for encoding information related to academic offerings, there are vocabularies such as the Academic Institution Internal Structure Ontology (AIISO), the Teaching Core Vocabulary (TEACH) (See http://linkedscience.org/teach /ns/), Metadata for Learning Opportunities (MLO), the XCRI Course Advertising Profile (XCRI-CAP), the Dublin Core metadata terms, the Friend of a Friend -vocabulary (FOAF), and the Open Provenance Model Vocabulary. These and other novel vocabularies should be further developed, evaluated, and taken into efficient use for increasing the linkage between scientific resources.

The LSC naturally faces the same challenges that have prevented the adoption of many other ontologies and vocabularies. Our aim is to tackle these challenges by (1) providing a lightweight structure, including only core terms and predicates and, therefore, minimizing ontological commitment; (2) providing clear examples of the use; (3) enabling the LSC to be technically accessible as Linked Data; and (4) involving the community to further develop and extend it.

15.5 CONCLUSIONS

In this chapter, we have laid the foundations for interconnecting scientific resources to increase the transparency, openness, and reproducibility of science. We first explained the concept of Linked Science, that is, publishing data using web techniques, opening and running methods in clouds for reproducing scientific processes, and explicating copyright and license issues. We also gave an overview of the technologies required for Linked Science. This includes publishing of data, methods, resources, results, license, and provenance information together with the documentation, and to help to establish trust related to them.

Linked Science, therefore, seeks to help authors, reviewers, publishers, and the whole scientific community in their challenging tasks, and be the key if not the future form of academic publishing. We gave an example of how scientific knowledge can be encoded as Linked Data by using a combination of existing and developed vocabularies and ontologies. We showed how the approach works with a set of publications related to research about deforestation in the Brazilian Amazon Rainforest. Formalization of the scientific results aimed at sharing the knowledge of the findings. Through visualizations one may actually see what periods and spatial regions are covered by the knowledge resulting from the research. Results may, therefore, be used to get an overview of the research in a certain domain, and further to find out potential gaps that need to be filled by new research settings.

15.6 GLOSSARY

Linked Data: A way to publish data using web standards and techniques. In the Linked Data approach, all resources are identified and resolvable via HTTP URIs, and there is a linkage between the resources.

Linked Science: An approach where scientific resources—for example, processes, models, data, methods, and evaluation metrics—are semantically linked.

LODUM: The Linked Open Data University of Münster project that works on opening up and linking research and educational data.

Semantic Web: A vision of a web where the meanings of things are explicitly defined such that autonomous agents can share these meanings, connect to and combine results of various semantic services, and perform useful actions for a user.

ACKNOWLEDGMENTS

This research has been partially funded by the International Research Training Group on *Semantic Integration of Geospatial Information* (DFG GRK 1498), the Linked Open Data University of Münster (LODUM) project, and by the German Academic Exchange Service (DAAD). In addition, the authors thank the insightful comments obtained through the anonymous peer-review process.

REFERENCES

1. S. Bechhofer, J. Ainsworth, J. Bhagat, I. Buchan, P. Couch, D. Cruickshank, D. D. Roure, M. Delderfield, I. Dunlop, M. Gamble, C. Goble, D. Michaelides, P. Missier, S. Owen, D. Newman, and S. Sufi. Why linked data is not enough for scientists. In *e-Science (e-Science), 2010 IEEE Sixth International Conference on e-Science*, Brisbane, Australia, pp. 300–307. IEEE, December 2010.
2. B. Mons, H. van Haagen, C. Chichester, P.-B. t'Hoen, J. T. den Dunnen, G. van Ommen, E. van Mulligen, B. Singh, R. Hooft, M. Roos, J. Hammond, B. Kiesel, B. Giardine, J. Velterop, P. Groth, and E. Schultes. The value of data. *Nature Genetics*, 43(4):281–283, 2011.
3. T. Kauppinen and G. M. de Espindola. Linked open science—communicating, sharing and evaluating data, methods and results for executable papers. In *The Executable Paper Grand Challenge, Proceedings of the International Conference on Computational Science (ICCS 2011)*, Elsevier Procedia Computer Science series, Singapore, 2011.

4. P. Groth, A. Gibson, and J. Velterop. The anatomy of a nanopublication. *Information Services and Use*, 30(1):51–56, 2010.

5. A. Passant, P. Ciccarese, J. Breslin, and T. Clark. SWAN/SIOC: aligning scientific discourse representation and social semantics. In *Workshop on Semantic Web Applications in Scientific Discourse*. The 8th International Semantic Web Conference (ISWC 2009), Chantilly, VA, USA, October 25–29, 2009.

6. T. Berners-Lee. Linked Data. Personal view available from http://www .w3.org/DesignIssues/LinkedData.html (accessed February 4, 2013), 2009.

7. C. Bizer, T. Heath, and T. Berners-Lee. Linked data—the story so far. *International Journal on Semantic Web and Information Systems*, 5(3):1–22, 2009.

8. T. Berners-Lee, J. Hendler, and O. Lassila. The semantic web. *Scientific American*, 284(5):34–43, 2001.

9. T. R. Gruber. A translation approach to portable ontology specifications. *Knowledge Acquisition*, 5:199–220, 1993.

10. Y. Bishr. Overcoming the semantic and other barriers to GIS interoperability. *International Journal of Geographical Information Science*, 12(4):299–314, 1998.

11. P. M. Fearnside and R. I. Barbosa. Accelerating deforestation in Brazilian Amazonia: towards answering open questions. *Environmental Conservation*, 31(01):7–10, 2004.

12. D. Clodoveu, G. Câmara, and F. Fonseca. Beyond SDI: integrating science and communities to create environmental policies for the sustainability of the Amazon. *International Journal of Spatial Data Infrastructure Research (IJSDIR)*, 4:156–174, 2009.

13. T. Kauppinen, J. Väätäinen, and E. Hyvönen. Creating and using geospatial ontology time series in a semantic cultural heritage portal. In S. Bechhofer et al. (Eds.) *Proceedings of the 5th European Semantic Web Conference 2008 ESWC 2008, LNCS 5021*, Tenerife, Spain, pp. 110–123, 2008.

14. O. Hartig. Provenance information in the web of data. In *Proceedings of the Linked Data on the Web (LDOW) Workshop at the World Wide Web Conference (WWW)*, Madrid, Spain, April 2009.

15. O. Hartig and J. Zhao. Publishing and consuming provenance metadata on the web of linked data. In *Proceedings of the Third International Provenance and Annotation Workshop*, Troy, NY, 2010.

16. J. Zhao, G. Klyne, and D. Shotton. Provenance and linked data in biological data webs. In *The 17th International World Wide Web Conference (LDOW2008)*, Beijing, China, 2008.

17. W. R. van Hage and T. Kauppinen. SPARQL client for R. Available from http:// cran.r-project.org/package=SPARQL (accessed February 4, 2013), 2011.

18. W. F. Laurance, A. K. M. Albernaz, and C. D. Costa. Is deforestation accelerating in the Brazilian Amazon? *Environmental Conservation*, 28(04): 305–311, 2001.

19. D. Shotton. CiTO, the Citation Typing Ontology. *Journal of Biomedical Semantics*, 1(Suppl 1):S6+, 2010.

CHAPTER 16

Summary and Conclusions

Terence Critchlow and Kerstin Kleese van Dam

Due to the evolution of both experimental and computational capabilities, science today is data intensive. This book has presented a variety of perspectives on the challenges that science is facing due to this rapid increase in data volumes and complexities, as well as discussions of leaders in the field of data-intensive science (DIS) on current leading edge technologies and their vision for the future.

As shown in the three science-focused chapters (Chapters 3 through 5), the nature of scientific collaboration is changing in the face of DIS. To address today's grand challenge problems, projects are bringing together geographically dispersed teams of scientists with multidisciplinary backgrounds. The nature of challenges such as developing effective treatment for diseases, understanding the universe, or designing materials with specific, complex properties requires the participation of experts from various domains to formulate solutions. However, this new cross-disciplinary collaborative work also increases the communication challenges within these teams. These challenges are further heightened by the amount of distributed, heterogeneous information that must be effectively analyzed to gain new insights. Within a single scientific domain, the amount of information being produced is staggering, but having to combine information from multiple domains brings differences such as terminology and scale to the forefront and renders many traditional technology approaches obsolete. Furthermore, this integration needs to happen within a framework that supports data privacy and security—without these assurances, science teams will not be able to effectively share their information. There is currently no easy way to sort through this data to identify the relevant information and disseminate it across the team in a way that enables the key scientific insights required to make breakthroughs. However, given the complexity of these scientific grand challenges, it is clear that the teams that can effectively transform data into knowledge are going to be the leaders in the future of science.

As the chapters describing current technology approaches (Chapters 6 through 8) show, technology is quickly adapting to these new requirements. Capabilities are constantly being provided or enhanced, and the amount of information that can be effectively analyzed by science teams continues to increase. Despite this progress, however, enabling the full potential of DIS remains an elusive goal.

There are ongoing debates in the technical community about the best approach to data analysis, some of which have been reflected in the preceding chapters. For example, cloud computing, supercomputers, and Hadoop/MapReduce clusters are viewed as potential solutions to certain DIS challenges—cloud computing holds the promise of scalability, can be accessed by familiar tools, and brings the data and compute resources close together; supercomputers support scientific breakthroughs by providing an environment in which a class of applications, including data-intensive applications, can be executed in parallel on hundreds of thousands of cores; Hadoop provides a scalable analysis technology that is appropriate for many queries. Unfortunately, none of these technologies is a clear winner because they all have inherent drawbacks as well. Cloud computing can be expensive when data are not "born in the cloud" but rather must be transferred to a cloud storage mechanism from somewhere else, such as an experimental facility or a supercomputing center. Furthermore, the abstractions that enable the cloud—both compute and storage virtualization—can have a significant, negative impact on performance when dealing with information that is spread across multiple physical locations. Finally, cloud solutions are not currently easily transferred across vendors, locking science teams into specific services even as new solutions become available. Supercomputers require expert users to develop programs that will run efficiently. These experts are constantly required to update the underlying code base as machines evolve and new platforms are rolled out. Typically, these machines are oversubscribed, limited in the amount of time they give each science team, and not available to a majority of scientists. Finally, because they are, by definition, pushing technology, they tend to be extremely expensive to both build and maintain. Hadoop's shared-nothing approach provides a different hurdle. While it works well for many types of analysis, it is not efficient when algorithms require significant coordination or data sharing among parallel processes. It is also not clear how a Hadoop approach would be used to combine heterogeneous datasets, where the merge algorithms would need to adapt to the semantics of the data instead of being uniformly applied, and the results would be semantically incomparable. All of these approaches continue to evolve and adapt to overcome their shortcomings. Whether one will eventually evolve into a comprehensive solution for DIS or a new approach, not yet developed, will ultimately address these challenges is yet to be seen.

The heart of this book (Chapters 9–15), highlights some of the primary areas of research focused on providing solutions to DIS. These chapters were selected to show the breadth of work that remains and to provide some insight into how these challenges may be overcome. To that end, each chapter focused on a specific research domain, the key challenges facing that domain, and how solutions to those challenges will impact DIS:

- Chapter 9 focused on the network underpinning all DIS. While the network layer is simply assumed as given by most science teams, and even most programmers, ensuring it continues to scale will be critical to the success of future DIS efforts.

- Chapter 10 focused on understanding the semantics of the data, specifically how using the data to define the ontology provides an effective approach for defining the underlying concepts and relationships.

- Chapter 11 focused on the practical challenges involved in combining data from heterogeneous sources. In highly dynamic scientific domains, the idealized approach to data integration, where schema are precisely mapped between source and targets, is not as useful as a less precise but more flexible approach of simple to create, data transformation pipelines based on algebraic functions using cloud computing.

- Chapter 12 focused on understanding where information comes from. Having a clear, understandable, data provenance chain is critical to establishing the trust required to effectively share information between science teams. Unfortunately, that is not always as simple as it sounds.

- Chapter 13 focused on a new frontier for data analysis, in situ computing. For computational science, where simulations are run on the fastest available supercomputers, the disparity between the information that can be produced and what can be saved is increasing. This difference is forcing teams to analyze data as they are created, and store the analytical results instead of the original data.

- Chapter 14 focused on techniques for exploring large datasets. In an environment where the scientists do not know exactly what they are looking for, interactively manipulating the data is the key to gaining the critical insights required to advance science. New, scalable approaches to data exploration, such as multiresolution and topological analysis, are presented.

- Chapter 15 focused on how to establish the appropriate context around the information being produced. Information does not exist in a vacuum; to effectively understand it, we need to understand its relationship to other information through semantically meaningful connections. This is leading to new paradigms in data publishing and archival.

Currently, much of this research is disjointed, with advances being made in individual research communities but little in the way of an effort to pull together all of these advances into an overarching DIS solution that would be adopted and customized by science teams. Even in cases where there is clear overlap of capability, for example, the provenance of in situ data analysis results is critical to their acceptance, there is little cooperation between the leading researchers in each field. We believe this will change over the coming years as the current research teams address their domain-specific challenges and begin to face the collective challenges of DIS.

While the challenges facing DIS are significant, there has already been substantial progress in transforming its promise into reality. The next few years will be exciting, both to watch and to participate in, as new technologies will enable even more impressive scientific breakthroughs.

Index